# U.S. Marine Corps Close Combat

# Plus

# Ranger Skills Handbook

## Department of Defense

Pentagon
Publishing

DEPARTMENT OF THE NAVY
Headquarters United States Marine Corps
Washington, D.C. 20380-1775

18 February 1999

# FOREWORD

## 1. PURPOSE

Today's Marines operate within a continuum of force where conflict may change from low intensity to high intensity over a matter of hours. Marines are also engaged in many military operations other than war, such as peacekeeping missions or noncombatant evacuation operations, where deadly force may not be authorized. During noncombative engagements, Marines must determine if a situation warrants applying deadly force. Sometimes Marines must decide in a matter of seconds because their lives or the lives of others depend on their actions. To make the right decision, Marines must understand both the lethal and nonlethal close combat techniques needed to handle the situation responsibly without escalating the violence unnecessarily. Marine Corps Reference Publication (MCRP) 3-02B, *Close Combat*, provides the tactics, techniques, and procedures of Marine Corps close combat. It also provides the doctrinal basis for the Marine Corps Close Combat Training Program (MCCCTP).

## 2. SCOPE

This publication guides individual Marines, unit leaders, and close combat instructors in the proper tactics, techniques, and procedures for close combat. MCRP 3-02B is not intended to replace supervision by appropriate unit leaders and close combat instruction by qualified instructors. Its role is to ensure standardization and execution of tactics, techniques, and procedures throughout the Marine Corps.

## 3. SUPERSESSION

MCRP 3-02B supersedes Fleet Marine Force Manual (FMFM) 0-7, *Close Combat*, dated 9 July 1993. There are significant differences between the two publications. MCRP 3-02B should be reviewed in its entirety.

4. WARNING

Techniques described in this publication can cause serious injury or death. Practical application in the training of these techniques will be conducted in strict accordance with approved Entry Level Close Combat, Close Combat Instructor (CCI), and Close Combat Instructor Trainer (CCIT) lesson plans. Where serious danger exists, the reac er is alerted by the following:

WARNING

5. CERTIFICATION

Reviewed and approved this date.

BY DIRECTION OF THE COMMANDANT OF THE MARINE CORPS

J. E. RHODES
Lieutenant General, U.S. Marine Corps
Commanding General
Marine Corps Combat Development Command

DISTRIBUTION: 14400006600

# OVERVIEW OF CLOSE COMBAT

## 1. Purpose of Close Combat

Close combat is the physical confrontation between two or more opponents. It involves armed and unarmed and lethal and nonlethal fighting techniques that range from enforced compliance to deadly force. The purpose of close combat is to execute armed and unarmed techniques to produce both lethal and nonlethal results. Unarmed techniques include hand-to-hand combat and defense against hand-held weapons. Armed techniques include techniques applied with a rifle, bayonet, knife, baton, or any weapon of opportunity.

## 2. Continuum of Force

Marines will find themselves in both combative and noncombative situations. The threat level in these situations can rise and fall several times based on the actions of both Marines and the people involved. The escalation of force stops when one person complies with the demands imposed by another person. This range of actions is known as a continuum of force. Continuum of force is the concept that there is a wide range of possible actions, ranging from voice commands to application of deadly force, that may be used to gain and maintain control of a potentially dangerous situation (MCO 5500.6_, *Arming of Security and Law Enforcement [EE] Personnel and the Use of Force)*. Continuum of force consists of five levels that correspond to the behavior of the people involved and the actions Marines use to handle the situation (see the table below). Close combat techniques are executed in levels three, four, and five.

### Level One: Compliant(Cooperative)

The subject complies with verbal commands. Close combat techniques do not apply.

### Level Two: Resistant (Passive)

The subject resists verbal commands but complies immediately to any contact controls. Close combat techniques do not apply.

| Continuum of Force | | |
|---|---|---|
| **Level** | **Description** | **Actions** |
| 1 | Compliant (Cooperative) | Verbal commands |
| 2 | Resistant (Passive) | Contact controls |
| 3 | Resistant (Active) | Compliance techniques |
| 4 | Assaultive (Bodily Harm) | Defensive tactics |
| 5 | Assaultive (Serious Bodily Harm/ Death) | Deadly force |
| Note: Shading indicates levels in which Marines use close combat techniques. | | |

### Level Three: Resistant (Active)

The subject initially demonstrates physical resistance. Marines use compliance techniques to control the situation. Level three incorporates close combat techniques to physically force a subject to comply. Techniques include—

- Come-along holds.
- Soft-handed stunning blows.
- Pain compliance through joint manipulation and the use of pressure points.

### Level Four: Assaultive (Bodily Harm)

The subject may physically attack Marines, but he does not use a weapon. Marines use defensive tactics to neutralize the threat. Defensive tactics include the following close combat techniques:

- Blocks.
- Strikes.
- Kicks.
- Enhanced pain compliance procedures.
- Nightstick blocks and blows.

### Level Five: Assaultive (Serious Bodily Harm/Death)

The subject usually has a weapon and will either kill or seriously injure someone if he is not stopped immediately and brought under control. Typically, to control the subject, Marines apply deadly force through the use of a firearm, but they may also use armed and unarmed close combat techniques.

## 3. Marine Corps Tactical Concepts

Close combat techniques support the following key Marine Corps tactical concepts. The concepts are not standalone ideas but are to be combined to achieve an effect that is greater than their separate sum.

### Achieving a Decision

Achieving a decision is important in close combat. An indecisive fight wastes energy and possibly Marines' lives. Whether the intent is to control an opponent through restraint or defend themselves in war, Marines must have a clear purpose before engaging in close combat and act decisively once engaged.

### Gaining an Advantage

A basic principle of martial arts is to use the opponent' s strength and momentum against him to gain more leverage than one's own muscles alone can generate, thereby gaining an advantage. In close combat, Marines must exploit every advantage over an opponent to ensure a successful outcome. This can include employing various weapons and close combat techniques that will present a dilemma to an opponent. Achieving surprise can also greatly increase leverage. Marines try to achieve surprise through deception, stealth, and ambiguity.

### Speed

Marines use speed to gain the initiative and advantage over the enemy. In close combat, the speed and violence of the attack against an opponent provides Marines with a distinct advantage. Marines must know and understand the basics of close combat so they can act instinctively with speed to execute close combat techniques.

### Adapting

Close combat can be characterized by friction, uncertainty, disorder, and rapid change. Each situation is a unique combination of shifting factors that cannot be controlled with precision or certainty. For example, a crowd control mission may call for Marines to employ various techniques ranging from nonlethal restraint to more forceful applications. Marines who adapt quickly will have a significant advantage.

## Exploiting Success

Typically, an enemy will not normally surrender simply because he was placed at a disadvantage. Marines cannot be satisfied with gaining an advantage in a close combat situation. They must exploit any advantage aggressively and ruthlessly until an opportunity arises to completely dominate the opponent. Marines must exploit success by using every advantage that can be gained.

(reverse blank)

# Close Combat

## Table of Contents

Page

CHAPTER 1

# FUNDAMENTALS OF CLOSE COMBAT

***This chapter describes all techniques for a right-handed person. However, all techniques can be executed from eittwr side.***

***The Marine is depicted in camouflage utilities. The opponent is depicted without camouflage.***

The fundamentals of close combat include ranges, weapons of the body, target areas of the body, and pressure points of the body. These fundamentals form the basis for all close combat techniques. They provide Marines with a common framework regardless of the type of confrontation or the techniques used. If Marines apply these fundamentals properly in a close combat situation, they may save their lives or the lives of fellow Marines.

## 1. Ranges of Close Combat

Close combat engagements occur within three ranges: long range, midrange, and close range. During any engagement, these ranges may blur together or may rapidly transition from one to another until either the opponent is defeated or the conflict is resolved.

### Long Range

During long range engagements, combatants engage each other with rifles, bayonets, sticks, or entrenching tools. See figure below.

### Midrange

During midrange engagements, combatants engage each other with knives, punches, or kicks.

### Close Range

During close range engagements, combatants grab each other. Close range engagements also involve elbow strikes, knee strikes, and grappling.

## 2. Weapons of the Body

### Hands and Arms

The hands, forearms, and elbows are the arm's individual weapons. The hands consist of several areas that can be used as weapons: fists, edges of hands, palms, and fingers.

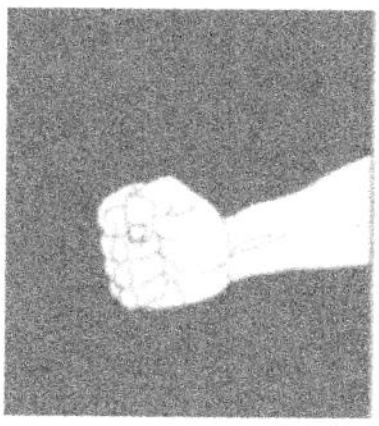

**Fists.** To minimize injury to the fists, Marines use their fists as weapons to target soft tissue areas such as the throat. The fists' striking surfaces are the first two knuckles of the hands or the meaty portions of the hands below the little fingers.

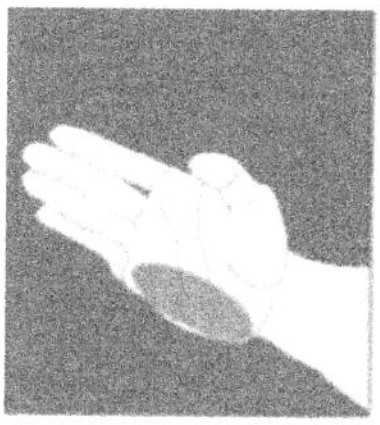

**Edge of Hand.** Marines use the edge of the hand (knife edge) as a weapon. Marines use the edge of the hand to strike soft tissue areas.

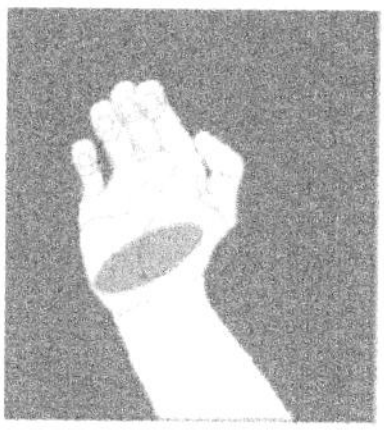

**Palms.** Because of the palm's padding, Marines use the heels of the palms to strike, parry, and/or block.

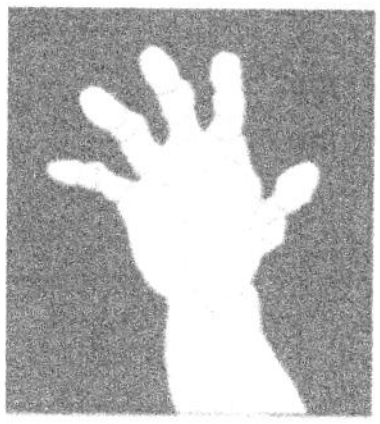

**Fingers.** Marines use the fingers to gouge, rip, and tear soft tissue areas (e.g., eyes, throat, groin).

**Forearms.** Marines use the forearms as a defensive tool to deflect or block attacks. Forearms can also be used as striking weapons to damage or break an opponent's joints and limbs. Marines sustain less self-injury when strikes are conducted with the forearms than when strikes are conducted with fists and fingers.

**Elbows.** Marines use the elbows as striking weapons. Because of the short distance needed to generate power, elbows are excellent weapons for striking during the close range of close combat.

## Legs

The legs are more powerful than any other weapon of the body, and they are less prone to injury when striking. The feet are protected by boots and are the preferred choice for striking.

**Feet.** Marines use the balls of the feet, the insteps, and the toes to kick an opponent. Marines use the cutting edge of the heels and the heels to stomp on an opponent. Marines must be wearing boots when striking with the toes.

**Knees.** Like elbows, knees are excellent weapons in the close range of close combat. Knee strikes are most effective while fighting close to an opponent where kicks are impractical. The opponent's groin area is an ideal target for the knee strike if he is standing upright. Knee strikes can deliver a devastating secondary attack to an opponent's face following an initial attack that caused him to bend at the waist.

# 3. Target Areas of the Body

During close combat, Marines strive to attack the accessible target areas of an opponent's body. The readily accessible areas will vary with each situation and throughout the engagement. The target areas are divided into five major groups: head, neck, torso, groin, and extremities. The figure below illustrates target areas of the body.

## Head

The vulnerable regions of the head are the eyes, temple, nose, ears, and jaw. Massive damage to the head kills an opponent.

**Eyes.** The eyes are excellent targets because they are soft tissue areas that are not protected by bone or muscle. Attacks to this area may cause the opponent to protect the area with his hands, allowing Marines to execute a secondary attack to other

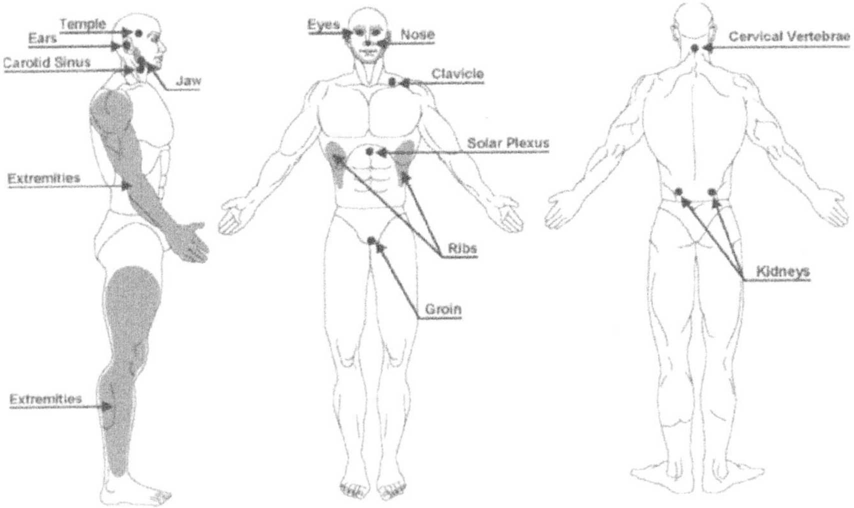

target areas while the opponent uses his hands to protect his eyes.

**Temple.** The temple is one of the most fragile areas of the head. Powerful strikes to the opponent's temple cause permanent damage and death.

**Nose.** The nose is very sensitive and easily broken. An attack to this area causes involuntary watering and closing of the opponent's eyes, rendering him vulnerable to secondary attacks. However, through training, individuals can condition themselves to withstand attacks to the nose. Therefore, any attack to the nose must be powerfully delivered.

**Ears.** Attacks to the ears may cause the eardrum to rupture. But this may not stop or even distract an opponent unless Marines powerfully deliver the strike.

**Jaw.** The jaw region, when struck forcefully, renders the opponent unconscious. Strikes to the jaw cause painful injuries to the teeth and surrounding tissues (e.g., lips, tongue), but the risk of self-injury is great unless Marines deliver strikes with a hard object such as a helmet, rifle butt, or boot heel.

## Neck

The front of the neck, or throat area, is a soft tissue area that is not covered by natural protection. Damage to this region causes the opponent's trachea to swell, closing his airway, which can lead to death.

**Carotid Sinus.** The carotid sinus is located on both sides of the neck just below the jaw. Strikes to the carotid sinus restrict blood flow to the brain, causing loss of consciousness or death.

**Cervical Vertebrae.** The cervical vertebrae on the back of the neck, from the base of the skull to the top of the shoulders, contains the spinal cord, which is the nervous system's link to the brain. The weight of the head and the lack of large muscle mass allow damage to the cervical vertebrae and spinal cord. Excessive damage to this area causes pain, paralysis, or death.

## Torso

**Clavicle.** The opponent's clavicle (or collar bone) can be easily fractured, causing immobilization of the arm.

**Solar Plexus.** Attacks to the opponent's solar plexus or center of the chest can knock the breath out of him and immobilize him.

**Ribs.** Damage to the opponent's ribs immobilizes him. It may also cause internal trauma.

**Kidneys.** Powerful attacks to the opponent's kidneys cause immobilization, permanent damage, or death.

## Groin

The groin area is another soft tissue area not covered by natural protection. Any damage to this area causes the opponent to involuntarily protect his injured area, usually with his hands or legs. In male opponents, the scrotum is the main target since even a near miss causes severe pain, contraction of the lower abdominal muscles, deterioration of his stance, and possible internal trauma.

## Extremities

Rarely will an attack to the opponent's extremities (arms and legs) cause death, but they are important target areas in close combat. Damage to an opponent's joints causes immobilization.

# 4. Pressure Points of the Body

There are nerves in the human body that, when pressure is applied or when they are struck, allow Marines to control a subject through pain compliance. Marines use pressure points to control an opponent when deadly force is not authorized. They also use pressure points to soften or distract an opponent so a lethal or nonlethal technique can be employed. The figure on page 1-5 illustrates the body's pressure points. Marines execute attacks to pressure points by—

- Rapidly kicking or striking pressure points.
- Slowly applying steady pressure to pressure points.

## Infraorbital Nerve

The infraorbital nerve is just below the nose. Marines apply pressure to this nerve with an index finger to control the opponent.

## Mastoid Process

The mastoid process is behind the base of the ear and beneath the edge of the jaw. Marines apply inward and upward pressure to this pressure point with the fingers to distract and control the opponent.

## Jugular Notch

The jugular notch is at the base of the neck in the notch formed at the center of the clavicle. Marines apply pressure in a quick, stabbing motion with the index finger. Strikes to the jugular notch cause serious damage.

## Brachial Plexus (Tie In)

The brachial plexus (tie in) is on the front of the shoulder at the joint. Strikes and pressure applied with the hand are effective on this nerve.

## Radial Nerves

Radial nerves are on the inside of the forearms along the radius bones. Strikes and pressure applied with the hand to the radial nerve serve as a softening technique.

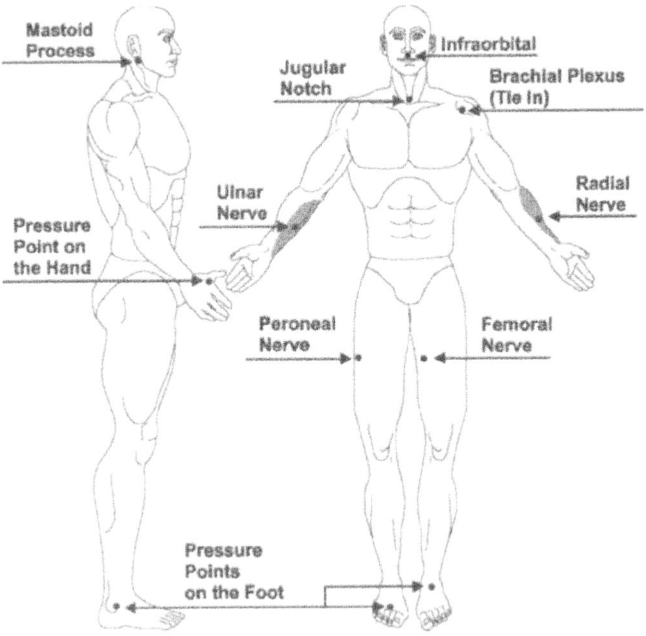

### Ulnar Nerve

Ulnar nerves are on the outside of the forearms along the ulnar bones. Strikes and pressure applied with the hand to the ulnar nerve serve as a softening technique.

### Pressure Point on the Hand

The hands contain a pressure point on the webbing between the thumbs and index fingers where the two bones of the fingers meet.

To force an opponent to soften or release his grip, Marines apply pressure with their index fingers to this pressure point or strike this pressure point with their fists.

### Femoral Nerves

Femoral nerves are on the inside of the thighs along the femur bones. Strikes to the femoral nerve serve as a softening technique.

### Peroneal Nerves

Peroneal nerves are on the outside of the thighs along the femur bones. Strikes to the peroneal nerve serve as a softening technique.

### Pressure Points on the Feet

There are pressure points on the feet that, when pressure is applied or when they are struck, serve to soften or distract the opponent. Marines apply pressure with the toe, edge, or heel of their boots to the following points:

- The notch below the ball of the ankle.
- The top center of the foot, above the toes.
- The top of the foot where the leg and foot meet.

## 5. Basic Warrior Stance

Achieving a solid stance is critical to stability and movement throughout any close combat confrontation. The basic warrior stance provides the foundation for stability and movement that is needed to execute close combat techniques. To execute the basic warrior stance, Marines put their feet apart, hands up, elbows in, and chin down.

### Feet Apart

Place feet shoulder-width apart.

Keep the head forward and eyes on the opponent, take a half step forward with the left foot, and pivot on the heels so the hips and shoulders are at approximately a 45-degree angle to the right.

Distribute body weight evenly on both legs. Bend the knees slightly.

### Hands Up

Curl the fingers naturally into the palm of the hand. Position the thumb across the index and middle fingers. Do not clench the fists. Clenching the fists increases muscular tension in the forearms and decreases speed and reaction time.

Bring the hands up to the face at chin level, with the palms facing each other. Hold the fists up high enough to protect the head, but not so high that they block the field of vision. Ensure continuous eye contact with the opponent.

### Elbows In

Tuck the elbows in close to protect the body.

### Chin Down

Tuck the chin down to take advantage of the natural protection provided by the shoulders.

## 6. Angles of Approach and Movement

Marines use movement to control a confrontation and to retain a tactical advantage. Movement increases power and maximizes momentum. By moving around the opponent, Marines gain access to different target areas of the opponent's body. Movement allows Marines to use different weapons of the body and different close combat techniques to attack specific target areas.

### Angles of Approach

Marines move anywhere within a 360-degree circle around the opponent to gain a tactical advantage. This circle provides access to different target areas of the opponent's body.

When facing an opponent, Marines move in a 45-degree angle to either side of the opponent. Moving at a 45-degree angle avoids an opponent's strike and puts Marines in the best position to attack the opponent. Marines should avoid being directly in front of an opponent during a confrontation. If a Marine is directly in front of an opponent, the opponent can rely on his forward momentum and linear power to seize the tactical advantage.

### Movement

Marines must know how to move in all directions while maintaining the basic warrior stance. During any movement, the legs or feet should not be crossed. Once a movement is completed, the basic warrior stance should be resumed. Maintaining the basic warrior stance protects Marines and puts them in the proper position to launch an attack against an opponent.

*Note: Before body movement begins. Marines turn their heads quickly to the new direction. The faster the head turns, the faster the body moves, and the quicker Marines attain visual contact with the opponent.*

**Forward to the Left.** To move forward to the left from the basic warrior stance, Marines-

: Move the left foot forward at a 45-degree angle from the body (approximately 12 to 15 inches), keeping the toe pointed toward the opponent.

Bring the right foot behind the left foot as soon as the left foot is in place. This returns Marines to the basic warrior stance.

**Forward to the Right.** To move forward to the right from the basic warrior stance, Marines-

Move the right foot forward at a 45-degree angle from the body (approximately 12 to 15 inches).

Bring the left foot, toe pointing toward the opponent, in front of the right foot as soon as the right foot is in place. This returns Marines to the basic warrior stance.

**Backward to the Left.** To move backward to the left from the basic warrior stance, Marines execute the forward movement in reverse. Marines-

Move the left foot backward at a 45-degree angle from the body (approximately 12 to 15 inches), keeping the toe pointed toward the opponent.

Bring the right foot behind the left foot as soon as the left foot is in place. This returns Marines to the basic warrior stance.

**Backward to the Right.** To move backward to the right from the basic warrior stance, Marines execute the forward movement in reverse. Marines-

Move the right foot backward at a 45-degree angle from the body (approximately 12 to 15 inches).

Bring the left foot, toe pointing toward the opponent, in front of the right foot as soon as the right foot is in place. This returns Marines to the basic warrior stance.

# 7. Balance and Off-Balancing

## Balance

In any close combat situation, Marines must strive to maintain balance. The last place to be in a close combat situation is on the ground. Marines must maintain a strong base and a low center of balance, their feet must be a shoulder-width apart, and they must stay on their toes to enable quick movement.

## Off-balancing

Marines use off-balancing techniques to control an opponent. These techniques are used to throw an opponent to the ground while Marines remain standing, or they are used to put Marines in a position for an offensive attack.

Off-balancing techniques use the opponent's momentum to move or throw him. For example, if the opponent is charging a Marine, the Marine pulls the opponent to drive him to the ground. Likewise, if the opponent is pulling a Marine, the Marine pushes the opponent to drive him to the ground.

Off-balancing techniques also rely on the power generated by the opponent. For example, during combat a Marine may be tired or outnumbered. Depending on the generated energy and momentum of the opponent, the Marine employs off-balancing techniques with very little effort and still provides effective results.

Because off-balancing techniques rely on the momentum and power generated by the opponent, these techniques are particularly effective for Marines who may be outsized by their opponent or lack their opponent's strength.

## Angles of Off-balancing

There are eight angles or directions in which an opponent can be off-balanced: forward, rear, right, left, forward right, forward left, rear right, and rear left.

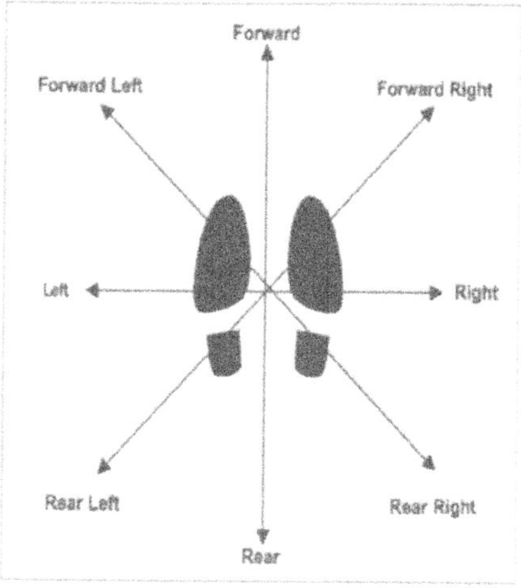


*Note: The angles correspond to the Marine's perspective, not the opponent's. Forward, rear, right, and left are straight angles. Forward right, forward left, rear right, and rear left are considered quadrants that are at a **45**-degree angle in either direction to the front or rear.*

## Off-balancing Techniques

Marines off-balance an opponent by pushing, pulling, or bumping the opponent with their hands, arms, or bodies.

To pull, Marines grab an opponent with their hands and drive him forcefully to one of the rear quadrants or to the right or left.

To push, Marines grab the opponent with their hands and drive him forcefully into one of the front quadrants or to the right or left.

Marines execute bumping in the same manner as pushing, but use their shoulders, hips, and legs instead of their hands to off-balance the opponent.

## 8. Falls

Marines may lose their balance or be thrown to the ground during encounters with an opponent. Marines use falling techniques to absorb the impact of a fall and to quickly return to their feet following an opponent's attack.

Whether falling or being thrown by an opponent, Marines strive to reduce the force of the impact, to prevent serious personal injury, and to increase the chances of survival. Falling techniques use the body's large muscles (back, thighs, buttocks) to protect vital organs and bones from injury and immobilization.

### Front Fall

Marines execute a front fall to break a fall on the front. To execute the front fall, Marines-

Bend the elbows and place the palms facing out in a position to spread and absorb the impact of the fall.

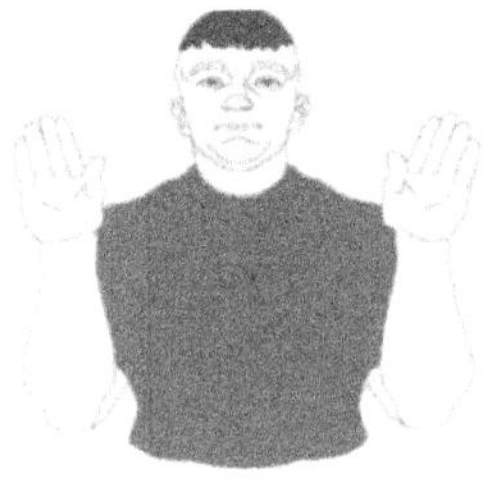

Fall forward, breaking the fall with the forearms and palms. The forearms and hands, down to the fingertips, should strike the ground simultaneously-

Offer resistance with the forearms and hands to keep the head raised off the ground.

### Side Fall

Marines execute a side fall to break a fall on the side. To execute the side fall, Marines-

Bring the right arm across the body so the hand is next to the left shoulder with the palm facing inboard.

Fall to the side, breaking the fall with the right arm by slapping the ground and making contact from the shoulder or forearm down to the hand. At the same time, tuck the chin and keep the head raised off the ground. The chin should be tucked to the chest at all times to prevent whiplash.

Stretch the right leg out to make contact with the ground and to distribute and absorb the impact. Bend the left leg, allowing the foot to make contact with the ground.

## Back Fall

Marines execute a back fall to break the fall when being thrown or falling backward. To execute the backfall, Marines-

Cross the hands in front of the chest and tuck the chin.

Fall backward and slap the ground with the forearms and hands to absorb the impact of the fall and keep the head off the ground.

## Forward Shoulder Roll

Marines use the forward shoulder roll to break a fall from an opponent's attack and to use the momentum of the fall to get back on their feet quickly. Ideally, Marines execute the forward shoulder roll to a standing position so they can continue fighting. To execute the forward shoulder roll, Marines-

Contact the ground with the back of the right forearm and upper arm. Tuck the chin into the chest.

Roll onto the right shoulder, rolling diagonally across the back to land on the left hip.

Slap the ground with the left arm, absorbing the impact from the shoulder to the hand, palm down.

Keep the left leg straight to absorb as much of the impact as possible. The right leg is bent and the foot hits flat on the ground.

Bend the left leg upon impact to push off with the left knee and leg to a squatting and then a standing position. Forward momentum should carry the Marine to a standing position.

(reverse blank)

CHAPTER 2

# LETHAL AND NONLETHAL WEAPONS TECHNIQUES

*This chapter describes all techniques for a right-handed person. However, all techniques can be executed from either side.*

*In drawings, the Marine is depicted in woodland camouflage utilities; the opponent is depicted without camouflage. In photographs, the Marine is depicted in woodland camouflage utilities; the opponent is depicted in desert camouflage utilities.*

## 1. Bayonet Techniques

WARNING

**During training, Marines must have bayonets sheathed. Marines use bayonet dummies to practice bayonet techniques. When practicing offensive and defensive bayonet techniques student-on-student, Marines use pugil sticks.**

All Marines armed with a rifle carry a bayonet. The bayonet is an effective weapon if Marines are properly trained in offensive and defensive bayonet techniques. An offensive attack, such as a thrust, is a devastating attack that can quickly end a fight. Defensive techniques, such as the block and parry, can deter the opponent's attack and allow Marines to regain the initiative. Through proper training, Marines develop the courage and confidence required to effectively use a bayonet to protect themselves and destroy the enemy. In situations where friendly and enemy troops are closely mingled and rifle fire and grenades are impractical, the bayonet becomes the weapon of choice.

### Holding the Rifle

To execute bayonet techniques, Marines hold the rifle in a modified basic warrior stance. All movement begins and ends with the basic warrior stance. To hold the rifle, Marines—

Use an overhanded grasp to grab the small of the rifle's stock. Use an underhanded grasp to grab the handguards of the rifle.

Lock the buttstock of the rifle against the hip with the right forearm.

Orient the blade end of the rifle toward the opponent.

### Offensive Bayonet Techniques

**Straight Thrust.** Marines use the straight thrust to disable or kill an opponent. It is the most deadly offensive technique because it causes the most trauma to an opponent. Target areas are the opponent's throat, groin, or face. The opponent's chest and stomach are also excellent target areas if not protected by body armor or combat equipment. To execute the straight thrust, Marines—

Lift the left leg and lunge forward off the ball of the right foot while thrusting the blade end of the weapon forward, directly toward the opponent.

Retract the weapon and return to the basic warrior stance.

**Slash.** Marines use the slash to kill an opponent or to create an opening in his defense. The target area is the opponent's neck. To execute the slash, Marines—

Extend the left hand back toward the left shoulder.

Thrust the left hand forward and swing it to the right, bring the right hand back toward the hip, and turn the cutting edge of the blade toward the opponent's neck. The movement is a slashing motion so the blade cuts across the opponent's neck.

**Horizontal Buttstroke.** Marines use the horizontal buttstroke to weaken an opponent's defenses, to cause serious injury, or to set him up for a killing blow. Target areas are the opponent's head, neck, and legs. To execute the horizontal buttstroke, Marines—

Step forward with the right foot and drive the right hand forward. Rotate the hips and shoulders into the strike. Move the left hand back toward the left shoulder.

Strike the opponent with the butt of the weapon.

**Vertical Buttstroke.** Marines use the vertical buttstroke to weaken an opponent's defenses, to cause serious injury, or to set him up for a killing blow. Target areas are the opponent's groin and face. To execute the vertical buttstroke, Marines—

Step forward with the right foot and drive the right hand straight up.

Pull the left hand back over the left shoulder.

Strike the opponent with the butt of the weapon.

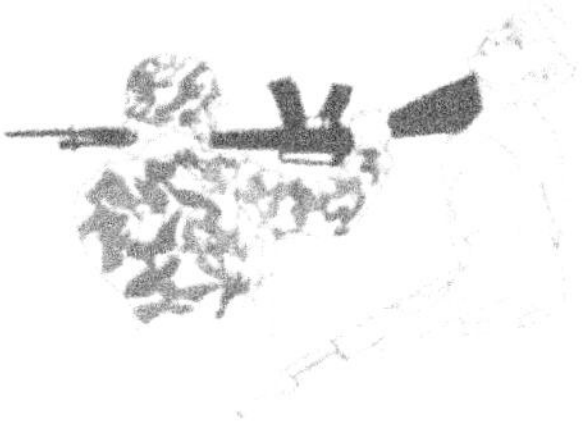

**Smash.** Marines use the smash as a follow-up technique to the horizontal or vertical buttstroke, particularly if they missed the target. The target area is the opponent's head. To execute the smash following a buttstroke, Marines-

Step forward with the right foot and place the blade end of the weapon over the left shoulder and elevate the right elbow above the shoulders.

Drive the arms straight forward, striking the opponent with the butt of the weapon.

## Defensive Bayonet Techniques

**Parry.** Marines use a parry as a defensive technique to redirect or deflect an attack. A parry is a slight redirection of a linear attack by an opponent; e.g., a straight thrust or a smash. To execute the parry, Marines-

Use the bayonet end of the rifle to redirect the barrel or bayonet of the opponent's weapon.

Lock the weapon against the hip with the right forearm.

Rotate to the right or left, moving the bayonet end of the rifle to parry the opponent's attack. Rotation is generated from the hips.

Redirect or guide the opponent's weapon away from the body by exerting pressure against the opponent's weapon.

**High Block.** Marines execute a high block against a vertical attack coming from high to low. To execute the high block, Marines—

Thrust the arms up forcefully at approximately a 45-degree angle from the body. The weapon should be over the top of the head and parallel to the ground. The elbows are bent, but there is enough muscular tension in the arms to absorb the impact and deter the attack.

**Low Block.** Marines execute the low block

against a vertical attack coming from low to high. To execute the low block, Marines—

Thrust the arms down forcefully at approximately a 45-degree angle from the body. The weapon

should be at or below the waist and parallel to the ground. The elbows are bent, but there is enough muscular tension in the arms to absorb the impact and deter the attack.

**Left and Right Block.** Marines execute a left or right block against a horizontal buttstroke or a slash. To execute the left or right block, Marines-

Thrust the arms forcefully to the right or left, holding the rifle vertically in the direction of the attack. The elbows are bent, but there is enough muscular tension in the arms to absorb the impact and deter the attack.

**Counter Action Following the Block.** After deflecting an opponent's attack with a block, Marines counter with a slash or a horizontal buttstroke to regain the initiative. However, the objective in any bayonet fight is to thrust forward with the blade end of the weapon to immediately end the fight.

## Group Strategy

On occasion, Marines may engage an opponent as a member of a group or numerous opponents by one's self or as a member of a group. By combining bayonet fighting movements and simple strategies, Marines can effectively overcome their opponent or opponents.

**Offensive Strategy: Two Against One.** If two bayonet fighters engage one opponent, the fighters advance together.

Fighter 1 engages the opponent while fighter 2 swiftly and aggressively attacks the opponent's exposed flank and destroys the opponent.

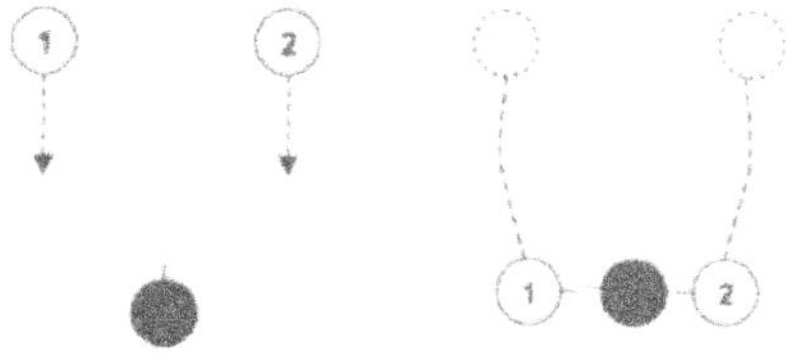


**Offensive Strategy: Three Against Two.** If three bayonet fighters engage two opponents, the fighters advance together keeping their opponents to the inside.

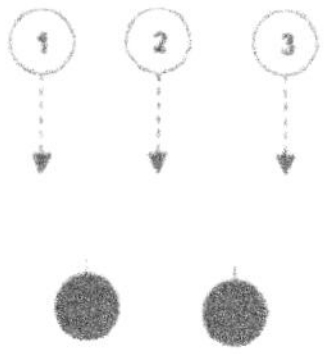


Fighters 1 and 3 engage opponents. Fighter 2 attacks the opponent's exposed flank engaged by fighter 1 and destroys the opponent.

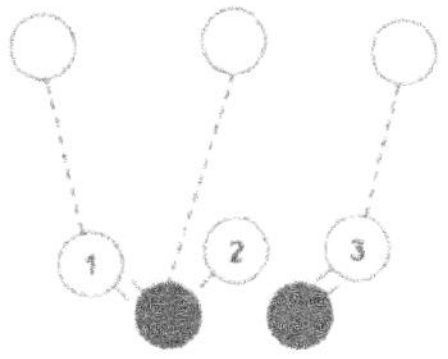


Fighters 1 and 2 turn and attack the exposed flank of the opponent engaged by fighter 3 and destroy the opponent.

**Defensive Strategy: One Against Two.** If a fighter is attacked by two opponents, the fighter immediately positions himself at the flank of the nearest opponent and keeps that opponent between himself and the other opponent.

Using the first opponent's body as a shield against the second opponent, the fighter destroys the first opponent quickly before the second opponent moves to assist.

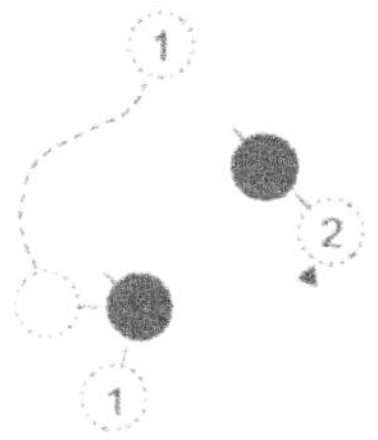


Then, the fighter engages and destroys the second opponent.

**Defensive Strategy: Two Against Three.** If two fighters are attacked by three opponents, the fighters immediately move to the opponent's flanks.

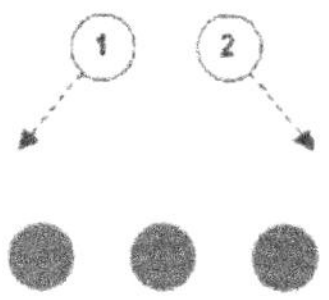

Fighters 1 and 2 quickly attack and destroy their opponents before the third opponent closes in.

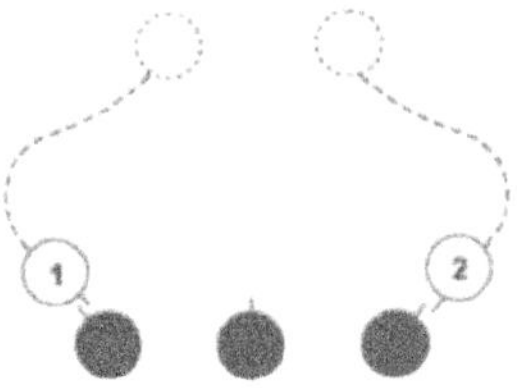


Fighter 1 engages the third opponent while fighter 2 attacks the opponent's exposed flank and destroys the opponent.

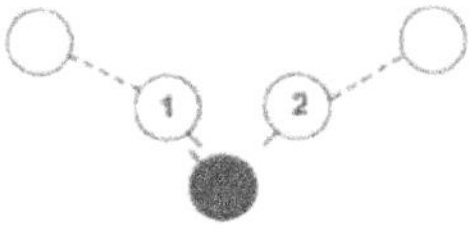


## 2. Nonlethal Rifle and Shotgun Retention Techniques

Most Marines are armed with the M16A2 service rifle. Marines are taught to keep their weapon with them at all times. Marines must be constantly alert to their surroundings and the people moving in and around their environment. Marines may be confronted by an individual who tries to take theii weapons. If this happens, Marines should not struggle with the individual. To retain positive control of their weapons, Marines must understand and apply weapons retention techniques, otherwise known as armed manipulation. The fol lowing techniques can be used with either the rifle or the shotgun.

### Blocking Technique

To execute a blocking technique, Marines-

Stand in a defensive posture.

Use the weapon to block the opponent by thrusting it out firmly, with the elbows bent. Do not try to hit the opponent with the rifle. The rifle is used as a barrier.

### Technique if Opponent Grabs Weapon Underhanded

Usually an opponent will try to grab the weapon or block it as an instinctive action. If the opponent uses an underhand grab to seize the handguards of the weapon, Marines—

Trap the opponent's closest finger(s) above the knuckle with the thumb so he cannot release his grip-

Apply bone pressure on the opponent's finger to initiate pain compliance.

Rotate the barrel of the weapon up or down quickly while maintaining pressure on the opponent's hand. At the same time, quickly pivot to off-balance the opponent.

## Technique if Opponent Grabs Weapon Overhanded

If the opponent uses an overhand grab to seize the weapon overhanded, Marines—

Trap the opponent's finger to hold his hand in place.

Rotate the barrel to place it across the opponent's forearm and apply downward pressure. This action is similar to an armbar.

## Technique if the Opponent Grabs the Muzzle

If the opponent grabs the muzzle of the weapon, Marines—

Rotate the muzzle quickly in a circle motion. Slash downward with the muzzle to release the opponent's grip.

## Butt Strikes

Strikes with the butt of the weapon control or ward off an attacker. During any of the retention techniques, Marines use the heel or cutting edge of the weapon to deliver butt strikes to the inside or the outside of the opponent's thigh. The inside butt strike targets the femoral nerve as illustrated in the photo to the left.

The outside butt strike targets the peroneal nerve as illustrated in the photo to the left. Strikes can be made to the outside or inside of the thighs. If a strike to one side of the thigh misses, Marines follow back through with the butt of the weapon on the other side of the thigh.

## Off-balancing Techniques

Marines apply off-balancing techniques to throw an opponent to the ground and retain possession of the weapon.

If the opponent grabs the weapon and pushes, Marines should not push on the weapon. They should—

Move with the momentum and movement of the

opponent by pivoting in the direction of the movement by stepping back.

Throw the opponent to the ground with a quick jerking movement by lowering the mu/./le and swinging the butt of the weapon.

If the opponent grabs the weapon and pulls, Marines—

Step on the opponent's foot and push forward to off-balance him and drive him to the ground.

Sweep the opponent's feet out from under him by hooking his leg with the leg and kicking backward.

## 3. Nonlethal Handgun Retention Techniques

Many Marines are armed with the M9 service pistol. Marines must keep their weapons in their possession at all times. Marines must be constantly alert to their surroundings and the people moving in and around the environment. Marines may be confronted by an individual who tries to take their weapons. To retain positive control of the weapon, Marines must understand and apply handgun retention techniques.

### Blocking Technique

Marines perform the following blocking techniques if an opponent attempts to grab their pistol in the holster. Marines—

Place the body between the weapon and the opponent by immediately pivoting so the weapon is away from the opponent.

Step back and away from the opponent while placing the hand on the pistol grip.

Extend the left hand and block, deflect, or strike the opponent's arm.

### Armbar Technique

Marines use the armbar technique when an opponent uses his right hand to grab their pistol in the holster. To execute the armbar technique, Marines—

Trap the opponent's right hand by grasping the opponent's wrist or hand with the right hand and applying pressure against the body.

Step back with the right foot and pivot sharply to the right to be next to the opponent. Always pivot in a direction that keeps the weapon away from the opponent.

Straighten the opponent's arm to apply an armbar. The arm should be straight across the torso.

Continue pivoting to the right while pulling back on the opponent's shoulder. This action may break the opponent's arm.

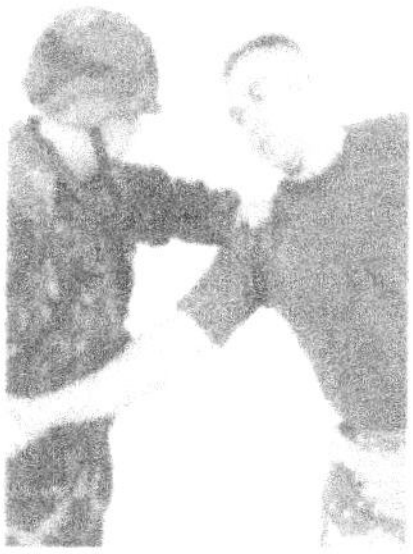

Place the left hand across the opponent's face and apply pressure back and down to take the oppo-

nent to the ground. Pressure applied beneath the nose or on the trachea is equally effective.

## Wristlock Technique

Marines use the wristlock technique when an opponent grabs their pistol while it is in the holster with his right hand. To execute the w ristlock technique, Marines—

Grasp the opponent's wrist or hand with the right hand and apply pressure against the body.

Step back and away from the opponent and pivot to the right so the weapon is away from the opponent. Always pivot in a direction that keeps the weapon away from the opponent.

Reach over the opponent's arm with the left hand and grab his hand, applying pressure against his arm with the left forearm. Execute a wristlock.

Incorporate the second hand into the wristlock and, stepping back with the left foot, pivot to the left.

Execute a two-handed wristlock by exerting downward pressure with the thumb and rotating his hand to the left.

Continue to pivot to off-balance the opponent and drive him to the ground.

*Note: If the opponent grabs the pistol with his left hand, Marines execute the wristlock with one hand and step in toward the opponent, rather than*

*away from the opponent.*

## Softening Techniques

Handgun retention techniques use softening techniques applied to pressure points. Bone pressure

and strikes with the hands (i.e., hammer fist), knees, and feet are also effective softening techniques.

**Pressure Points.** Marines use pressure point techniques to get the opponent to loosen his grip. Marines use their finger tips to apply pressure to the webbing between the index finger and thumb, the jugular notch, and the brachial plexus tie in. The following figure illustrates pressure applied to the brachial plexus tie in.

**Bone Pressure.** Bone pressure is the application of pressure on a bone against a hard object to initiate pain compliance. To apply bone pressure, Marines use their hand to trap the opponent's hand on the weapon. Marines apply a slow, steady pressure to the opponent's hand and fingers until his grip is softened or he releases his hold.

**Strikes.** If it is difficult to apply a retention technique, Marines employ strikes or kicks to force the opponent to loosen his grip. Strikes to the

eyes, the arms (radial nerve), or shoulder (brachial plexus tie in) soften the opponent's grip on the weapon.

Kicks and knee strikes to the peroneal nerve, the femoral nerve, or the groin are effective because the opponent is typically unprepared to counter the strike.

Stomping on the top of the opponent's foot may distract him or loosen his grip on the weapon.

# 4. Firearm Disarmament Techniques

Marines use firearm disarmament techniques during a close-range confrontation if they are un-

armed and the opponent has a firearm (pistol). These techniques are equally effective if Marines are armed but do not have time to withdraw and present the weapon. The goal of firearm disarmament techniques is to gain control of the situation

so Marines gain the tactical advantage. The goal is not necessarily to get control of the opponent's weapon.

## Pistol to the Front

This technique is used when Marines are unarmed and the opponent has a pistol pointing at their front (e.g., head, chest). The technique is the same if the opponent sticks the pistol under the Marine's chin. To execute the counter when an opponent is pointing a pistol toward the front of a Marine, Marines—

Place the hands close to the weapon, about chest high, palms out.

Use the left hand to grab the opponent's forearm and push the opponent's hand with the pistol away to clear the body from in front of the weapon. At the same time, rotate the right shoulder back to clear the body from the weapon.

Maintain control of the opponent's arm.

Grasp the weapon with the right hand by placing the thumb underneath the pistol and the fingers over top of the pistol.

Keep the right hand wrapped tightly around the muzzle and quickly rotate the pistol in the opponent's hand so the muzzle is facing the opponent.

Grasp and pull the opponent's wrist or forearm away from the body while rotating the weapon.

Rotate the weapon toward the opponent while pulling it up and back and out of the opponent's grasp.

## Pistol to the Rear

This technique is performed when Marines are unarmed and the opponent has a pistol pointing to the back of the Marine's head. To execute the counter when the opponent is pointing a pistol to the rear, Marines—

Place the hands close together about chest high, palms out.

Step back with the left foot, pivoting on the right foot so the side is against the opponent's front. This action clears the body from the weapon's line of fire. Keep the left hand up.

Pivot on the left foot to face the opponent and, at the same time, raise the left elbow and reach over the top of the opponent's arm with the left arm.

Wrap the left arm tightly around the opponent's arm above his elbow to control it.

Push on the opponent's shoulder with the right hand while pulling up with the left arm to achieve an armbar. This action releases the opponent's grip on the weapon. If necessary, the opponent can be taken to the ground with a leg sweep.

*Note: To execute this technique, the weapon must be close to or touching the rear of the Marine. If the weapon is too far away from the body, this technique would be difficult to execute or it would be ineffective.*

# CHAPTER 3

# HAND-HELD WEAPONS

***This chapter describes all techniques for a right-handed person. However, all techniques can be executed from either side.***

***In drawings, the Marine is depicted in woodland camouflage utilities; the opponent is depicted without camouflage. In photographs, the Marine is depicted in woodland camouflage utilities; the opponent is depicted in desert camouflage utilities.***

Marines must know how to defend against attacks when an opponent is either unarmed or armed with a held-held weapon. This chapter addresses the combative use of knives, specific weapons of opportunity, and sticks. However, virtually anything can be used as a hand-held weapon.

## 1. Fundamentals of Knife Fighting

Marines must be trained in knife fighting techniques. Marines experienced in offensive knife techniques can cause enough damage and massive trauma to stop an opponent. When engaged against each other, experienced knife fighters employ various maneuvers and techniques that are specific to knife fighting. Seldom, if ever, will Marines engage an opponent in a classical knife fight.

*Note: When armed with a rifle, Marines are issued a bayonet. When armed with a pistol, Marine are issued a combat knife.*

### Angles of Attack

There are six angles from which an attack with a knife can be launched:

- Vertical strike coming straight down on an opponent.
- Forward diagonal strike coming in at a 45-degree angle to the opponent.
- Reverse diagonal strike coming in at a 45-degree angle to the opponent.
- Forward horizontal strike coming in parallel to the ground.
- Reverse horizontal strike coming in parallel to the ground.
- Forward thrust coming in a straight line to the opponent.

### Target Areas of the Body

During any confrontation, the parts of the opponent's body that are exposed or readily accessible will vary. The goal in a knife fight is to attack the body's soft, vital target areas that are readily accessible (e.g. the face, the sides and front of the neck, the lower abdomen [or groin]).

**Neck.** Carotid arteries, located on either side of the neck, are good target areas because they are not covered by body armor or natural protection.

**Lower Abdomen (or Groin).** The lower abdomen (or groin region) is a good target area because it is not covered by body armor.

**Heart.** The heart, if not covered by body armor, is an excellent target which, if struck, can prove fatal in a matter of seconds or minutes.

**Secondary Targets.** There are secondary target areas that will cause substantial bleeding if an artery is severed. These target areas are not immediately fatal, but can become fatal if left unattended. Attacks to—

The legs can cause a great deal of trauma and prove fatal. For example, the femoral artery located on the inside of the thigh is a large artery which, if cut, will cause extensive blood loss.

The brachial artery, located between the biceps and triceps on the inside of the arm, can cause extensive bleeding and damage.

The arm's radial and ulnar nerves can cause extensive bleeding and damage.

## Movement

Marines can move anywhere within a 360-degree circle around the opponent. This allows accessibility to different target areas of the opponent's body. Marines should avoid being directly in front of an opponent because the opponent can rely on his forward momentum to seize the tactical advantage. If Marines face an opponent, movement is made in a 45-degree angle to either side of the opponent. This angle avoids an opponent's strike and places Marines in the best position to attack an opponent.

## Wearing the Combat Knife

Marines must wear the combat knife where it is easily accessible and where it can best be retained. It is recommended the combat knife be worn on the weak side hip, blade down, sharp edge facing forward. Marines can place it behind the magazine pouch where it is easily accessible to them, but not easily grabbed by an opponent.

## Grip

The grip on the knife should be natural. Marines grasp the knife's grip with the fingers wrapped around the grip naturally as it is pulled out of its sheath. This is commonly known as a hammer grip. The blade end of the knife is always facing the opponent.

## Stance

Marines use the basic warrior stance as the foundation for knife techniques. The left hand is a vertical shield that protects either the ribs or the head and neck. The right elbow is bent with the blade pointing forward toward the opponent' s head. This position serves as an index point, where all techniques are initiated.

## Principles of Knife Fighting

The following are key principles of knife fighting:

Execute movements with the knife blade within a box, shoulder-width across from the neck down to the waistline. The opponent has a greater chance of blocking an attack if the blade is brought in a wide, sweeping movement to the opponent.
Close with the opponent, coming straight to the target.
Move with the knife in straight lines.
Point the knife's blade tip forward and toward the opponent.
Apply full body weight and power in each of the knife techniques. Full body weight should be put into the attack in the direction of the blade's movement (slash or thrust).
Apply constant forward pressure with the body and blade to keep the opponent off-balanced.

# 2. Knife Fighting Techniques

## Slashing Techniques

Marines use slashing techniques to close with an enemy. Slashing techniques distract the opponent or damage the opponent so Marines can close in. Typically, Marines target the opponent's limbs, but any portion of the body that is presented can become a target.

**Vertical Slash Technique.** The vertical slash follows a vertical line straight down through the target. To execute the vertical slash, Marines—

Thrust the right hand out and bring the weapon straight down on the opponent, continuing to drag the knife down through the opponent's body.

Maintain contact on the opponent's body with the blade of the knife.

**Forward Slash Technique.** The forward slash follows a straight line in a forehand stroke, across the target areas of either the neck (high diagonal slash) or abdominal region (low horizontal slash). To execute the forward slash, Marines-

Extend the right hand while simultaneously rotating the palm up until the knife blade makes contact with the opponent.

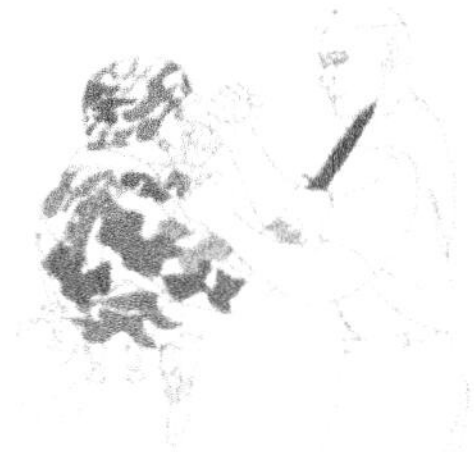

Snap or rotate the wrist through the slashing motion to maximize blade's contact with the opponent.

Drag the knife across the opponent's body, from right to left, in a forehand stroke. The movement ends with the forearm against the body and the knife at the left hip with its blade oriented toward the opponent.

**Reverse Slash Technique.** The reverse slash is a follow-up technique to a forward attack. It allows Marines both a secondary attack and the ability to resume the basic warrior stance. The reverse slash follows a straight line in a backhand stroke, across the target areas of either the neck (high diagonal slash) or abdominal region (low horizontal slash). To execute the reverse slash, Marines—

Extend the right hand while simultaneously rotating the palm down until the knife blade makes contact with the opponent.

Snap or rotate the wrist through the slashing motion to maximize the blade's contact with the opponent.

Drag the knife across the opponent's body, from left to right, in a backhand stroke. Maintain contact on the opponent's body with the blade of the knife.

## Thrusting Techniques

The primary objective of knife fighting is to insert the blade into an opponent to cause extensive damage and trauma. This is done with a thrusting technique. Thrusting techniques are more effective than slashing techniques because of the damage they can inflict. However, Marines use slashing techniques to close with the enemy so that they are closer to the opponent, which allows them to use the thrusting technique.

**Vertical Thrust.** The thrusting motion follows a vertical line straight up through the target (low into the abdomen region or high into the neck). To execute the vertical thrust, Marines—

Thrust the right hand toward the target, inserting the knife blade straight into the opponent.

Pull the knife out of the opponent.

**Forward Thrust.** The forward thrust follows a straight line straight into the opponent's neck (high thrust) or abdominal region (low thrust). To execute the forward thrust, Marines—

Thrust the right hand, palm down, toward the target, inserting the knife blade straight into the opponent.

Rotate the palm up once the knife is inserted to twist the blade.

Drop the right elbow and bring the knife to the opposite side of the opponent's body from where it was inserted. At the same time, rotate the hips and shoulders downward to bring body weight to bear on the attack.

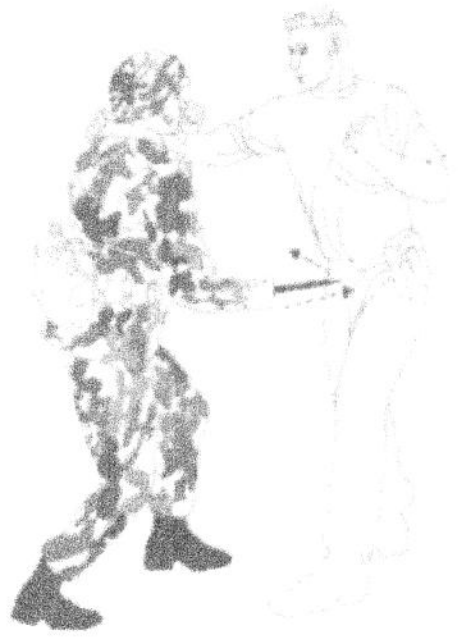

**Reverse Thrust.** The reverse thrust is a follow-up technique to a forward attack. It allows Marines both a secondary attack and the ability to resume the basic warrior stance. The reverse thrust follows a horizontal line straight into the opponent's neck (high thrust) or abdominal region (low thrust). To execute the reverse thrust, Marines-

Bend the right arm, crossing the arm to the left side of the body.

Thrust the right hand, palm up, toward the target, inserting the knife blade straight into the opponent.

Rotate the palm down to twist the blade once the knife is inserted.

Bring the knife to the opposite side of the opponent's body from where it was inserted.

## 3. Weapons of Opportunity

During an unarmed close combat situation, Marines use their bodies as weapons, but they should be ready and able to use anything around them as a weapon. For example, Marines could throw sand or liquid in an opponent's eyes to temporarily impair his vision or smash the opponent's head with a rock or helmet. Marines must use whatever means are available and do whatever it takes to take control of the situation and to win, or they face the possibility of losing their lives. Some weapons of opportunity are discussed in the following subparagraphs.

### Entrenching Tool

An entrenching tool (E-tool) is commonly carried by Marines. It can be an excellent weapon, especially when sharpened. Marines can use the E-tool to block, slash, and thrust at an opponent.

### Tent Pole and Pins

Marines can use tent poles and pins to block, strike, or thrust at an opponent.

### Web Belt

Marines can stretch a web belt between their hands to block attacks by an opponent.

### Battlefield Debris

Marines can use debris on the battlefield (e.g., sticks, glass, a sharp piece of metal) to cut, slash, or stab an opponent. They can also use other types of debris such as shovel or ax handles, boards, metal pipes, or broken rifles to strike an opponent or apply a choke.

### Helmet

A helmet can be used to strike an opponent on an unprotected area like the head and face. Grasp the rim of the helmet and thrust the arms forward, striking the opponent with the top of the helmet.

## 4. Fundamentals of Combative Stick

On the battlefield, Marines must be ready and able to use anything as a weapon. They must learn and be able to use techniques that can be employed with most weapons of opportunity. Among these techniques are combative stick techniques. Combative stick techniques can be used with a stick, a club, a broken rifle, an E-tool, or even a web belt.

## Angles of Attack

There are six angles from which an attack with a hand-held weapon can be launched:

Vertical strike coming straight down on an opponent.
Forward diagonal strike coming in at a 45-degree angle to the opponent.
Reverse diagonal strike coming in at a 45-degree angle to the opponent.
Forward horizontal strike coming in parallel to the ground.
Reverse horizontal strike coming in parallel to the ground.
Forward thrust coming in a straight line to the opponent.

## Grip

Grasp the stick about 2 inches from its base.

## Stance

The basic warrior stance serves as the foundation for combative stick techniques. The left hand becomes a vertical shield that protects the ribs or the head and neck. Depending on how heavy the weapon is, it should be held at a level approximately shoulder height.

## Movement

Movement during combative stick techniques is the same as it is for other close combat techniques. Marines can move anywhere within a 360-degree circle around the opponent. This allows accessibility to different target areas of the opponent's body and gains a tactical advantage. Marines should avoid being directly in front of an opponent because the opponent can rely on his forward momentum to seize the tactical advantage. If Marines face an opponent, movement is made in a 45-degree angle to either side of the opponent. This angle avoids an opponent's strike and places Marines in the best position to attack an opponent.

# 5. Combative Stick Techniques

## Strikes

Strikes are intended to inflict as much damage on an opponent as possible. Striking techniques apply to a weapon of opportunity such as a stick, a tent pole, a club, a broken rifle, an E-tool, or a Pipe-

**Vertical Strike.** To execute the vertical strike, Marines—

Bend the right arm, extending the weapon over the back of the right shoulder.

Rotate the forearm straight down off the elbow to bring the weapon down on the opponent.

Rotate the hips and shoulders forcefully toward the opponent.

Follow through with the strike by allowing the weight of the weapon to go through the target area of the body.

**Forward Strike.** To execute a forward strike, Marines—

Step forward with the left foot in the direction of the strike.

Bend the right arm with the elbow extending out to the right and the weapon extended over the right shoulder.

Rotate the forearm to the right of the elbow to bring the weapon down onto the opponent. At the same time, forcefully rotate the hips and shoulders toward the opponent.

Follow through with the strike by allowing the weight of the weapon to go through the target area of the body.

**Reverse Strike.** The reverse strike is a follow-up technique to a forward strike. It allows Marines both a secondary attack and the ability to resume the basic warrior stance.To execute a reverse strike, Marines—

Step forward with the right foot in the direction of the strike.

Bend the right arm with the hand near the left shoulder. The weapon is extended over the left shoulder.

Rotate the forearm to the right of the elbow to bring the weapon down onto the opponent. At the same time, forcefully rotate the hips and shoulders toward the opponent.

Follow through with the strike by allowing the weight of the weapon to go through the target area of the body.

**Forward Thrust.** To execute the forward thrust, Marines—

Grasp the stick with the left hand, palm up, in a position where the stick can be controlled with two hands.

Lift the left leg and lunge forward off the ball of the right foot. At the same time, thrust the end of the weapon directly toward the opponent by thrusting both hands forward in a straight line.

## 6. Blocking Techniques

A block is meant to deter or deflect an attack by an opponent. A block sets up Marines for a follow-on attack against the opponent. Blocks are executed by deflecting, rather than hitting or following through like a strike.

### Blocks Against Unarmed Attacks

To block against an unarmed attack, Marines—

Step forward at a 45-degree angle with either the right or left foot. This moves the body out of the line of attack.

Raise the left arm and block the strike with the meaty portion of the forearm.

Employ the stick by one of two techniques:

- Strike the opponent with the stick.
- Use the stick to block on two points of contact. When the stick is used to block, it serves as an incidental strike. This technique is only used if Marines have closed with the opponent and are inside his strike.

## Blocks Against Armed Attacks

**Block for a Vertical Strike.** To execute the block for a vertical strike, Marines—

Step forward with the left foot at a 45-degree angle to the left. This moves the body out of the line of attack.

Block on two points of contact to disperse the impact of the attack:

- Block the opponent's stick by positioning the stick so it is perpendicular to the opponent's stick. If the stick is not perpendicular to the opponent's stick, the stick can slide through and make contact on the Marine.
- Block the opponent's wrist or forearm with the back of the left forearm.

*Note: Use the stick to block the opponent's arm if closer to the opponent. It is the same movement as blocking with the arm, except the opponent's arm is blocked with both the stick and the arm.*

**Block for a Forward Strike.** To execute the block for a forward strike, Marines-

Step forward with the left foot at a 45-degree angle to the left. This moves the body out of the line of attack and inside the opponent's strike.

Block on two points of contact to disperse the impact of the attack:

- Block the opponent's wrist or forearm with the meaty portion of the back of the left forearm.
- Strike the opponent's attacking biceps with the stick.

**Block for a Reverse Strike.** To execute the block for a reverse strike, Marines—

Step forward at a 45-degree angle with the right foot to the right. This moves the body out of the line of attack and inside the opponent's strike.

Block on two points of contact to disperse the impact of the attack:

Block the opponent's stick by positioning the stick so it is perpendicular to the opponent's stick.

Block the opponent's forearm with the meaty portion of the left forearm.

*Note: If closer to the opponent, block his triceps with the back of the left forearm and strike his forearm with the stick.*

## 7. Unarmed Against Hand-Held Weapons

If Marines are engaged against an opponent with a knife, a stick, or some other weapon of opportunity, they must establish and maintain an offensive mindset, not a defensive mindset. Their survival depends on it. They cannot afford to think about getting cut or hurt.

### Angles of Attack

Before Marines can learn to block or counter an attack with a hand-held weapon (e.g., knife, stick), they must know from what angle the opponent is attacking. There are six angles from which an opponent will typically attack with a hand-held weapon:

Vertical strike coming in straight down on the Marine.
Forward diagonal strike coming in at a 45-degree angle to the Marine.
Reverse diagonal strike coming in at a 45-degree angle to the Marine.
Forward horizontal strike coming in parallel to the ground.
Reverse horizontal strike coming in parallel to the ground.
Forward thrust coming in a straight line to the Marine.

### Blocks

If the opponent has a hand-held weapon (e.g., knife or stick), Marines parry the opponent's hand or arm to block the attack.

**Basic Block.** To execute the basic block technique, Marines-

Step forward at a 45-degree angle to move out of the line of the attack. Always step in the direction of the strike.

Thrust the forearms forward, hands up, against the opponent's attacking arm. Contact is made on the opponent's arm with the backs of the fore-arms.

**Block for a Vertical Strike.** To execute the block against a vertical strike, Marines—

Step forward with the left foot at a 45-degree angle to the left to move out of the line of attack.

Thrust the forearms forward, hands up, against the outside of the opponent's attacking arm.

**Block for a Forward Strike.** To execute the block against a forward diagonal or forward horizontal strike, Marines-

Step forward with the left fool inside the opponent's attacking arm.

Block the attack with both arms bent so the forearms make contact with the opponent's biceps and forearm.

**Block for a Reverse Strike.** To execute the block against a reverse diagonal or reverse horizontal strike, Marines—

Step forward with the right foot to the outside of the opponent's attacking arm.

Block the attack with both arms bent so the forearms make contact with the opponent's triceps and forearm.

**Block for a Forward Thrust.** To execute the block against a forward thrust, Marines—

Bend at the waist, move the hips backward, and jump backward with both feet to move away from the attack. This action is known as "hollowing out."

Hollow out and block the attack with the arms bent and hands together on top of the attacking arm.

Overlap the hands slightly so one thumb is on top of the other hand's index finger. The other thumb should be under the other hand's index finger.

## 8. Counters to Hand-Held Weapon Attacks

### Principles of Counters

A counter is used to control the situation to regain the tactical advantage and end the fight. Regardless of the type of weapon or angle of attack, the following principles apply to countering the attack with a hand-held weapon:

Move out of the line of attack. Movement is executed in a 45-degree angle forward to the left or right.

- Block the attack.

*Note: The first two actions are taken simultaneously.*

- Take control of the weapon by controlling the hand or arm that is holding the weapon. Never attempt to grab the opponent's weapon.
- Execute the appropriate follow-up to end the fight; e.g., strikes, joint manipulations, throws, or takedowns (see chap. 8). Marines should continue to attack the opponent until the fight ends.

### Counter Techniques

There are two techniques that can be used to counter any armed attack: forward armbar counter and reverse armbar counter. These techniques can be used to counter a vertical attack, a forward diagonal strike, or a forward horizontal strike. With minor variations, the same techniques are used to counter reverse strikes. A third technique, the bent armbar counter, is used to counter a vertical attack.

**Forward Armbar Counter.** To execute the forward armbar counter to an attack coming from a forward strike, Marines-

Step forward with the left foot inside the opponent's attacking arm.

Block the attack with both arms bent so the forearms make contact with the opponent's biceps and forearm.

Slide the left arm over the opponent's forearm and wrap the arm tightly around his arm, trapping the opponent's attacking arm between the biceps and torso.

Place the right hand on the opponent's shoulder or upper arm to further control his arm and to effect an armbar.

Execute an armbar and continue to exert steady pressure against the arm.

**Reverse Armbar Counter.** To execute the reverse armbar counter to an attack coming from a forward strike, Marines-

Step forward with the left foot inside the opponent' s attacking arm.

Block the attack with both arms bent so the forearms make contact with the opponent's biceps and forearm.

Control the opponent's arm with the left hand and pivot to the right so the back is against the opponent's side. Immediately slide the right arm over the opponent's biceps and wrap the arm tightly around his arm, trapping the opponent's attacking arm between the biceps and torso.

Grasp the opponent's wrist with the left hand and twist his thumb away from the body.

Take the opponent to the ground with an armbar takedown if he does not drop the weapon from the wristlock.

**Bent Armbar Counter.** This counter is particularly effective against a vertical attack. To execute the bent armbar counter, Marines—

Step forward with the left foot inside the opponent's attacking arm.

Block the attack with both arms bent so the forearms make contact with the opponent's biceps and forearm.

Grasp the opponent's forearm with the left hand. At the same time, slide the right arm underneath the opponent's triceps and grasp the opponent's forearm or wrist with the right hand.

Apply downward pressure with the hands against the opponent's forearm to off-balance the opponent.

# CHAPTER 4

# STRIKES

***This chapter describes all techniques for a right-lmnded person. However, all techniques can be executed from either side.***

***In drawings, the Marine is depicted in woodland camouflage utilities; the opponent is depicted without camouflage. In photographs, the Marine is depicted in woodland camouflage utilities; the opponent is depicted in desert camouflage utilities.***

Strikes are unarmed individual hitting techniques. Strikes use the hands, elbows, knees, feet, and, in some instances, other parts of the body as personal weapons. Marines must know how to execute strikes effectively. They must also know how to counter strikes from an opponent.

## 1. Principles of Punches

### Muscle Relaxation

Muscle relaxation is crucial when executing punches. The natural tendency in a fight is to tense up, which results in rapid fatigue and decreased power generation. Marines who remain relaxed during a close combat situation generate greater speed, which results in greater generation of power. Relaxing the forearms generates speed and improves reaction time. At the point of impact, Marines clench the fist to cause damage to the opponent and avoid injury to the wrist and hand.

### Weight Transfer

Weight transfer is necessary to generate power in a punch. Marines accomplish this by-

Rotating their hips and shoulders into the attack.

Moving their body mass straight forward or backward in a straight line.

Dropping their body weight into an opponent. Body mass can be transferred into an attack from high to low or from low to high.

### Rapid Retraction

When Marines deliver a punch, rapid retraction of the fist is important. Once the hand has made contact with the target, Marines quickly return to the basic warrior stance. Rapid retraction-

Returns the hand and arm to the protection afforded by the basic warrior stance.

Prevents the opponent from being able to grab the hand or arm.

Permits the hand and arm to be "chambered" or "re-cocked" in preparation for delivering a subsequent punch.

### Telegraphing

Telegraphing a strike occurs when body movements inform the opponent of the intention to launch an attack. Staying relaxed helps to reduce telegraphing.

Often, an untrained fighter telegraphs his intention to attack by drawing his hand back in view of his opponent, changing facial expression, tensing neck muscles, or twitching. These movements, however small, immediately indicate an attack is about to be delivered. If the opponent is a trained fighter, he may be able to evade or counter the attack. If the opponent is an untrained fighter, he may still be able to minimize the effect of an attack.

## 2. Punches

Punches may be thrown during any hand-to-hand confrontation. Most people resort to punching because it is a natural reaction to a threat. The purpose of a punch is to stun the opponent or to set him up for a follow-up finishing technique. However, punches should only be executed to the soft tissue areas of an opponent. A correctly delivered punch maximizes the damage to an opponent while minimizing the risk of injury to Marines.

### Basic Fist

Punches are executed using the basic fist. To make the basic fist, the fingers are curled naturally into the palm of the hand and the thumb is placed across the index and middle fingers. Do not clench the fist until movement has begun. This reduces muscular tension in the forearm and increases speed and reaction time. Just before impact, Marines exert muscular tension on the hand and forearm to maximize damage to the opponent and reduce their chances of injury. Contact should be made with the knuckles of the index and middle fingers.

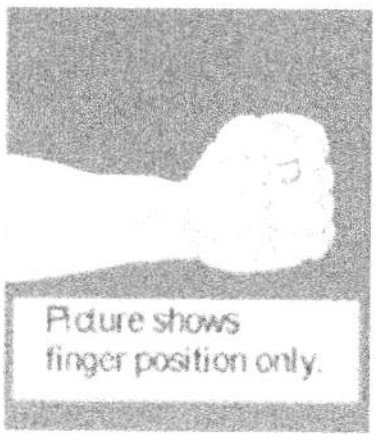
Picture shows finger position only.

When striking with the basic fist, Marines must keep the hand straight, or in line, with the wrist to avoid injury to the wrist.

### Lead Hand Punch

The lead hand punch is a snapping straight punch executed by the forward or lead hand. It is a fast punch designed to keep the opponent away and to set up an attack. A lead hand punch conceals movement and allows Marines to get close to the opponent. Lead hand punches should strike soft tissue areas, if possible. To execute the lead hand punch, Marines-

Snap the lead hand out to nearly full extension, while rotating the palm to the ground.

Contact the opponent with the first two knuckles of the fist.

Retract the hand immediately, resuming the basic warrior stance.

### Rear Hand Punch

The rear hand punch is a snapping punch executed by the rear (right) hand. It is a power punch designed to inflict maximum damage on the opponent. Its power comes from pushing off the rear leg and rotating the hips and shoulders. To execute the rear hand punch, Marines —

Rotate the hips and shoulders forcefully toward the opponent and thrust the rear hand straight out, rotating the palm down, to nearly full extension.

Shift body weight to the lead foot while pushing off on the ball of the rear foot.

Contact the opponent with the first two knuckles of the fist.

Retract the hand immediately.

## Uppercut

The uppercut is a powerful punch originating below the opponent's line of vision. It is executed in an upward motion traveling up the centerline of the opponent's body. It is delivered in close and usually follows a preparatory strike that leaves the target area unprotected. When delivered to the chin or jaw, the uppercut can render an opponent unconscious, cause extensive damage to the neck, or sever the tongue. To execute the uppercut, Marines—

Bend the arms, rotating the palm inboard. The distance the arms bend depends on how close the opponent is.

Rotate the hips and shoulders forcefully toward the opponent, thrusting the fist straight up toward the opponent's chin or jaw.

Contact the opponent with the first two knuckles of the fist.

Retract the hand immediately.

## Hook

The hook is a powerful punch that is executed close in and is usually preceded by a preparatory strike. To execute the hook, Marines—

Thrust the right arm in a hooking motion toward the opponent, keeping the elbow bent while forcefully rotating the right shoulder and hip toward the opponent.

Contact the opponent with the first two knuckles of the fist. Continue rotating the shoulder and hip, following through with the fist to the target.

Retract the hand immediately.

## 3. Strikes with the Upper Body

Strikes stun the opponent or set him up for a follow-up finishing technique. The hands, forearms, and elbows are individual weapons of the arms that can be used to execute strikes including the hammer fist, chin jab, knife hand, eye gouge, and elbow strikes. These strikes provide a variety of techniques that can be used in any type of close combat situation.

### Principles of Execution

There are several principles of execution that ensure a strike's effectiveness.

**Generating Power.** Weight transfer is necessary to generate maximum power in a punch. Marines accomplish this by-

- Rotating the hips and shoulders into the attack.
- Moving the body mass straight forward or backward in a straight line.
- Dropping body weight into an opponent. Body mass can be transferred into an attack from high to low or from low to high.

**Muscular Tension.** The arms are relaxed until the moment of impact. At the point of impact, Marines apply muscular tension in the hand and forearm to maximize damage to the opponent and to avoid injury to the hand. The arms are relaxed until the moment of impact.

**Hit and Stick and Follow-Through.** A strike should be delivered so that the weapon (e.g., hand, elbow) hits and remains on the impact site (target) and follows through the target. This technique inflicts maximum damage on the opponent.

Strikes with the arms are executed with "heavy hands"; i.e., the strike is executed by driving through with the strike to allow the weight of the hand to go through the target area of the body. Contact with the opponent should be made with the arm slightly bent, and the arm extends as it moves through the target. This technique allows Marines to deliver strikes effectively without executing full force.

**Movement.** Movement puts Marines in the proper position for launching an attack against an opponent as well as providing protection.

Movement is initiated from the basic warrior stance and ends by resuming the basic warrior stance. Strikes can be performed with either the left or right arm depending upon—

The angle of attack.
The position of the opponent.
The opponent's available, vulnerable target areas.

## Target Areas of the Body

For each strike, there are target areas of the body that, when struck, maximize damage to an opponent. Strikes use gross motor skills as opposed to fine motor skills. The target areas of the body are just that—areas. Pinpoint accuracy on a specific nerve is not needed for the strike to be effective.

## Hammer Fist

Striking with the hammer fist concentrates power in a small part of the hand, which, when transferred to the target, can have a devastating effect. The striking surface of the hammer fist is the meaty portion of the hand below the little finger. To execute the hammer fist strike, Marines-

Make a fist and bend the arm at approximately a 45- to 90-degree angle. At the same time, rotate the right hip and right shoulder backward.

Thrust the fist forward onto the opponent while rotating the right hip and shoulder forward.

Rotate the wrist so the hammer fist makes contact on the opponent.

## Chin Jab

The chin jab can immediately render an opponent unconscious and cause extensive damage to the neck and spine. The striking surface is the heel of the palm of the hand. To execute the chin jab, Marines—

Bend the right wrist back at a 90-degree angle with the palm facing the opponent and the fingers pointing up.

Keep the right arm bent and close to the body. Extend the hand into a concave position with the fingers slightly spread apart.

Step forward with the left foot toward the opponent, keeping the feet approximately shoulder-width apart and the knees bent. This is done to close with the opponent.

Keep the right arm bent and close to the side. Thrust the palm of the hand directly up under the opponent's chin. At the same time, rotate the right hip forward to drive the body weight into the attack to increase the power of the strike. The attack should travel up the centerline of the opponent's chest to his chin.

## Knife Hand

The knife hand is one of the most versatile and devastating strikes. The striking surface is the cutting edge of the hand, which is the meaty portion of the hand below the little finger extending to the top of the wrist. The striking surface is narrow, allowing strikes on the neck between the opponent's body armor and helmet. The knife hand strike is executed from one of three angles: outside, inside, and vertical.

**Outside Knife Hand.** To execute the outside knife hand strike, Marines—

Execute a knife hand by extending and joining the fingers of the right hand and placing the thumb next to the forefinger (like saluting).

Retract the right hand. At the same time, rotate the right hip and right shoulder backward.

Thrust the knife hand forward (horizontally) onto the opponent while rotating the right hip and shoulder forward.

**Inside Knife Hand.** To execute the inside knife hand strike, Marines—

Execute a knife hand.

Bring the right hand over the left shoulder. At the same time, rotate the right shoulder forward and the left hip forward.

Thrust the knife hand forward (horizontally) onto the opponent while rotating the right hip and shoulder forward and the left shoulder backward.

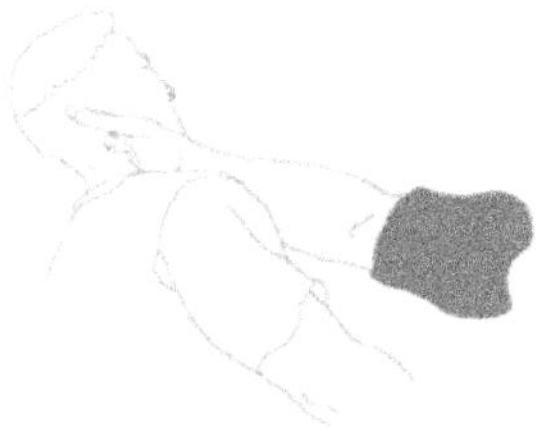

**Vertical Knife Hand.** When thrown vertically, the knife hand strike comes straight down in a straight line.

## Eye Gouge

The eye gouge is used to attack an opponent's eyes, blinding him so follow-up strikes can be executed. The striking surface is the tips of the fingers or thumb. To execute the eye gouge, Marines—

Extend the right hand with the fingers slightly spread apart to allow entry into the eye sockets.

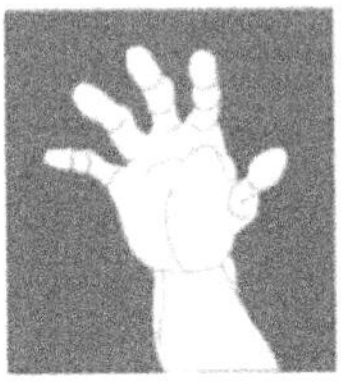

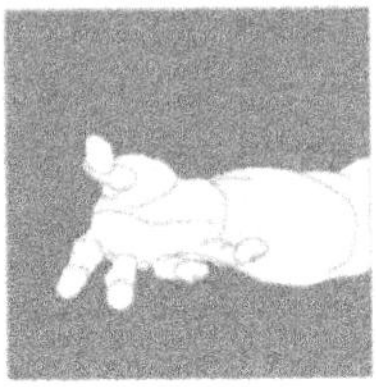

Place the palm of the hand either toward the ground or toward the sky and thrust the right hand forward into the opponent's eyes.

Thrust the hand forward at the opponent's nose level so the fingers or thumb slide naturally into the grooves of the opponent's eye sockets.

## Elbow Strikes

The elbow is a powerful weapon that can be used in several different ways to attack virtually any part of an opponent's body. Elbow strikes can be performed either vertically (upward or downward) or horizontally (forward or reverse). The striking surface is 2 inches above or below the point of the elbow, depending upon the angle of attack, the opponent's attack angle, and the position of the opponent.

Above Elbow

Below Elbow

**Vertical Elbow Strike (Up).** To execute an upward vertical elbow strike, Marines-

Bend the right elbow, keeping the fist close to the body. The fist is at shoulder level and the elbow is next to the torso.

Thrust the elbow vertically upward toward the opponent while rotating the right shoulder and hip forward to generate additional power.

Contact the opponent with the right forearm 2 inches above the point of the elbow.

**Vertical Elbow Strike (Down).** To execute a downward vertical elbow strike, Marines—

Bend the right elbow, keeping the fist close to the body. The fist is on the shoulder and the elbow is raised well above the shoulder.

Thrust the elbow vertically downward toward the opponent while dropping body weight into the attack to generate additional power.

Contact the opponent with the right triceps 2 inches above the point of the elbow.

**Horizontal Elbow Strike (Forward).** To execute the forward horizontal elbow strike, Marines-

Tuck the right fist near the chest with the palm heel facing the ground.

Thrust the right elbow horizontally forward toward the opponent. The forearm is parallel to the ground.

Rotate the right shoulder and hip forward to generate additional power.

Contact the opponent with the right forearm 2 inches below the point of the elbow.

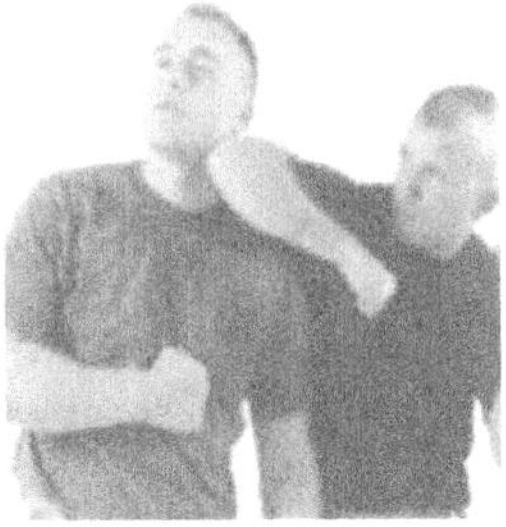

**Horizontal Elbow Strike (Rear).** To execute the rear horizontal elbow strike, Marines-

Tuck the right fist near the left shoulder with the palm heel facing the ground. At the same time, rotate the right shoulder forward and the left hip forward.

Thrust the right elbow horizontally rearward toward the opponent. The forearm is parallel to the ground and the hand moves toward the direction of the attack.

Rotate the right hip back and the right shoulder backward to generate additional power.

Contact the opponent with the right triceps 2 inches above the point of the elbow.

## 4. Strikes with the Lower Body

The legs are the body's most powerful weapons because they use the largest muscle groups to generate a strike. Legs are also less prone to injury. The feet are the preferred choice for striking because they are protected by boots. Marines use their feet, heels, and knees to execute kicks, knee strikes, and stomps.

### Kicks

The purpose of a kick is to stop an opponent's attack or to create an opening in his defense in order to launch an attack. Kicks can be performed with the left (lead) leg or the right (rear) leg. Kicks with the rear leg have greater power because the hips are rotated into the attack. However, the rear leg is further away from the opponent so a strike with the rear leg will not contact the opponent as quickly as a strike with the lead leg.

**Front Kick.** The front kick is executed when the opponent is in front of the Marine. The front kick, delivered with either the rear or lead leg, is effective for striking below the waist. Attempting to kick higher results in diminished balance and provides the enemy with a greater opportunity to grab the leg or foot. The striking surfaces are the toe of the boot or the bootlaces. To execute the front kick, Marines-

Raise the left knee waist high, pivot the hips into the attack, and thrust the left foot forward toward the opponent.

Contact the opponent with the toe of the left boot or bootlaces.

Return to the basic warrior stance.

**Side Kick.** The side kick, delivered with the lead leg, is effective for striking the knees. The side kick is executed when the opponent is to the side of the Marine. The striking surface is the outside cutting edge of the boot near the heel. To execute the side kick, Marines—

Raise the right knee waist high and rotate the right hip forward.

Thrust the right foot to the right side toward the opponent, turning the foot at a 90-degree angle to maximize the striking surface on the opponent.

Contact the opponent with the cutting edge of the right boot.

Return to the basic warrior stance.

## Knee Strikes

Knee strikes are excellent weapons during the close range of close combat fighting. The knee strike is generally delivered in close.

**Vertical Knee Strike.** The striking surface is the thigh, slightly above the knee. To execute the vertical knee strike, Marines-

Raise the right knee and drive it up forcefully into the opponent. Power is generated by thrusting the leg upward.

Contact the opponent 2 inches above the right knee.

**Horizontal Knee Strike.** The horizontal knee strike is executed with the leg generally parallel with the ground while rotating the hips to generate power. It is often delivered to the peroneal nerve. The striking surface is the front of the leg, slightly above or below the knee. To execute the horizontal knee strike, Marines-

Raise the right knee, rotate the right hip forward while pivoting on the left foot, and drive the knee horizontally into the opponent.

Contact the opponent 2 inches above the right knee.

## Stomps

Stomps are delivered with the feet, usually when the opponent is down. Remember, when the opponent is down, Marines take whatever target is available.

**Vertical Stomp.** The vertical stomp allows Marines to remain upright and balanced, to rapidly deliver multiple blows with either foot, and to quickly and accurately attack the target. It is the preferred stomp. The striking surface is the flat bottom of the boot or the cutting edge of the heel. To execute the vertical stomp, Marines—

Raise the right knee above the waist with the right leg bent at approximately a 90-degree angle.

Drive the flat bottom of the right boot or the cutting edge of the right heel down onto the opponent forcefully. At the same time, bend the left knee slightly to drop the body weight into the strike.

**Ax Stomp.** The striking surface of the ax stomp is the cutting edge of the heel. To execute the ax stomp, Marines-

Raise the right heel above the waist, keeping the right leg straight.

Drive the cutting edge of the right heel down onto the opponent forcefully. At the same time, bend the left knee slightly to drop the body weight into the strike.

# 5. Counters to Strikes

In a close combat situation, an opponent will attempt to strike Marines with punches and kicks. When an opponent uses a strike, Marines must first avoid the strike. This is accomplished by moving quickly and blocking. Next, Marines must get into an offensive position. This allows Marines to use offensive strikes to attack the opponent. Regardless of the strike, the counter to a strike requires Marines to move, block, and strike.

## Move

The first step in countering a strike is to move out of the way of the strike's impact. Movement removes Marines from the opponent's intended strike point and positions Marines to attack. Movement is executed at approximately a 45-degree angle to the front or rear. Movement is always initiated from the basic warrior stance. Return to the basic warrior stance with the toe of the lead foot pointing toward the opponent once the movement is complete.

## Block

Different blocks are executed based on the strike. These will be covered by individual counters.

## Strike

Any of the upper body or lower body strikes can be executed as a follow-on attack or part of the counter to an opponent's strike. The follow-on strike is determined by the angle to the opponent, the position of the opponent, and the opponent's available vulnerable target areas.

## Counters to Punches

**Counter to a Lead Hand Punch.** This counter is used when the opponent throws a lead hand punch. To execute the counter to the lead hand punch, Marines-

Step forward and to the left at approximately a 45-degree angle, moving in to the outside of the opponent's attacking arm.

Raise the left arm and block or deflect the opponent's lead hand with the palm of the hand or the meaty portion of the forearm.

Hit and stick by leaving the left arm against the opponent's right arm while stepping forward and to the left at approximately a 45-degree angle to close with the opponent.

Execute a finishing technique, such as a strike or a kick, to the opponent's exposed target areas.

**Counter to a Rear Hand Punch.** This counter is used when the opponent throws a rear hand punch. To execute the counter to the rear hand punch, Marines—

Step forward and to the left at approximately a 45-degree angle, moving in to the outside of the opponent's attacking arm.

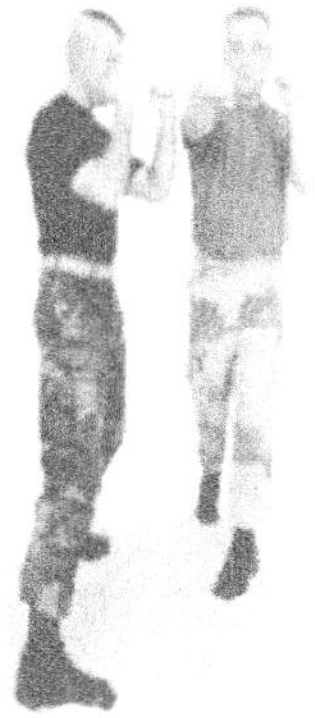

Raise the left arm and block or deflect the opponent's rear hand with the palm of the hand or the meaty portion of the forearm.

Hit and stick by leaving the left arm against the opponent's right arm while stepping forward and to the right at approximately a 45-degree angle to close with the opponent.

Execute a finishing technique, such as a strike or a kick, to the opponent's exposed target areas.

## Counters to Kicks

**Counter to a Front Kick (Left or Lead Leg).** This counter is used when the opponent executes a front kick with his left leg. To execute the counter to a front kick, Marines-

Step forward and to the right at approximately a 45-degree angle, moving in to the outside of the opponent's striking leg.

Lower the left arm and block or deflect the opponent's leg with the palm of the hand or the meaty portion of the forearm.

Hit and stick by leaving the left arm against the opponent's leg while stepping forward and to the left at approximately a 45-degree angle to close with the opponent.

Execute a finishing technique, such as a strike or a kick, to the opponent's exposed target areas.

**Counter to a Front Kick (Right or Rear Leg).** This counter is used when the opponent executes a front kick with his right leg. To execute the counter to a front kick, Marines—

Step forward and to the left at approximately a 45-degree angle, moving in to the outside of the opponent's striking leg.

Lower the left arm and block or deflect the opponent's leg with the palm of the hand or the meaty portion of the forearm.

Hit and stick by leaving the left arm against the opponent's leg while stepping forward and to the right at approximately a 45-degree angle to close with the opponent.

# CHAPTER 5

# THROWS

***This chapter describes all techniques for a right-handed person. However, all techniques can be executed from either side.***

***The Marine is depicted in camouflage utilities. The opponent is depicted without camouflage.***

Marines use throwing techniques to maintain the tactical advantage and to throw the opponent to the ground during close combat. Throws apply the principles of balance, leverage, timing, and body position to upset an opponent's balance and to gain control by forcing the opponent to the ground. Throwing techniques are effective because they are size- and gender-neutral, and they rely on the momentum and power generated by the opponent rather than the strength or size of the Marine. Marines also execute a throw as a devastating attack against an opponent, possibly causing unconsciousness or broken limbs. When Marines execute throws, they must maintain balance and, simultaneously, prevent the opponent from countering a throw or escaping after being forced to the ground.

## 1. Turning Throw

Marines use a turning throw to take the opponent to the ground while they remain standing. A turning throw can also be executed from a stationary position. It is particularly effective if Marines and the opponent are wearing gear. To execute the turning throw, Marines-

Grasp the opponent's right wrist with the left hand.

Step forward with the right foot, place it against the outside of the opponent's right foot, and pivot so the back of the heel is next to the middle of the opponent's foot.

Hook the opponent's right arm with the right arm and pinch his arm between the biceps and forearm, touching the opponent with the body.

Pull the opponent's wrist downward, keeping it close to the body.

Pivot to the left on the ball of the foot and continue pulling downward on the opponent's wrist while rotating the wrist outward to off-balance him.

## 2. Hip Throw

Marines use a hip throw to take the opponent to the ground while they remain standing. A hip throw is particularly effective if the opponent is moving forward or pushing. Marines use the opponent' s forward momentum to execute the hip throw. To execute the hip throw, Marines-Grasp the opponent's right wrist with the left hand.

Step forward with the right foot and place it against the outside of the opponent's right foot.

The back of the heel should be next to the opponent' s foot.

Step back with the left foot and rotate on the ball of the foot. The back of the heel is next to the opponent's toe. The knees are bent.

Rotate at the waist and hook the right hand around the back of the opponent's body (anywhere from his waist to his head). The side and hip should be against the opponent.

Rotate the hip against the opponent. The hips must be lower than the opponent's.

Use the right hand to pull the opponent up on the hip to maximize contact.

Pull the opponent's arm across the body and, at the same time, lift the opponent off the ground slightly by bending at the waist, straightening the legs, and rotating the body to the left.

## 3. Leg Sweep

Marines use a leg sweep to take the opponent to the ground while they remain standing. A leg sweep is particularly effective if the opponent is already off-balanced and moving backward or pulling on the Marine. To execute the leg sweep, Marines—

Grab the opponent's right wrist with the left hand and grab the opponent's left shoulder with the right hand.

*Note: Grab the opponent's clothing or gear if his wrist and shoulder cannot be grabbed.*

Step forward with the left foot and place it on the outside of the opponent's right foot.

Pull the opponent's wrist downward, close to the body, and push his shoulder backward to off-balance him.

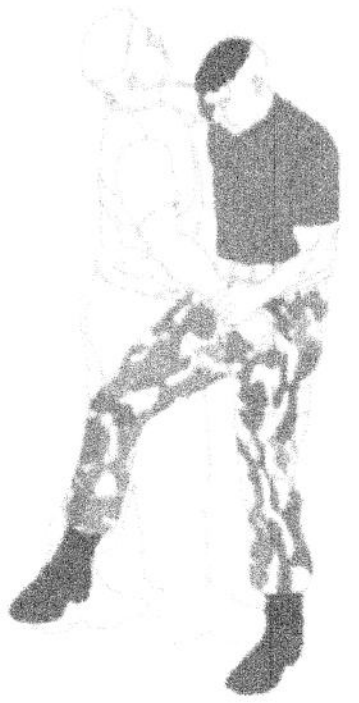

Raise the right knee no higher than the waist.

Kick the foot past the opponent's right leg.

Use the heel of the boot to make contact with the opponent's calf (anywhere from the top of the calf down to the Achilles tendon or on the inside of the calf).

Sweep the opponent to the ground.

# CHAPTER 6

# CHOKES AND HOLDS

*This chapter describes all techniques for a right-handed person. However, all techniques can be executed from either side.*

*The Marine is depicted in camouflage utilities. The opponent is depicted without camouflage.*

WARNING

**During training, never execute a choke at full force or full speed and never hold a choke for more than 5 seconds.**

When Marines correctly perform a choke, they render an opponent unconscious in as little as 8 to 13 seconds. Chokes are easily performed regardless of size or gender. Marines must know how to apply chokes and how to counter a choke or a hold executed by an opponent.

## 1. Types of Chokes

There are two types of chokes: an air choke and a blood choke. An air choke closes off the airway to the lungs, thereby preventing oxygen from reaching the heart. A blood choke cuts off the blood flow to the brain. Both types can result in unconsciousness and eventual death for an opponent.

### Air Choke

An air choke is performed on the opponent's windpipe (or trachea), cutting off the air to the lungs and heart. If Marines execute the air choke properly, the opponent loses consciousness within 2 to 3 minutes. Due to the length of time it takes to immobilize the opponent, air chokes are not recommended.

### Blood Choke

A blood choke is performed on the opponent's carotid artery, which carries oxygen-enriched blood from the heart to the brain. The carotid artery is located on both sides of the neck. If Marines execute a blood choke properly, the opponent will lose consciousness within 8 to 13 seconds. The blood choke is the preferred choke because its intended effect (i.e., the opponent losing consciousness) can be executed quickly, ending the fight.

## 2. Chokes

### Front Choke

Marines execute a front choke when they are facing the opponent. Marines use the opponent's lapels or collar to execute a front choke. To execute the front choke, Marines-

Grab the opponent's right lapel with the right hand, making certain that the knuckles or the back of the hand are against the opponent's carotid artery on the right side of his neck.

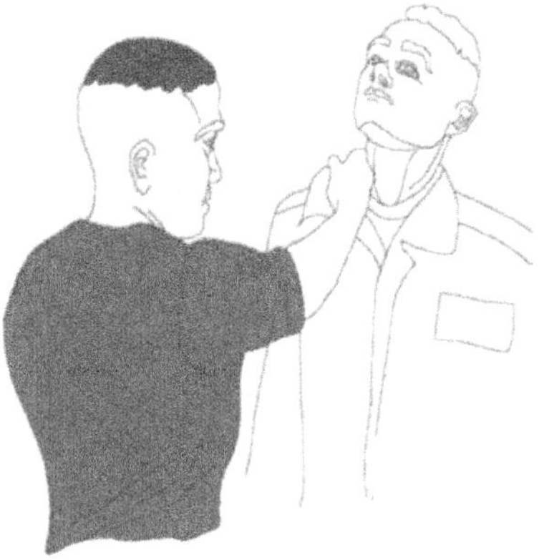

Keep the right hand against the opponent's neck, reach under the opponent's right arm with the left hand, grab the opponent's left lapel, and form an X with the wrists.

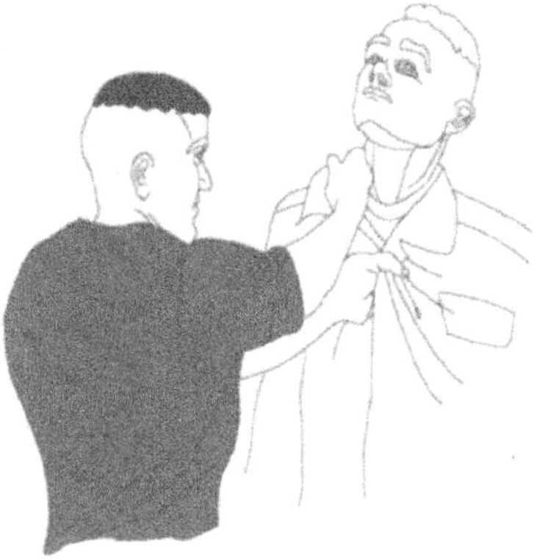

Press the right hand against the opponent's carotid artery.

Pull his lapel to the left with the left hand.

## Side Choke

Marines execute a side choke when they are facing the opponent. The side choke is particularly effective when deflecting a punch thrown by an opponent. To execute the side choke, Marines—

Use the left hand to parry the opponent's arm inboard (to the inside of the opponent's reach).

Bring the right arm underneath the opponent's arm and up around the front of his neck.

Extend the fingers, place the back of the forearm against the opponent's neck just below his ear, and press the carotid artery on his neck.

*Note: If Marines are unable to place their fingers against the opponent's neck, they may use the back of the thumb or wrist.*

Reach, with the left hand, around the back of the opponent's neck and clasp the hands together.

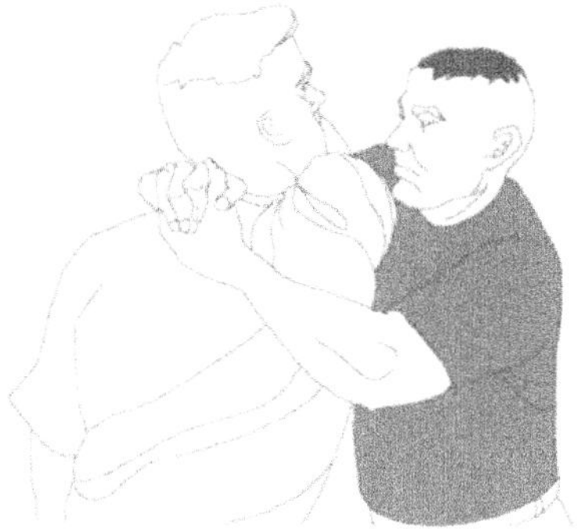

Pull the opponent toward the chest by pulling the clasped hands toward the chest.

Exert pressure on the side of the opponent's neck with the forearm.

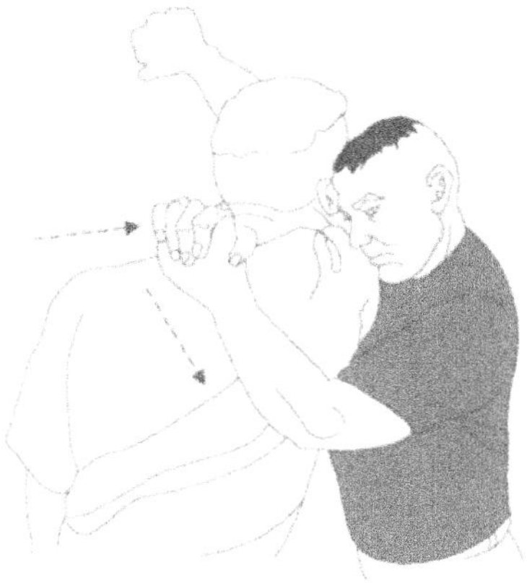

## Rear Choke

Marines execute a rear choke when they are behind the opponent, the opponent is on the ground, or when they are taking the opponent to the ground. To execute the rear choke, Marines—

Reach, with the right arm, over the opponent's right shoulder and hook the bend of the arm around his neck.

Clasp both hands together.

Exert pressure with the biceps and forearms on both sides of the opponent's neck on his carotid arteries.

Maintain pressure with the biceps and forearms on both sides of the neck and draw the opponent closer by drawing the right arm in.

## Figure-Four Choke

A variation of the rear choke is the figure-four choke. The figure-four choke allows Marines to gain more leverage than the rear choke. If the rear choke cannot be secured, the figure-four variation is employed to increase the pressure of the choke on the opponent. To execute the figure-four variation of the rear choke, Marines—

Apply a rear choke. The Marine's body should be against the opponent's body.

Grasp the left biceps with the right hand and place the left hand against the back of the opponent's head.

Push the opponent's head forward and down with the left hand.

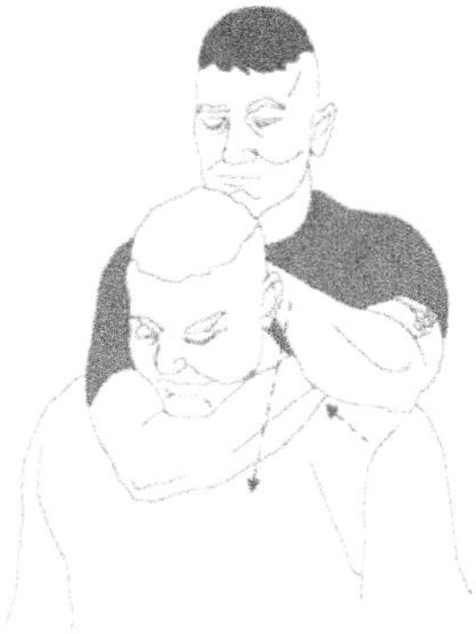

Draw the right arm in, maintaining pressure with the bicep and forearm on both sides of the opponent's neck.

## 3. Counters to Chokes and Holds

During a close combat situation, an opponent may apply a choke or hold on a Marine. If the opponent correctly applies a choke, a Marine quickly loses consciousness. If a choke is not executed properly, it often results in a hold, typically a bear hug or a headlock. A hold allows the opponent to control a Marine and removes the Marine's ability to attack. It is important for Marines to extract themselves from chokes and holds, regain the tactical advantage, and counter with strikes.

Although a choke causes unconsciousness in 8 to 13 seconds for a blood choke and 2 to 3 minutes for an air choke, the first movement in any counter to a choke is to clear the airway. Marines use softening techniques to loosen an opponent's grip and to clear their airway. Softening techniques are particularly effective if Marines lack the physical strength of their opponent. These techniques include groin strikes, eye gouges, foot stomps, etc. Softening techniques are not offensive; rather, they are used to loosen an opponent's hold.

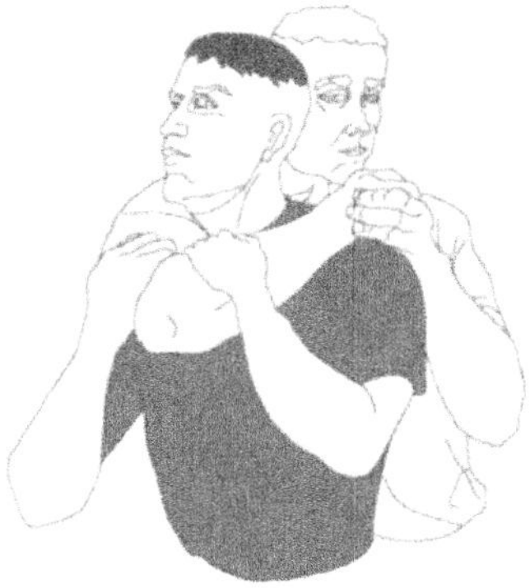

The second movement is to tuck the chin. Once the airway is clear, Marines tuck their chins to prevent the opponent from reapplying the choke.

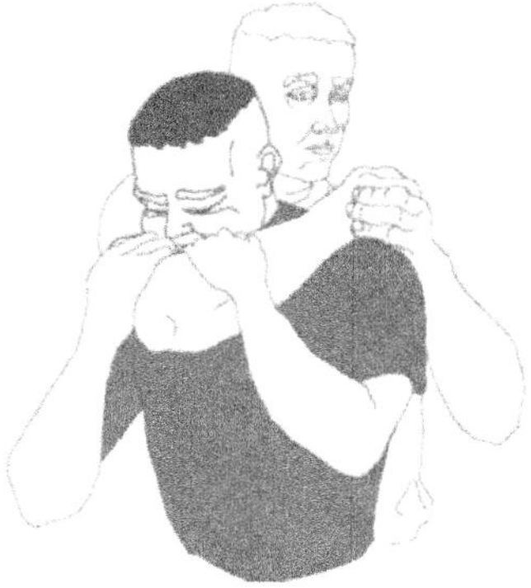

### Counter to a Front Choke

Marines use a counter to a front choke when the opponent approaches from the front and uses both hands to choke a Marine around the throat. To execute the counter to the front choke, Marines—

Grasp the opponent's right forearm (where the elbow bends) with the left hand and apply downward pressure on the opponent's radial nerve with the fingers.

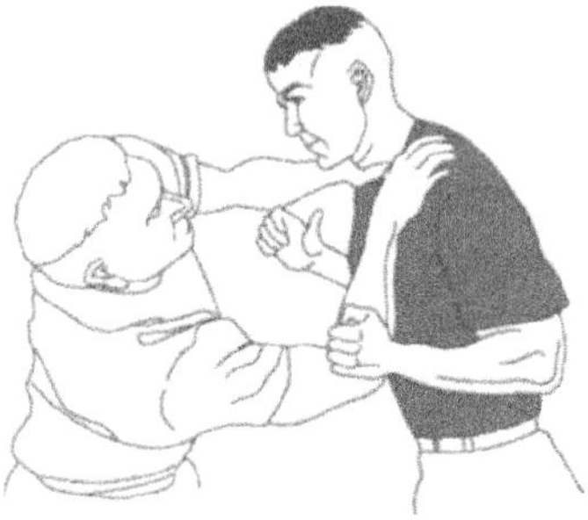

Execute a chin jab to the opponent's chin with the right hand. To generate power into the strike, bring the left foot to the outside of the opponent's right foot and rotate the hips into the strike.

## Counter to a Rear Choke

Marines execute a counter to a rear choke when the opponent approaches from the rear and puts his right arm around a Marine's throat. To execute the counter to the rear choke, Marines-

Grasp the opponent's forearm (at the radial nerve) and bicep with both hands and pull down just enough to clear the airway. Once the airway is clear, tuck the chin to protect the airway and to prevent the opponent from reapplying the choke.

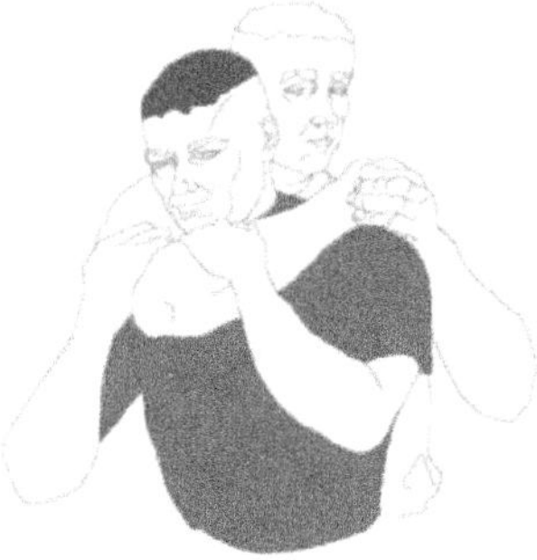

Step behind the opponent's right leg with the left foot, keeping both legs bent (almost in a squatting position).

Strike and drive the left elbow into the opponent's torso while rotating the hips and pivoting to the left, throwing the opponent back and to the ground.

## Counter to a Front Headlock

Marines use a counter to a front headlock when the opponent approaches from the front and puts his right arm around the Marine's neck, bends the Marine forward, and locks the Marine's head against his hip. To execute the counter to a front headlock, Marines—

Grasp the opponent's wrist and forearm with both hands and pull down to clear the airway. Maintain control of the opponent's wrist throughout the move. Once the airway is clear, tuck the chin to protect the airway and to prevent the opponent from reapplying the choke.

Move the right hand and arm across the opponent's torso.

Step forward and to the left with the left foot at a 45-degree angle.

Execute a sweep with the right foot against the opponent's right leg. At the same time, push against the opponent's chest with the right arm and shoulder to generate power in the sweep.

## Counter to a Rear Headlock

Marines use a counter to a rear headlock when the opponent approaches from the rear and puts his right arm around the Marine's neck, bends the Marine forward, and locks the Marine's head against his hip. To execute the counter to the rear headlock, Marines-

Grasp the opponent's wrist and forearm with the right hand and pull down to clear the airway. Once the airway is clear, tuck the chin to protect the airway and to prevent the opponent from re-applying the choke.

Reach over the opponent's right shoulder with the left arm.

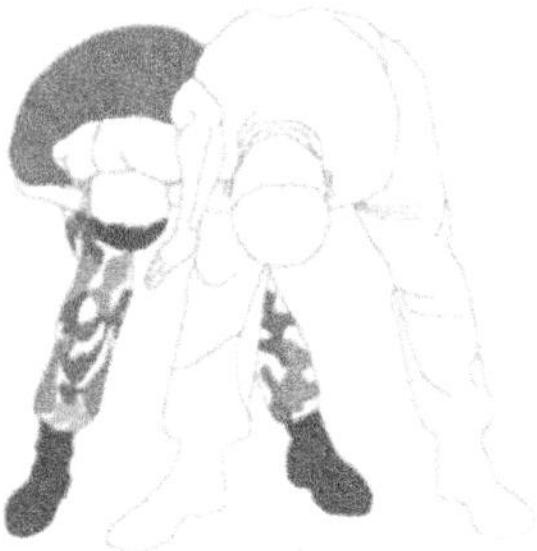

Grab any part of the opponent's face (chin, nose, eyes) and pull back while rising to a standing position.

Execute, with the right hand, a hammer fist strike to the opponent's exposed throat.

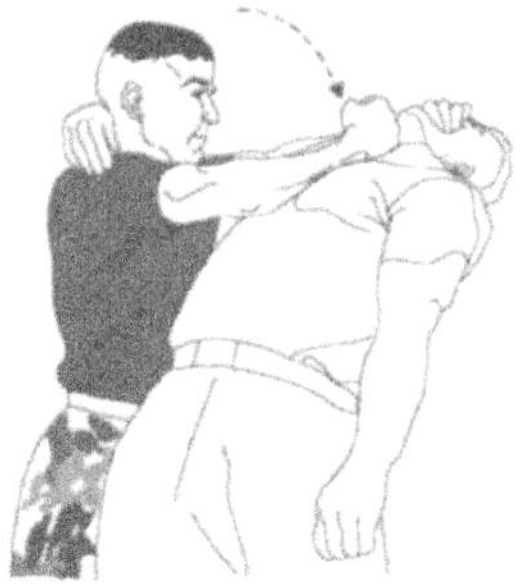

## Counter to a Front Bear Hug

Marines execute a counter to a front bear hug when the opponent approaches from the front and puts both of his arms around the Marine's body, trapping the Marine's arms to the sides. To execute the counter to a front bear hug, Marines—

Step forward and to the left with the left foot at a 45-degree angle to the outside of the opponent's right leg, keeping the left leg bent.

Grasp the opponent's torso or arms to gain balance and to assist in throwing the opponent. It may be helpful to hook the opponent's right arm with the left arm.

Drive the right arm and shoulder forward and, at the same time, bring the right leg forward and sweep the opponent's right leg, bringing him to the ground.

## Counter to a Rear Bear Hug

Marines use a counter to a rear bear hug when the opponent approaches from the rear and puts both of his arms around the Marine's body, trapping the Marine's arms to the sides. To execute the counter to a rear bear hug, Marines—

Step behind the opponent's right leg with the left foot, keeping both legs bent (almost in a squatting position). The left side of the body should be against the opponent's.

Pivot the hip, turning the body to the left and throwing the opponent back over the bent leg.

(reverse blank)

# CHAPTER 7

# GROUND FIGHTING

***This chapter describes all techniques for a right-handed person. However, all techniques can be executed from either side.***

***In drawings, the Marine is depicted in woodland camouflage utilities; the opponent is depicted witlwut camouflage. In photographs, the Marine is depicted in woodland camouflage utilities; the opponent is depicted in desert camouflage utilities.***

Marines should avoid being on the ground during a close combat situation because the battlefield may be covered with debris and there is an increased risk of injury. However, many close combat situations involve fighting on the ground. The priority in a ground fight is for Marines to get back on their feet as quickly as possible. In any ground fighting scenario, Marines will usually end up in one of four positions with the opponent. The offensive positions, in which Marines have a tactical advantage, are the guard and mount. The defensive positions, which are used as counters when the opponent has the tactical advantage, are the counter to the guard and the counter to the mount. Marines can also employ chokes during ground fighting to quickly end a fight.

## 1. Offensive Ground Fighting

### Guard Position

Marines execute a guard position when the opponent is on top and the legs are wrapped around the opponent's legs. If the opponent is on top, he may try to choke the Marine, but the Marine still has the tactical advantage because the Marine is in a position to control the situation. To execute the guard position technique, Marines—

Trap the opponent's hands on the chest by crossing the hands on the chest so the forearms are resting on the opponent's forearms. Apply pressure with the elbows or forearms.

Strike the outside of the opponent's thigh with the cutting edge of the right heel. This causes the opponent to jerk to that side.

Move the head quickly to the left and swivel the hips to the right. At the same time, bring up both of the legs. Both legs are on the right side of the opponent's body.

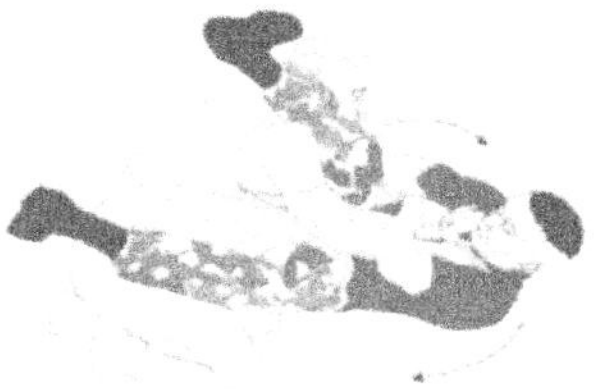

Bring the right leg down, hooking the opponent's neck and head. Exert downward pressure to roll him over on his back. Grasp and maintain control of the opponent's left arm. Upon completion of the move, the Marine is sitting up with legs bent over the opponent while maintaining control of his arm.

Keep legs and knees bent. Maintain pressure against the opponent's neck with the back of the right foot and against his side with the left foot underneath his armpit. Squeeze the knees together, locking the opponent's arm.

Pull the opponent's arm straight up and fall back sharply, pulling his arm to the side in the direction of his little finger.

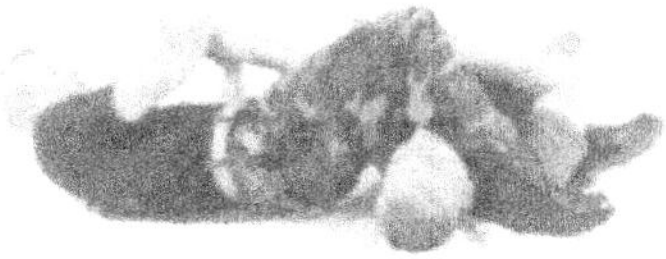

Return to standing.

## Mount Position

Marines execute the mount if the opponent is lying on his back on the ground and the Marine is on top with legs wrapped around the opponent's body. This position is an offensive position because Marines are in a better position to control the opponent and to execute ground fighting techniques. To execute the mount position technique, Marines—

Grab the opponent's wrists or forearms with the hands. Hold them tightly against the torso.

Maintain control of the opponent's arms with the left arm and apply pressure to the opponent's brachial plexus (tie in) with the right hand.

Bring the left foot close to the opponent's right armpit and swing the right leg across the opponent's head.

Keep the legs and knees bent. Maintain pressure against the opponent's neck with the back of the right foot and against his side with the left foot which is underneath his armpit. Squeeze the knees together, locking the opponent's arm.

Pull the opponent's arm straight up and fall back sharply, pulling his arm to the side in the direction of his little finger.

Return to standing.

## 2. Defensive Ground Fighting

### Counter to the Guard

Marines use this technique if the opponent is lying on his back on the ground and the Marine is kneeling on the ground between the opponent's legs. To execute the counter to the guard position, Marines—

Strike the opponent's femoral nerve, located on the inside of the thigh, with the elbows. This forces the opponent to separate his legs.

Strike the opponent's groin with the right fist.

Hook the left arm underneath the opponent's right knee from the inside. Throw the opponent's leg over the head.

Duck the head quickly and move to the left.

Return to standing.

## Counter to the Mount

Marines use this technique when they are lying on their back on the ground, the opponent is mounted on top of the Marine, and the opponent's legs are wrapped around the Marine. The opponent has the tactical advantage. To execute the counter to the mount position, Marines—

Grab the opponent's gear or clothing on his upper torso and pull him down close.

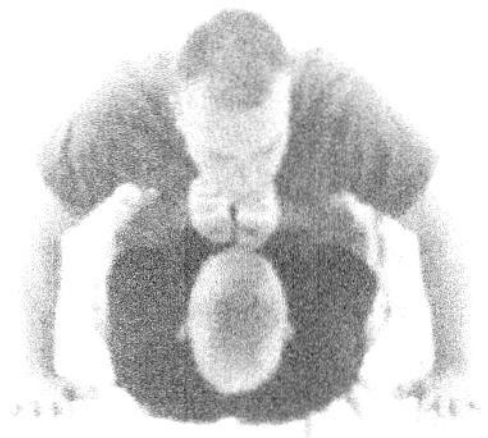

Use the right arm to hook the opponent's left arm, from the inside around the outside, above his elbow. With the right foot, hook the opponent's left leg or ankle.

Draw the elbow in to bend the opponent's elbow, bringing him down close. The arm must be hooked above the opponent's elbow in order to bend it.

Strike the opponent's side with the left hand.

Push the opponent over and roll him off to the right side.

Maintain control of the opponent's hooked arm and move to a kneeling position.

Move to a standing position while maintaining pressure on the opponent's arm and using the knee to apply pressure against the opponent's elbow.

## 3. Ground Fighting Chokes

WARNING

**During training, never execute a choke at full force or full speed. Never hold a choke for more than 5 seconds.**

The priority in ground fighting is for Marines to get back on their feet as quickly as possible. Sometimes, Marines can quickly end a ground fight by executing a choke on the opponent. When performed correctly, a choke can render an opponent unconscious in as little as 8 to 13 seconds. Chokes are easily performed regardless of size or gender. The chokes performed during ground fighting are the same as those performed while standing (see chap. 6).

### Ground Fighting Front Choke

The ground fighting front choke is a blood choke performed most effectively from the mount position. The front choke employs the opponent's lapels or collar to execute the choke. To execute the ground fighting front choke, Marines—

Grab the opponent's right lapel with the right hand, making certain that the knuckles or the back of the hand are against the carotid artery on the right side of the opponent's neck.

Keep the right hand pressed against the opponent's carotid artery, reach under the right arm with the left hand, and grab the opponent's left lapel, forming an X with the wrists.

Keep the right hand pressed against the opponent's carotid artery and pull the opponent's left lapel to the left with the left hand.

## Ground Fighting Side Choke

The ground fighting side choke is a blood choke performed from the mount position. The ground fighting side choke is particularly effective when the opponent raises his arms and places them on the Marine's chest or throat. To execute the ground fighting side choke, Marines-

Use the left hand to parry the opponent's right arm inboard (to the inside of the opponent's reach). Bring the right arm underneath the opponent's arm and up around the front of his neck.

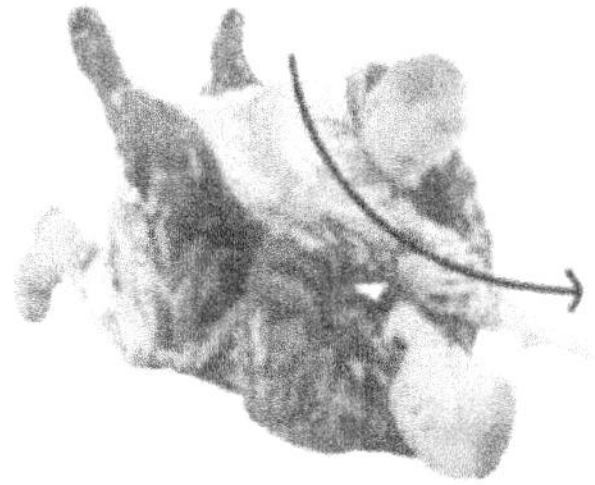

Keep the fingers extended, place the back of the forearm against the opponent's neck just below his ear and press the carotid artery on his neck. Depending on the Marine's or the opponent's body configuration, the back of the thumb or wrist may press against the opponent's neck.

Use the left hand to reach around the back of the opponent's neck and clasp the hands together.

Pull the opponent toward the chest. Use the forearm to exert pressure on the side of his neck. This is done by pulling the clasped hands toward the chest.

## Ground Fighting Rear Choke

The ground fighting rear choke is a blood choke performed when Marines are behind the opponent. To execute the ground fighting rear choke, Marines—

Wrap the legs around the opponent with the left foot against the inside of his left thigh and the right foot against the inside of his right thigh.

Use the right arm to reach over the opponent's right shoulder and hook the bend of the arm around his neck.

Use the left hand to clasp both hands together.

Exert pressure with the biceps and forearms on both sides of the opponent's neck. This applies pressure to the opponent's carotid arteries. While maintaining pressure with the biceps and forearms, draw the opponent closer by drawing the right arm in. To increase the effectiveness of the choke, lean backward by arching the back and pulling the opponent back. At the same time, push the feet against the opponent's thighs.

## Figure-Four Choke

A variation of the ground fighting rear choke is the figure-four choke. The figure-four choke allows Marines to gain more leverage on the rear choke. If the rear choke cannot be secured, the figure-four variation may be applied to increase the pressure of the choke on the opponent. To execute the figure-four variation of the rear choke, Marines—

Apply the ground fighting rear choke. The chest should be against the opponent's back.

Grasp the left biceps with the right hand and place the left hand against the back of the opponent's head.

Use the left hand to push the opponent's head forward and down.

Draw the right arm in while maintaining pressure with the biceps and forearm on both sides of the opponent's neck.

Increase the effectiveness of the choke by arching backward and pushing the feet against the opponent's thighs. Continue to pull the opponent back with the right arm while exerting pressure forward with the left hand against the opponent's head.

# CHAPTER 8

# NONLETHAL TECHNIQUES

*This chapter describes all techniques for a riglit-handed person. However, all techniques can be executed from either side.*

*In drawings, the Marine is depicted in woodland camouflage utilities; the opponent is depicted without camouflage. In photographs, the Marine is depicted in woodland camouflage utilities; the opponent is depicted in desert camouflage utilities.*

The Marine Corps' involvement in military operations other than war—e.g., humanitarian, peacekeeping, or evacuation missions—has greatly increased. These missions require skills that span the spectrum of conflict and support operations within a continuum of force. But the Corps' day-to-day existence also demands a responsible use of force. Nonlethal techniques are among the skills Marines use to apply a responsible use of force.

## 1. Unarmed Restraints and Manipulation

Marines operate within a continuum of force, particularly in support of peacekeeping or humanitarian types of missions. In these situations, Marines must act responsibly to handle a situation without resorting to deadly force. Unarmed restraints and manipulation techniques including joint manipulation, come-alongs, and takedowns can be used to control a subject without resorting to deadly force. Marines must train to become proficient in nonlethal techniques and to respond in a responsible manner. These techniques are referred to as compliance techniques, and they are applied in the third level in the continuum of force.

**WARNING**

**During training, never apply the techniques for unarmed restraints and manipulation at full force or full speed. Use a slow and steady pressure to avoid injury.**

### Compliance Techniques

Compliance techniques are unarmed restraint and manipulation techniques used to physically force a subject or opponent to comply. Compliance can be achieved through the close combat techniques of-

Pain compliance using joint manipulation and pressure points. (Pain compliance is the initiation of pain to get compliance on the part of the subject.)

Come-along holds.

### Principles of Joint Manipulation

Joint manipulation is used to initiate pain compliance and gain control of a subject. It involves the application of pressure on thejoints (elbow, wrist, shoulder, knee, ankle, and fingers). Pressure is applied in two ways:

In the direction in which the joint will not bend. For example, joints such as the knees and elbows only bend in one direction and when pressure is applied in the opposite direction, pain compliance can be achieved. Beyond the point where the joint stops naturally in its range of movement (i.e., it no longer bends).

Since each joint has a breaking point, Marines should apply slow steady pressure only until pain compliance is reached. Continued pressure will break the joint and may escalate the violence of the situation.

Joint manipulation also uses the principle of off-balancing. A subject can be better controlled when he is knocked off balance.

## Wristlocks

A wristlock is a joint manipulation that can be applied in a number of ways to achieve pain compliance. The wrist rotates in a number of directions and will bend in a single direction until its movement stops naturally. In a wristlock, pressure is exerted beyond that point by bending or twisting the joint. A wristlock is executed when an opponent tries to grab Marines or is successful in grabbing Marines or their equipment. A wristlock can also be performed by Marines if they wish to initiate control of an opponent.

**Basic Wristlock.** A basic wristlock is executed when Marines grab the opponent's left hand with the right hand. To execute the basic wristlock, Marines—

Use the right hand to grab the opponent's left hand by placing the thumb on the back of the opponent's hand so that the Marine's knuckles are facing to the left.

Hook the fingers across the fleshy part of the opponent's palm below the thumb. The fingers are used to anchor the hand so leverage can be applied to twist and bend the joint.

Exert downward pressure with the thumb to bend the opponent's joint. Rotate the opponent's hand to the right to twist the joint.

Step in to the opponent to keep the opponent's hand in close to the body to control him and provide more leverage on the wristlock.

Use the left hand to further control the opponent.

*Note: When executing the basic wristlock with the left hand, the Marine grabs the opponent's hand so that the Marine's knuckles are facing to the right, and then rotates and twists the opponent's hand to the left.*

**Reverse Wristlock.** A reverse wristlock is executed when Marines grab the opponent's right hand with the right hand. To execute the reverse wristlock, Marines—

Place the right palm on the back of the opponent's right hand and wrap the fingers across the fleshy part of his palm below his little finger.

Twist the opponent's hand to the right while stepping in to place his hand against the chest. Apply downward pressure on the opponent's hand against the chest. Leave the opponent's hand on the chest to fully control the subject and to gain leverage.

Lean forward to use body weight to add additional pressure to the joint.

Use the left hand to further control the opponent.

**Two-Handed Wristlock.** Both hands can be used in the wristlock to maximize the leverage and pressure needed to bend and twist the joint. To execute the two-handed wristlock, Marines—

Place both thumbs on the back of the opponent's hand, thumbs crossed.

Hook the fingers of both hands around the fleshy part of the opponent's palm on both sides of his hand.

Step into the opponent and apply pressure downward on the back of his hand to bend the joint and rotate his wrist away from the body to twist the joint.

**Enhanced Pain Compliance on Wristlock.** Enhanced pain compliance techniques are applied in the third and fourth levels in the continuum of force. Additional pain be applied to a wristlock by-

Adding downward pressure to the elbow with the other hand or elbow by using the fingers to pull in on the opponent's radial nerve located on the inside of the forearm. When pressure is added to the opponent's radial nerve, his direction can be controlled.

Applying pressure against the opponent's finger joint to bend it in a direction it cannot bend (i.e., splitting the fingers).

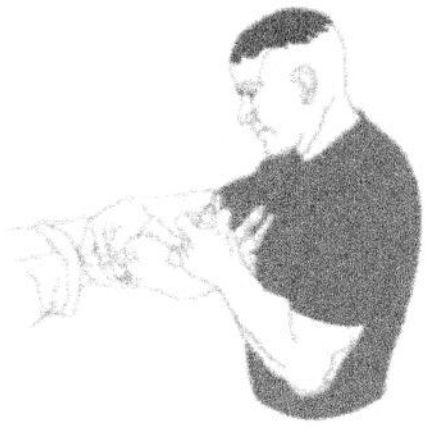

## Come-Along Holds

Marines use a come-along hold to control and move an opponent.

**Escort Position.** A common come-along hold is the escort position. To execute the escort position, Marines—

Face the opponent. Use the left foot to step forward at a 45-degree angle. Turn to face the right side of the opponent.

Use the right hand to firmly grasp the opponent's right wrist. With the left hand, firmly grasp the opponent's right triceps.

Position the opponent's controlled arm diagonally across the torso, keeping his wrist against the right hip. The Marine should be standing to the right of and behind the opponent.

*Note: This technique works well when escorting an opponent on either the right or left side. Take caution when escorting an opponent by ensuring his controlled hand is not in a position to grab the holstered weapon. The preferred escort position is from the Marine's left side, so that the opponent is kept further away from the weapon.*

**Wristlock Come-Along.** To execute the wristlock come-along, Marines-

Use the left hand to execute a basic wristlock. Incorporate the right hand in a two-handed wristlock for more control.

Maintain pressure on the opponent's wrist with the right hand, step forward, and pivot around to stand next to the opponent.

Release the left hand, quickly reach under the opponent's arm from behind, and grab his hand.

Use the left hand and apply downward pressure on the opponent's wrist.

**Controlling Technique.** The following controlling technique is used when an opponent grabs the Marine's wrist. To execute the technique, Marines—

Trap the opponent's hand with the palm of the other hand.

Rotate the opponent's trapped hand up and on his forearm while maintaining downward pressure on his trapped hand.

Apply downward pressure with both hands until the opponent is taken to the ground.

## Armbars

An armbar is a joint manipulation in which pressure is applied on a locked elbow, just above the joint, in the direction the joint will not bend. An armbar has to be locked in quickly, but still requires a slow, steady pressure to gain compliance.

**Basic Armbar.** To execute a basic armbar, Marines-

Use the right hand to grab the opponent's right wrist.

Bring the left hand down on or above the opponent's elbow joint. To gain additional leverage, pivot to face the opponent.

Use the left hand to apply downward pressure on or above the opponent's elbow joint while pulling up on his wrist.

**Armbar from a Wristlock.** To execute an armbar from a wristlock, Marines-

Use the right hand to grab the opponent's right hand and execute a reverse wristlock.

Bring the left hand down on or above the opponent's elbow joint.

Use the left hand to apply downward pressure on or above the opponent's elbow joint while pulling up on his wrist.

## Takedowns

A takedown is used to bring an opponent to the ground to further control him.

**Takedown From a Wristlock Come-Along.** To take the opponent to the ground from a wristlock come-along, Marines—

Use the right foot to push down on the opponent's calf or Achilles tendon.

Maintain control of the opponent's wrist and elbow and apply a slow, steady pressure to bring him to the ground.

**Armbar to a Takedown.** This technique is used to take a noncompliant opponent to the ground from an armbar. To execute the armbar takedown, Marines—

Use the right hand to execute a reverse wristlock.

Bring the left hand or forearm down on or above the opponent's elbow joint.

Pivot so the back is facing the opponent and, at the same time, lift the left elbow and slide the body so it is against the opponent, placing the armpit high above the opponent's elbow joint.

Lean back, placing the body weight on the opponent's arm until he complies or is taken to the ground.

*Note: This technique may break the opponent's arm. Therefore, this technique should not be employed if the objective is a nonlethal takedown.*

**Wristlock Takedown.** This technique is used to take a noncompliant opponent to the ground from a basic wristlock and to put him in a position where he can be handcuffed, if necessary. To execute the takedown, Marines-

Use the right hand to execute a basic wristlock.

Incorporate the left hand in a two-handed wristlock.

Apply downward pressure on the wristlock, pivot on the ball of the right foot, and quickly turn to the right to take the opponent to the ground.

Continue to apply pressure on the wrist joint as the opponent lands on his back with his arm straight in the air.

Slide the left foot under the opponent's back, underneath his armpit.

Apply pressure with the knee against the opponent's triceps while pulling back on his arm.

Continue applying pressure with the knee on the opponent's arm. Pivot and step around the opponent's arm to roll him on his stomach.

Kneel down with one knee on the opponent's back. The other knee is placed on the opponent's neck and shoulder on either side of his arm. Apply inward pressure with the knees to lock his arm in place.

Tell the opponent to put his other hand in the middle of his back. Bring the opponent's controlled hand to the center of his back.

**Escort Position Takedown.** This technique can be used to control a noncompliant opponent from the escort position. To execute the escort position takedown, Marines—

Lock the opponent's arm straight across the body while rotating his wrist away from the body.

Use the left hand or forearm and apply downward pressure above the opponent's elbow where the triceps meet.

Step back with the right foot and, keeping the opponent's hand controlled against the hip, pivot to the right while continuing to apply downward pressure on his arm to bring him to the ground.

*Note: This technique works well when escorting a subject on the Marine's right or left side. When taking down a subject from the right side, step back and pivot to the left.*

## 2. Nonlethal Baton

A baton or nightstick can be an effective compliance tool when used correctly. Batons can be used defensively (blocking), offensively (striking), and as a restraining device when needed. In the fourth level of the continuum of force (assaultive [bodily harm]), defensive tactics include baton or nightstick blocks and blows. Blows to the head or other bony parts of the body are considered deadly force. When deadly force is not authorized, Marines must be able to employ blocks, strikes, and restraints with the baton with the minimum of force.

### Grips

**One-Handed Grip.** To execute the one-handed grip, Marines—

Use the right hand to grasp the lower end of the baton, about 2 inches from the end.

Wrap the thumb and index finger around the baton so they are touching one another. The grip on the baton should be firm, but natural.

**Two-Handed Grip.** To execute the two-handed grip, Marines-

Use the right hand to grasp the lower end of the baton, about 2 inches from the end. Wrap the thumb and index finger around the baton so they touch one another.

Use the left hand to grasp the upper end of the baton, palm down, about 2 inches from the end. The hands should be approximately 10 to 12 inches apart.

## Stance and Method of Carry

The basic warrior stance serves as the foundation for initiating nonlethal baton techniques. The method of cany provides effective defensive positions with a wide range of options to control a combative opponent.

**One-Handed Carry.** To execute the one-handed carry, Marines—

Grip the baton using the one-handed grip.

Elevate the baton, with the gripping hand at a level between the belt and shoulder.

Keep the left hand in the position of the basic warrior stance.

**Two-Handed Carry.** This carry is effective for blocks. To execute the two-handed carry, Marines-

Grip the baton using the two-handed grip.

Elevate the baton, with the left hand higher than the right hand.

Orient the weapon toward the subject.

## Movement

In a nonlethal confrontation, movement may be made to create distance between Marines and the opponent or to close the gap to control him. When facing an opponent, Marines move in a 45-degree angle to either side of the opponent. Moving at a 45-degree angle is the best way to both avoid an opponent's strike and to put Marines in the best position to control the opponent.

## Blocking Techniques

One-handed blocks are used when carrying the baton in the one-handed carry. The same one-handed blocks used in combative stick techniques (see page 3-6) apply in nonlethal baton engagements. The blocks include blocks for a vertical strike, a forward strike, and a reverse strike. Because the baton is often carried with two hands, there are also two-handed blocks that are effectively used from this carry. Two-handed blocks are discussed in the following subparagraphs.

**High Block.** Marines execute a high block to deter a downward vertical attack directed at the head and shoulders. To execute the high block, Marines—

Raise the baton up to a level even with or above the head. The baton should be in a horizontal position to block the blow. The fingers of the left hand should be open and behind the baton.

Place the baton perpendicular to the opponent's striking surface to absorb the impact of the blow.

Bend the elbows slightly to help absorb the impact of the blow. The arms should give with the strike of the blow.

**Low Block.** Marines execute a low block to deter an upward vertical attack directed at the abdomen, groin, or torso. The opponent's blow can be delivered by a foot, knee, or fist. To execute the low block, Marines—

Lower the baton to a level even with or below the groin. The baton should be in a horizontal position to block the blow. The fingers of the left hand should be open and behind the baton.

Place the baton perpendicular to the opponent's striking surface to absorb the impact of the blow.

Bend the elbows slightly to help absorb the impact of the blow. The arms should give with the strike of the blow.

**Right Block.** Marines execute a right block to deter an attack directed at the head, neck, flank, or hip. The opponent's blow can be delivered by a foot, knee, fist, or elbow. To execute the right block, Marines-

Thrust the baton in a vertical position to the right side.

Pivot to the right by stepping with the left foot and pivoting off the ball of the right foot. Rotate the hips and shoulders into the direction of the block. The fingers of the left hand should be open and behind the baton.

Place the baton perpendicular to the opponent's striking surface to absorb the impact of the blow.

Bend the elbows slightly to help absorb the impact of the blow. The arms should give with the strike of the blow.

**Left Block.** Marines execute a left block to deter an attack directed at the head, neck, flank, or hip. The opponent's blow can be delivered by a foot, knee, fist, or elbow. To execute the left block, Marines—

Thrust the baton in a vertical position to the left side.

Pivot to the left by stepping with the right foot and pivoting off the ball of the left foot. Rotate the hips and shoulders into the direction of the block. The fingers of the left hand should be open and behind the baton.

Place the baton perpendicular to the opponent's striking surface to absorb the impact of the blow.

Bend the elbows slightly to help absorb the impact of the blow. The arms should give with the strike of the blow.

**Middle Block.** Marines execute either a left or right block to deter an attack directed at the face, throat, chest, or abdomen. To execute the middle block, Marines—

Thrust the baton in a vertical position straight out in front of the body. The fingers of the left hand should be open and behind the baton.

Place the baton perpendicular to the opponent's striking surface to absorb the impact of the blow. The baton should be held with the left hand forward of the right.

Bend the elbows slightly to help absorb the impact of the blow. The arms should give with the strike of the blow.

## Restraining Technique

The strong-side armlock is used to restrain an opponent who is not compliant. To execute the strong-side armlock, Marines—

Use the right hand to run the baton up under the opponent's left armpit, parallel to the ground.

Use the right foot to step forward at a 45-degree angle to the left side of the opponent. The baton is across his forearm.

Use the right hand to drive the baton forward and up so the action bends the opponent's arm behind his back. At the same time, continue moving around him to get behind him.

Place the baton on the opponent's forearm with the thumb and/or fingers and apply pressure to his forearm. At the same time, grasp the other end of the baton with the left hand.

Pull up on the low end of the baton with the right hand. At the same time, push down on the top end of the baton with the left forearm, reaching around with the left hand to grasp the opponent's biceps or shoulder.

Continue exerting downward pressure with the left forearm while pulling back on the opponent's biceps with the left hand. This places the opponent in a position where he is controlled and can be moved.

Apply pressure with the foot against the bend in the opponent's leg above his calf. This lowers the opponent to the ground rather than throwing him to the ground and risking severe injury.

## Striking Target Areas

Marines must avoid striking an opponent in the head, neck, or other bony parts with the baton because this is considered deadly force and can lead to serious bodily injury or death. Instead, the legs, arms, and buttocks are target areas that are considered nonlethal.

**Legs.** Primary targets are the thighs and lower legs. Avoid striking the knee and ankle joints because this can cause permanent damage.

**Arms.** Primary targets are the upper arms. Avoid striking the shoulder and elbow and wrist joints because this can cause permanent damage.

**Buttocks.** Primary targets are the buttocks. Avoid striking any other part of the torso, including the chest, rib cage, spine, tail bone, and groin because strikes to these areas can cause permanent damage or death.

## One-Handed Striking Techniques

**One-Handed Forward Strike.** A forward strike follows either a horizontal line or a downward diagonal line using a forehand stroke. To execute the one-handed forward strike, Marines-

Stand facing the opponent with the baton carried in a one-handed cany.

Place the right hand palm up, swing the baton from right to left, and make contact with the opponent.

**One-Handed Reverse Strike.** A reverse strike follows either a horizontal line or a downward diagonal line using a backhand stroke. To execute the one-handed reverse strike, Marines —

Stand facing the opponent with the baton carried in a one-handed carry.

Bend the right arm and cross the arm to the left side of the body. The baton should be close to or over the left shoulder.

Place the right hand palm down, swing the baton from left to right, and make contact with the opponent.

## Two-Handed Striking Techniques

**Two-Handed Forward Strike.** This strike is an effective follow-up to a middle block or left block. To execute the two-handed forward strike, Marines—

Pull back with the left hand while driving the right hand forward toward the opponent. The baton should be horizontal to the ground. Power is generated by stepping forward with the right foot and rotating the right hip and shoulder into the strike.

Contact the opponent with the end of the baton.

**Two-Handed Reverse Strike.** This strike is an effective follow-up to a middle block or left block. To execute the two-handed reverse strike, Marines—

Pull back with the right hand while driving the left hand forward toward the opponent. The baton should be horizontal to the ground. Power is generated by stepping forward slightly with the left foot and rotating the left hip and shoulder into the strike.

Contact the subject with the end of the baton.

**Front Jab.** This strike is effective for countering a frontal attack. It can also be executed as a quick poke to keep a subject away. To execute the front jab, Marines thrust both hands forward in a quick jab. The baton is held either horizontal to the ground or at a slight downward angle.

**Rear Jab.** This strike is effective for countering a bear hug from the rear. To execute the rear jab, Marines thrust both hands rearward in a quick jab. The baton is held either horizontal to the ground or at a slight downward angle.

# APPENDIX A

# PUGIL STICK TRAINING

A pugil stick is a training device used to simulate a rifle bayonet so that effective, but safe, training can be conducted to build proficiency of rifle bayonet techniques. Pugil stick training builds on the techniques used to throw punches. Pugil stick training is the only "full contact" training provided to Marines in the Close Combat Program. Pugil stick training teaches Marines to function when faced with stress and violence, and it prepares them to deliver a blow and take a blow. It also provides them with the physical and mental skills vital to success on the battlefield.

## 1. Pugil Stick Training

### Design

A pugil stick consists of a stick wrapped in padding at both ends that can be gripped like a rifle. The pugil stick is approximately the same weight and length of an unloaded rifle with a bayonet attached.

### Holding the Pugil Stick

The pugil stick is held in the same manner as the sen ice rifle. All movements come from the basic warrior stance. To hold the pugil stick correctly, Marines—

Use the right hand to grasp the lower end of the pugil stick overhanded.
Use the left hand to grasp the upper end of the pugil stick underhanded.
Use the right forearm to lock the lower end of the pugil stick against the hip.
Orient the blade end of the pugil stick toward the opponent.

### Safety Equipment

The following safety equipment must be worn during any pugil stick bout.

**Groin Protection.** Groin protection protects the groin from an accidental blow. It should be pulled high around the waist to protect the groin area, with the concave portion against the body.

**Flak Jacket.** The flak jacket provides protection to the body. It is worn completely fastened.

**Neck Roll.** The neck roll prevents whiplash if Marines receive a blow to the head. The neck roll further supports the head and protects the neck from blows. The neck roll is worn above the flak jacket and below the helmet. The tied end of the neck roll faces the front.

**Helmet.** A regulation football helmet protects the face and head. The helmet must fit snugly and the chin straps must be adjusted and snapped.

**Mouthpiece.** The mouthpiece is worn on the upper teeth to protect the teeth.

**Gloves.** Marines wear gloves to protect the hands if pugil sticks do not have gloves built-in.

## Pugil Stick Screening

Prior to pugil stick training, instructors must ask participating Marines the following questions. If a Marine answers "yes" to any question, he must see the corpsman or a competent medical authority who evaluates him and determines whether or not the he can participate in the training.

Are you on light duty?
Are you restricted to running shoes by a corpsman or a doctor?
Have you fought in a pugil stick bout within the last 7 days?
Have you received a blow to the head within the last 7 days?
Have you had a concussion within the last 6 months?
Have you had dental surgery within the last 24 hours?
Do you have stitches or staples on your body?
Have you had a shoulder or head injury within the last 5 years?
Are you taking a prescription drug?
Do you have an ear infection or current sinus infection?
Have you had a broken bone within the last 6 months?

## Safety Personnel

The following safety personnel are required to conduct pugil stick training:

One close combat instructor must officiate the bout.
One close combat instructor trainer, commissioned officer, or staff noncommissioned officer must be in the training area to serve as range safety officer.
One corpsman must be in the training area.

## Training Area

To prevent injury, Marines train on areas with soft footing (i.e., sand or grass). Training mats are not recommended because feet can stick to the mats, prohibiting movement or causing joint injuries by twisting a knee. Bouts should not take place on a hard surface area; e.g., a flight deck or parking lot. A boxing ring may be used to conduct pugil stick bouts; ring dimensions can vary as long as there is ample room to execute the techniques in the training area.

## Second Impact Syndrome

Second Impact Syndrome occurs when a second blow to the head produces a second concussion that occurs within 1 week following a previous concussion (before recovery from the first concussion). Second Impact Syndrome causes rapid brain swelling and can cause death. Therefore, there must be 7 days between pugil stick bouts to reduce the risk of severe injury resulting from Second Impact Syndrome. The 7 day separation between pugil stick bouts significantly reduces the possibility of injury, particularly in someone who may have suffered a brain injury or concussion but shows no symptoms.

Any Marine who experiences headaches or the following symptoms after training must be examined by appropriate medical personnel:

Blurred vision.
Ringing in the ears.
Dilation of the pupils.
Slurred speech.
Bleeding from ears or mouth.
Swelling in head or neck area.
Any unnatural discoloration in head or neck area.

The Marine should not be allowed to participate in pugil stick training or any other activity where a heavy blow might be sustained for a minimum of 7 days after the headache or other symptoms have subsided.

## Safety Measures

The following safety measures must be followed:

Gear must be worn properly throughout training.
Contact lenses or glasses will not be worn.
False teeth will be removed from the mouth.
Nothing will be worn around the neck except the neck roll.

Competition among groups of Marines is authorized as long as it does not overshadow training objectives or compromise safety procedures.

Safety and proper techniques are paramount.

Safety is more important than competition.

## 2. General Rules and Regulations Governing Pugil Stick Bouts

### Instructors and Support Personnel Requirements

At a minimum, one close combat instructor observes a pugil stick bout. For safety purposes, it is better to have two instructors judging a bout because each instructor can fully observe each of the fighters. The best position for observation is to the right of a fighter. This allows the instructor to see the fighter's facial expression and body movement. The instructor's position must not interfere with the fight.

In addition, one commissioned or staff noncommissioned officer (to serve as range safety officer) and one corpsman will be in the training area.

### Prior to the Bout

Prior to the bout, Marines—

Are paired according to height, weight, and gender.

Wear the proper safety equipment.

Once Marines have properly donned the safety equipment, they wait for instructions from the close combat instructor. On the command of the close combat instructor, a pair of Marines will enter the ring. Once in the ring, the close combat instructor inspects each Marine for the proper safety equipment.

*Note: There should be enough gear so that when two Marines are training, two other Marines can be donning safety equipment for the next bout.*

### During the Bout

The bout begins when the close combat instructor blows the whistle. All strikes are directed above the waist. Upon hearing a whistle blast, all fighting immediately ceases.

### Stopping a Bout

There are two reasons for stopping a bout: delivery of a scoring blow or an unsafe condition. The close combat instructor trainer, close combat instructor, or the range safety officer may stop the bout at any time an unsafe condition is observed.

**Scoring Blow.** A scoring blow is an offensive technique delivered to a vulnerable area of an opponent with sufficient force and precision to be considered as a disabling or killing blow. Scoring blows are not judged solely on the degree of force with which a blow is delivered, but on the accuracy and techniques employed. A scoring blow is defined as—

A straight thrust with the blade end of the weapon (red end of the pugil stick) to the opponent's face mask or throat.

A slash to the side of the opponent's helmet (below the ear) or neck with the red end of the pugil stick.

A heavy blow to the opponent's head with an authorized technique (i.e., buttstroke, smash) using the butt (black end) of the pugil stick.

When a scoring blow is delivered, the close combat instructor blows the whistle to stop the bout.

**Unsafe Condition.** The bout will be stopped as soon as an unsafe condition exists. An unsafe condition exists when a Marine is unable to defend himself, loses his balance and falls down on one or two knees or falls down completely, shows instability (e.g., buckling at the knees), loses muscular tension in his neck and his head snaps back or to one side, or appears disoriented. An unsafe condition also exists if a Marine lets go of one end of the pugil stick, the equipment (e.g., helmet, neck roll) falls off, or a Marine fails to use the proper techniques.

If any of these conditions occur—

The close combat instructor trainer, close combat instructor, range safety officer, or anyone supervising training stops the bout and separates the two Marines.

The corpsman evaluates the possibly injured Marine to determine if the fight can continue. The corpsman—

- Ensures that he is alert and responsive, both verbally and physically.
- Talks to him to see if he is coherent.
- Makes certain he comprehends and replies to verbal questions.
- Makes certain his speech is not slurred.
- Checks his physical signs.
- Ensures his eyes are focused and not dazed or glazed.
- Ensures his legs are not wobbly or shaky.

The pugil stick is not used as a baseball bat. The use of unauthorized techniques will result in expulsion from the bout. Only techniques taught in bayonet training are authorized. These techniques include-

Straight thrust.

Buttstroke (horizontal and vertical).

Smash.

Slash.

Parry.

Blocks (low, high, left, right).

## After the Bout

A third whistle blast is used prior to identifying the Marine who delivered the killing blow.

A fourth whistle blast is used prior to demonstrating the killing technique used by the Marine who won the bout.

# 3. Directions for Making Pugil Sticks

Close combat instructor trainers are taught to make pugil sticks to standard specifications at school. It is recommended that all pugil sticks be made under the supervision of a close combat instructor trainer. Until the Marine Corps acquires a standardized pugil stick that units can order through the supply system, instructors use the following information and instructions to make pugil sticks locally.

## Supplies

The following supplies are needed to make one pugil stick:

One 45" by 2" circular, oak, wood dowel.

Four 9" wide by 27" long foam pad pieces to make two inner handguards of double thickness.

Four 11" wide by 19" long foam pad pieces to make two outer handguards of double thickness.

Canvas material to cover inner and outer handguards.

One foam pad cut 1/4" thick by 7" wide by 13" long to make the centerguard.

One 8" wide by 14"long piece of canvas material to make a pugil stick sock.

Canvas material to make two pugil stick end caps. Ideally, two different color fabrics should be used to simulate the bayonet and buttstock ends of a rifle. Common colors for end caps are black and red. Each end cap is a different color to help the officiating certified close combat instructor determine the scoring blow during the pugil stick bout.

Two 1/4" thick by 11" wide by 54" long foam pad pieces to make two ends.

Two foam pads cut 1/4" thick by 3" wide by 14 1/2" long to make two end plug inserts.

One can of 3M Photo MountQ spray adhesive (NSN 8040-01-120-4009)to adhere foam padding to the stick. (Spray adhesive is available from self-service.)

One roll of duct tape or riggers tape (NSN 7510-00-266-5016) to protect foam pad and canvas from wear and to reinforce gluing action of spray adhesive. (Duct tape or riggers tape is available from self-service.)

One roll para-cord or 5-50 cord to tie both the inner and outer handguards together.

One roll of thread strong enough to sew canvas material together.

Sixteen grommets, one for each of the four corners of the inner and outer handguards.

One foam sleeping mat (NSN 8465-01-109-3369). Use unserviceable foam sleeping mats obtained from Defense Reutilization and Marketing Office.

## Tools

The following tools are needed to make a pugil stick:

Tape measure, ruler, or yardstick.
Scissors.
Grommet machine.
Knife.
Sewing machine.

## Directions

To make a pugil stick, Marines perform the following steps.

### Preparing Materials

*Step 1.* Cut foam sleeping mat into the dimensions listed above. Cut mats to double thickness.

*Step 2.* Use spray adhesive to adhere the two 9" by 27" pieces of foam together. Use the spray adhesive to adhere the two 11" by 19" pieces of foam together. Let adhesive dry before using the double-thick foam for handguards.

*Step 3.* Cut canvas material to make two coverings for the inner handguards. Cut material to make two coverings for the outer handguards. Two inner handguards and two outer handguards are required for each pugil stick.

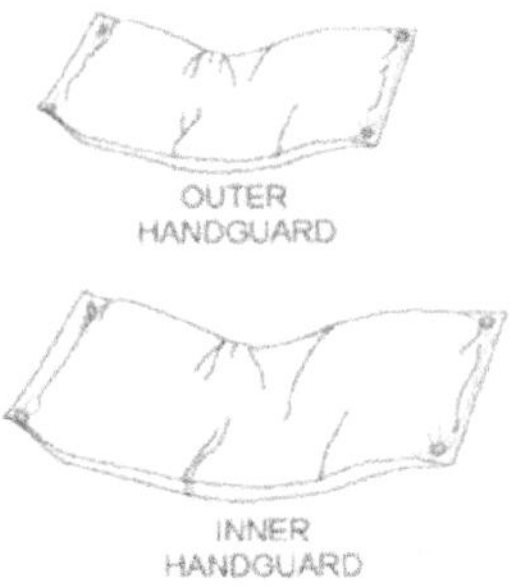


*Step 4.* Use a sewing machine to make a sleeve out of the 9" by 27" canvas. Sew down three sides. Insert foam pad inside the canvas sleeve and then sew down the fourth side. Use the grommet machine to create holes reinforced with metal grommets in each of the four corners. Once finished, the inner handguards are complete.

*Note: It is recommended that the Fabric Repair Shop be used to sew canvas sleeves and apply grommets.*

*Step 5.* Repeat step 4 to make the other inner handguard and both outer handguards.

### Constructing Ends, End Plugs, and End Caps

*Step 1.* Apply spray adhesive on approximately 8 inches of one end of the dowel. This adhesive serves to glue the foam pad to the wooden dowel.

*Step 2.* Apply spray adhesive on one side of the 11" by 54" foam pad. This causes the foam pad to stick to itself when it is rolled around the end of the wooden dowel. Ensure 3 inches of the rolled foam pad extends past the wooden dowel for stability.

*Step 3.* Apply spray adhesive on one side of 3" by 14 1/2" foam pad. Roll pad into a tight cylinder.

*Step 4.* Apply spray adhesive on the exposed end of the dowel and fill the hole in the end with the end plug. The end plug prevents the wooden dowel from protruding beyond the edge of the rolled foam padding. This is an important safety precaution to prevent injury.

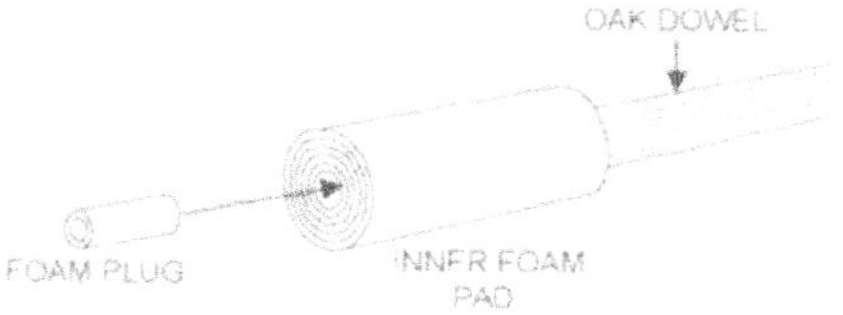


*Step 5.* Use duct tape or riggers tape to secure and reinforce the foam pad once the end plug is inserted in the end.

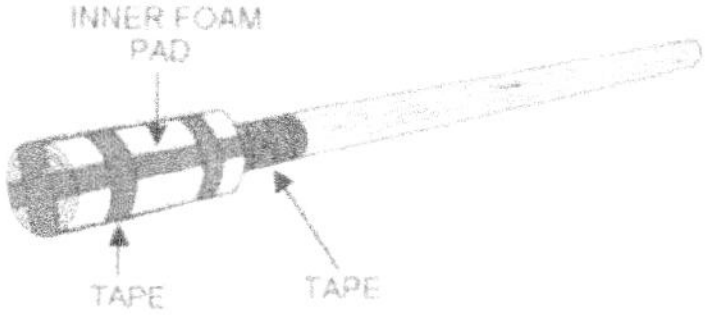


*Step 6.* Measure the dimensions of the ends (typically, the length of the cylinder should be 12 inches long to allow for taping). Cut and sew canvas material in the shape of a cylinder with one open end. Once complete, this becomes the end cap. The end cap is a colored sleeve that covers the rolled and taped foam pad at each end of the pugil stick.

*Note: It is recommended that the Fabric Repair Shop be used to make the end cap.*

*Step 7.* Slide one of the end caps over the rolled foam pad and tape it to the dowel.

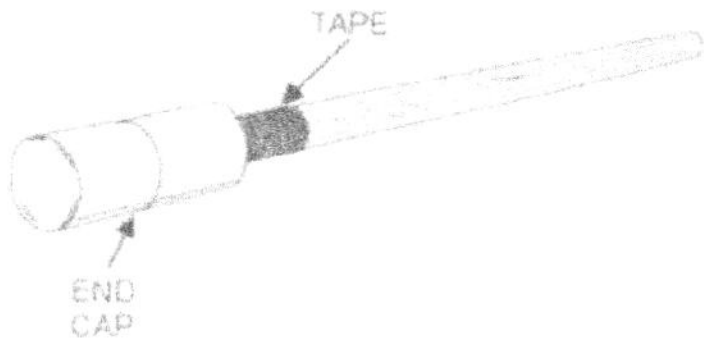


### Building the Centerguard

*Step 1.* Spray one side of the 7" by 13" foam pad with spray adhesive and wrap it tightly around the center of the dowel.

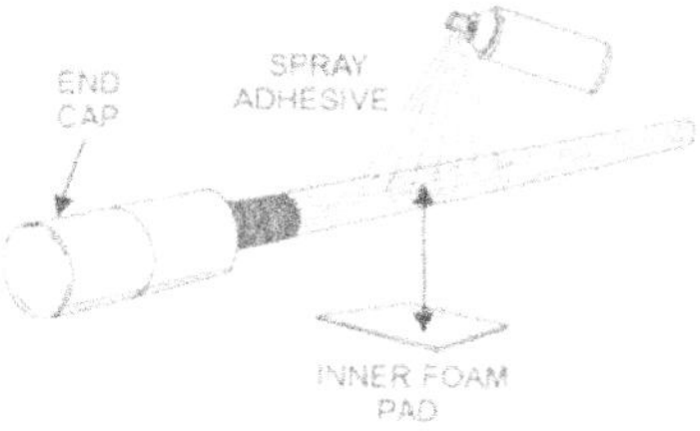

*Step 2.* Wrap duct tape or riggers tape around the foam pad to secure.

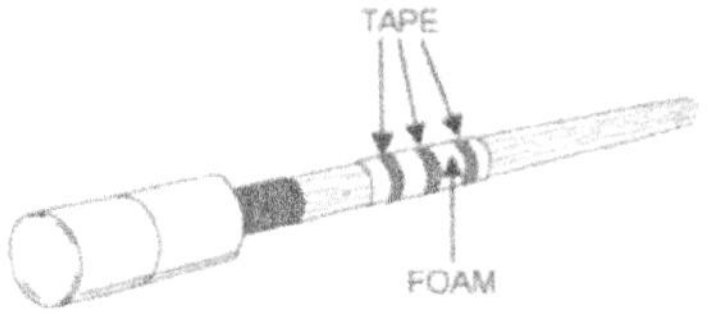


*Step 3.* Use a sewing machine to make a tubular pugil stick sock out of the 8" by 14" piece of canvas. Slide the pugil stick sock over the rolled foam pad and secure it with the duct tape or riggers tape. Duct tape or riggers tape protects the edges of the centerguard and reinforces the gluing effects of the spray adhesive.

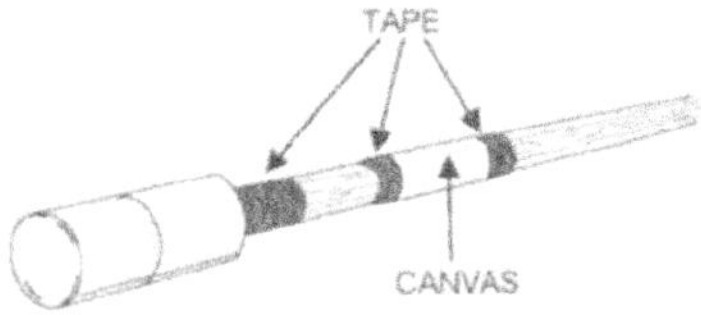


*Note: It is recommended that the Fabric Repair Shop be used to make the pugil stick sock.*

#### Finishing the Opposite End Cap

After one end cap is complete and the centerguard is finished, then repeat steps 1 through 7 in Constructing Ends, End Plugs, and End Caps for the other end of the pugil stick. Ensure that the opposite color end cap is used.

#### Fastening Handguards

*Step 1.* Place the inner handguards (longer piece) around the dowel. The inner handguard is placed on first because it protects the wrist. One inner handguard is needed for the left hand and one for the right hand.

*Step 2.* Fold the inner handguard around the pugil stick and run the para-cord or 5-50 cord through the metal grommets.

*Step 3.* Tighten down the corner ends of the inner handguards by pulling on the para-cord or 5-50 cord.

*Step 4.* Tie off the cord with a square knot above the metal grommets.

*Step 5.* Place the outer handguards (shorter piece) over the inner handguards (longer piece). Repeat steps 1-4 for the other handguard. The outer handguards provide additional protection to the knuckles and fingers and provides additional stability for the inner handguards.

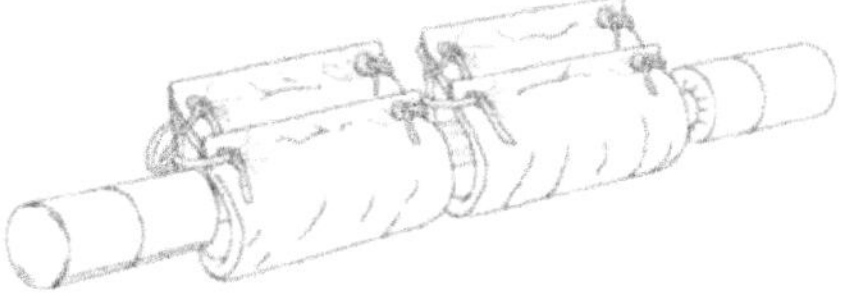

## Serviceability Inspection

Prior to any training, all pugil sticks are inspected to ensure they are fit for use. If any of the following conditions exist, the pugil stick is unsafe and will not be used:

Stick protrudes through the end caps.
Frayed or worn edges exist on canvas surfaces.
The stick can be felt through the pads or the padding is too soft.
End caps or handguards are not securely attached.

(reverse blank)

# APPENDIX B

# SAFETY PRECAUTIONS DURING TRAINING

## 1. General Safety Precautions

When training close combat techniques, certain safety precautions must be adhered to to prevent injuries.

Most training should be conducted on a training area with soft footing such as a sandy or grassy area. If training mats are available, they should be used. A hard surface area is not appropriate for close combat training.

All techniques should be executed slowly at first. Marines can increase the speed of execution as they become more proficient. Marine-on-Marine training that requires contact (chokes, throws, ground fighting, and unarmed restraints and manipulation) should not be executed at full force or full speed.

If a technique is applied to the point that a Marine is uncomfortable, the Marine must "tap out." This indicates immediate release of the pressure being applied or to immediately stop the technique. The Marine "taps out" by firmly tapping his hand several times on any part of the opponent's body that will get his attention or by saying *stop*.

Second Impact Syndrome occurs when a second concussion develops within hours, days, or weeks following a prior concussion (and before recovery from the first concussion). Second Impact Syndrome causes rapid brain swelling and can cause death. Marines who experience headaches or other symptoms following training must be examined by appropriate medical personnel. These symptoms can include, but are not limited to, blurred vision, ringing in the ears, dialation of the pupils, bleeding from the ears or mouth, slurred speech, swelling in head or neck area, or any unnatural discoloration of head or neck. They should not be allowed to participate in pugil stick training or any other activity where a heavy blow might be sustained for a minimum of 7 days after the symptoms have subsided.

## 2. Safety Precautions for Individual Techniques

### Falls

When training Marines to fall, they should progress from the ground, to a kneeling or squatting position, and then to a standing position. This ensures they are comfortable and understand the technique before progressing to executing falls from a higher profile. This instructional technique vastly reduces the risk of injury.

### Strikes and Punches

When training Marines to strike and punch, they begin by executing the techniques "in the air." As they become more proficient, they execute strikes on equipment (when available) such as an air shield or a heavy bag. At no time should they be permitted to execute strikes on another student.

### Chokes and Ground Fighting Chokes

When training Marines to execute chokes, they will not apply pressure to the opponent's throat during training because the trachea and windpipe can be crushed. During training, Marines should practice the proper procedures for blood chokes. They should never execute air chokes. At no time during training should a choke be applied at full force or full speed. No choke should be held for longer than 5 seconds.

## Unarmed Restraints and Manipulation

When training Marines to execute unarmed restraints and manipulation techniques, they utilize slow and steady pressure. Never apply these techniques at full force or at full speed.

## Pugil Stick Training

A blow to the head during training will bruise the brain. A second blow to the head can cause death. This is known as the Second Impact Syndrome. There must be 7 days between pugil stick training to prevent injury or death.

## RANGER CREED

**R**ecognizing that I volunteered as a Ranger, fully knowing the hazards of my chosen profession, I will always endeavor to uphold the prestige, honor, and high esprit de corps of the Rangers.

**A**cknowledging the fact that a Ranger is a more elite Soldier who arrives at the cutting edge of battle by land, sea, or air, I accept the fact that as a Ranger my country expects me to move further, faster, and fight harder than any other soldier.

**N**ever shall I fail my comrades I will always keep myself mentally alert, physically strong, and morally straight and I will shoulder more than my share of the task whatever it may be, one hundred percent and then some.

**G**allantly will I show the world that I am a specially selected and well trained Soldier. My courtesy to superior officers, neatness of dress, and care of equipment shall set the example for others to follow.

**E**nergetically will I meet the enemies of my country. I shall defeat them on the field of battle for I am better trained and will fight with all my might. Surrender is not a Ranger word. I will never leave a fallen comrade to fall into the hands of the enemy and under no circumstances will I ever embarrass my country.

**R**eadily will I display the intestinal fortitude required to fight on to the Ranger objective and complete the mission, though I be the lone survivor.

## STANDING ORDERS, ROGERS' RANGERS

1. Don't forget nothing.
2. Have your musket clean as a whistle, hatchet scoured, sixty rounds powder and ball, and be ready to march at a minute's warning.
3. When you're on the march, act the way you would if you was sneaking up on a deer. See the enemy first.
4. Tell the truth about what you see and what you do. There is an army depending on us for correct information. You can lie all you please when you tell other folks about the Rangers, but don't never lie to a Ranger or officer.
5. Don't never take a chance you don't have to.
6. When we're on the march we march single file, far enough apart so one shot can't go through two men.
7. If we strike swamps, or soft ground, we spread out abreast, so it's hard to track us.
8. When we march, we keep moving till dark, so as to give the enemy the least possible chance at us.
9. When we camp, half the party stays awake while the other half sleeps.
10. If we take prisoners, we keep' em separate till we have had time to examine them, so they can't cook up a story between' em.
11. Don't ever march home the same way. Take a different route so you won't be ambushed.
12. No matter whether we travel in big parties or little ones, each party has to keep a scout 20 yards ahead, 20 yards on each flank, and 20 yards in the rear so the main body can't be surprised and wiped out.
13. Every night you'll be told where to meet if surrounded by a superior force.
14. Don't sit down to eat without posting sentries.
15. Don't sleep beyond dawn. Dawn's when the French and Indians attack.
16. Don't cross a river by a regular ford.
17. If somebody's trailing you, make a circle, come back onto your own tracks, and ambush the folks that aim to ambush you.
18. Don't stand up when the enemy's coming against you. Kneel down, lie down, hide behind a tree.
19. Let the enemy come till he's almost close enough to touch, then let him have it and jump out and finish him up with your hatchet.

*--MAJOR ROBERT ROGERS, 1759*

# RANGER HISTORY

The history of the American Ranger is a long and colorful saga of courage, daring and outstanding leadership. It is a story of men whose skills in the art of fighting have seldom been surpassed. Only the highlights of their numerous exploits are told here.

Rangers primarily performed defensive missions until Benjamin Church's Company of Independent Rangers from Plymouth Colony proved successful in raiding hostile Indians during King Phillip's War in 1675. In 1756, Major Robert Rogers, a native of New Hampshire, recruited nine companies of American colonists to fight for the British during the French and Indian War. Ranger techniques and methods of operation were an inherent characteristic of the American frontiersmen; however, Major Rogers was the first to capitalize on them and incorporate them into the fighting doctrine of a permanently organized fighting force.

The method of fighting used by the first Rangers was further developed during the Revolutionary War by Colonel Daniel Morgan, who organized a unit known as "Morgan's Riflemen". According to General Burgoyne, Morgan's men were "....the most famous corps of the Continental Army, all of them crack shots."

Francis Marion, the "Swamp Fox," organized another famous Revolutionary War Ranger element known as "Marion's Partisans." Marion's Partisans, numbering anywhere from a handful to several hundred, operated both with and independent of other elements of General Washington's Army. Operating out of the Carolina swamps, they disrupted British communications and prevented the organization of loyalists to support the British cause, substantially contributing to the American victory.

The American Civil War was again the occasion for the creation of special units such as Rangers. John S. Mosby, a master of the prompt and skillful use of cavalry, was one of the most outstanding Confederate Rangers. He believed that by resorting to aggressive action he could compel his enemies to guard a hundred points. He would then attack one of the weakest points and be assured numerical superiority.

With America's entry into the Second World War, Rangers came forth to add to the pages of history. Major William O. Darby organized and activated the 1st Ranger Battalion on June19, 1942 at Carrickfergus, North Ireland. The members were all hand-picked volunteers; 50 participated in the gallant Dieppe Raid on the northern coast of France with British and Canadian commandos. The 1st, 3rd, and 4th Ranger Battalions participated with distinction in the North African, Sicilian and Italian campaigns. Darby's Ranger Battalions spearheaded the Seventh Army landing at Gela and Licata during the Sicilian invasion and played a key role in the subsequent campaign which culminated in the capture of Messina. They infiltrated German lines and mounted an attack against Cisterna, where they virtually annihilated an entire German parachute regiment during close in, night, bayonet, and hand-to-hand fighting.

The 2nd and 5th Ranger Battalions participated in the D-Day landings at Omaha Beach, Normandy. It was during the bitter fighting along the beach that the Rangers gained their official motto. As the situation became critical on Omaha Beach, the division commander of the 29th Infantry Division stated that the entire force must clear the beach and advance inland. He then turned to Lieutenant Colonel Max Schneider, Commander of the 5th Ranger Battalion, and said, "Rangers, lead the way." The 5th Ranger Battalion spearheaded the breakthrough and thus enabled the allies to drive inland away from the invasion beaches.

The 6th Ranger Battalion, operating in the Pacific, conducted Ranger type missions behind enemy lines which involved reconnaissance and hard-hitting, long-range raids. They were the first American contingent to return to the Philippines, destroying key coastal installations prior to the invasion. A reinforced company from the 6th Ranger Battalion formed the rescue force which liberated American and allied prisoners of war from the Japanese prison camp at Cabanatuan.

Another Ranger-type unit was the 5307th Composite Unit (Provisional), organized and trained as a long-range penetration unit for employment behind enemy lines in Japanese occupied Burma. The unit commander was Brigadier General (later Major General) Frank D. Merrill. Its 2,997 officers and men became popularly known as "Merrill's Marauders."

The men composing Merrill's Marauders were volunteers from the 5th, 154th, and 33rd Infantry Regiments and from other Infantry regiments engaged in combat in the southwest and South Pacific. These men responded to a call from then Chief of Staff, General George C. Marshall, for volunteers for a hazardous mission. These volunteers were to have a high state of physical ruggedness and stamina and were to come from jungle-trained and jungle-tested units.

Prior to their entry into the Northern Burma Campaign, Merrill's Marauders trained in India under the overall supervision of Major General Orde C. Wingate, British Army. There, they were trained from February to June 1943 in long-range penetration tactics and techniques of the type developed and first employed by General Wingate. The operations of the Marauders were closely coordinated with those of the Chinese 22nd and 38th Divisions in a drive to recover northern Burma and clear the way for the construction of Ledo Road, which was to link the Indian railhead at Ledo with the old Burma Road to China. The Marauders marched and fought through jungle and over mountains from Hukwang Valley in northwest Burma to Myitkyina and the Irrawaddy River. In 5 major and 30 minor engagements, they met and defeated the veteran soldiers of the Japanese 18th Division. Operating in the rear of the main force of the Japanese, they prepared the way for the southward advances of the Chinese by disorganizing supply lines and communications. The climax of the Marauder's operations was the capture of Myitkyina Airfield, the only all-weather strip in northern Burma. This was the final victory of "Merrill's Marauders," which disbanded in August 1944. Remaining personnel were consolidated into the 475th Infantry Regiment, which fought its last battle on February 3, and 4,1945, at Loi-Kang Ridge, China. This Infantry Regiment would serve as the forefather of today's 75th Ranger Regiment.

Shortly after the outbreak of the Korean War in June 1950, the 8th Army Ranger Company was formed of volunteers from American units in Japan. The Company was trained in Korea and distinguished itself in combat during the drive to the Yalu River, performing task force and spearhead operations. In November 1950 during the massive Chinese intervention, this small unit, though vastly outnumbered, withstood five enemy assaults on its position.

In September 1950, a Department of the Army message called for volunteers to be trained as Airborne Rangers. In the 82nd Airborne Division, five thousand regular Army paratroopers volunteered, and from that number nine hundred men were selected to form the initial eight Airborne Ranger Companies. An additional nine companies were formed from volunteers of regular Army and National Guard Infantry Divisions. These seventeen Airborne Ranger companies were activated and trained at Fort Benning, Georgia, with most receiving additional training in the mountains of Colorado.

In 1950 and 1951, some 700 men of the 1st, 2nd, 3rd, 4th, 5th. and 8th Airborne Ranger companies fought to the front of every American Infantry Division in Korea. Attacking by land, water, and air, these six Ranger companies conducted raids, deep penetrations, and ambush operations against North Korean and Chinese forces. They were the first Rangers in history to make a combat jump. After the Chinese intervention, these Rangers were the first Americans to re-cross the 38th parallel. The 2nd Airborne Ranger Company was the only African American Ranger unit in the history of the American Army. The men of the six Ranger companies who fought in Korea paid the bloody price of freedom. One in nine of this gallant brotherhood died on the battlefields of Korea.

Other Airborne Ranger companies led the way while serving with infantry divisions in the United States, Germany, and Japan. Men of these companies volunteered and fought as members of line infantry units in Korea. One Ranger, Donn Porter, would be posthumously awarded the Medal of Honor. Fourteen Korean War Rangers became general officers and dozens became colonels, senior noncommissioned officers, and leaders in civilian life. They volunteered for the Army, the Airborne, the Rangers, and for combat. The first men to earn and wear the coveted Ranger Tab, these men are the original Airborne Rangers.

In October 1951, the Army Chief of Staff, General J. Lawton Collins directed, "Ranger training be extended to all combat units in the Army." The Commandant of the Infantry School was directed to establish a Ranger Department. This new department would develop and conduct a Ranger course of instruction. The objective was to raise the standard of training in all combat units. This program was built upon what had been learned from the Ranger Battalions of World War II and the Airborne Ranger companies of the Korean conflict.

During the Vietnam Conflict, fourteen Ranger companies consisting of highly motivated volunteers served with distinction from the Mekong Delta to the DMZ. Assigned to separate brigade, division, and field force units, they conducted long-range reconnaissance and exploitation operations into enemy-held areas providing valuable combat intelligence. Initially designated at LRRP, then LRP companies, these units were later designated as C, D,E,F,G,H,I,K,L,M,N,O and P (Ranger) 75th Infantry.

Following Vietnam, recognizing the need for a highly trained and highly mobile reaction force, the Army Chief of Staff, General Abrams directed the activation of the first battalion-sized Ranger units since World War II, the 1st and 2nd Battalions (Ranger), 75th Infantry. The 1st Battalion was trained at Fort Benning, Georgia and was activated February 8, 1974 at Fort Stewart, Georgia with the 2nd Battalion being activated on October 3, 1974. The 1st Battalion is now located at Hunter Army Airfield, Georgia and the 2nd Battalion at Fort Lewis, Washington.

The farsightedness of General Abrams' decision, as well as the combat effectiveness of the Ranger battalions, was proven during the United States' invasion of the island of Grenada in October 1983 to protect American citizens there, and to restore democracy. As expected, Rangers led the way! During this operation, code named "Urgent Fury," the Ranger battalions conducted a daring, low level airborne assault (from 500 feet) to seize the airfield at Point Salines, and then continued operations for several days to eliminate pockets of resistance, and rescue American medical students.

As a result of the demonstrated effectiveness of the Ranger battalions, the Department of the Army announced in 1984 that it was increasing the strength of Ranger units to its highest level in 40 years. To do this, it activated another Ranger battalion as well as a Ranger Regimental Headquarters. These new units, the 3rd Battalion (Ranger), 75th Infantry, and Headquarters Company (Ranger), 75th Infantry, have increased the Ranger strength of the Army to over 2,000 soldiers actually assigned to Ranger units. On February 3, 1986, the 75th Infantry was renamed the 75th Ranger Regiment.

On December 20, 1989, the 75th Ranger Regiment was once again called upon to demonstrate its effectiveness in combat. For the first time since its reorganization in 1984, the Regimental Headquarters and all three Ranger battalions were deployed on Operation "Just Cause" in Panama. During this operation, the 75th Ranger Regiment spearheaded the assault into Panama by conducting airborne assaults onto Torrijos/Tocumen Airport and Rio Hato Airfield to facilitate the restoration of democracy in Panama, and protect the lives of American citizens. Between December 20, 1989 and January 7, 1990, numerous follow-on missions were performed in Panama by the Regiment.

Early in 1991, elements of the 75th Ranger Regiment deployed to Saudi Arabia in support of Operation Desert Storm.

In August 1993, elements of the 75th Ranger Regiment deployed to Somalia in support of Operation Restore Hope, and returned November 1993.

In 1994, elements of the 75th Ranger Regiment deployed to Haiti in support of Operation Uphold Democracy.

In 2000 – 2001, elements of the 75th Ranger Regiment deployed to Kosovo in support of Operation Joint Guardian.

Since September 11, 2001, the 75th Ranger Regiment has led the way in the Global War on Terrorism. In October 2001, elements of the 75th Ranger Regiment deployed to Afghanistan in support of Operation Enduring Freedom. In March 2003, elements of the Regiment deployed in support of Operation Iraqi Freedom.

The performance of these Rangers significantly contributed to the overall success of these operations and upheld the Ranger tradition of the past. As in the past, the Regiment stands ready to execute its mission to conduct special operations in support of the United States' policies and objectives.

## RANGER MEDAL OF HONOR WINNERS

| | | | |
|---|---|---|---|
| Millett, Lewis L. Sr.......... | Captain........................ | Feb 7 1951 ..................... | Co. E 2/27th Infantry |
| Porter, Donn F.*............. | Sergeant...................... | Sept 7 1952.................... | Co. G 2/14th Infantry |
| Mize, Ola L. ................... | Sergeant...................... | June 10-11 1953............ | Co. K 3/15th Infantry |
| Dolby, David C............... | Staff Sergeant ............. | May 21 1966................... | Co. B 1/8th (ABN) Calvary |
| Foley, Robert F............. | Captain........................ | Nov 5 1966 ..................... | Co. A 2/27th Infantry |
| Zabitosky, Fred M. ......... | Staff Sergeant ............. | Feb 19 1968 ................... | 5th Special Forces |
| Bucha, Paul W............... | Captain........................ | May 16-19 1968............. | Co. D 3/187 Infantry |
| Rabel, Laszlo*................ | Staff Sergeant ............. | Nov 13 1968 ................... | 74th Infantry (LRRP) |
| Howard, Robert L........... | Sergeant First Class.... | Dec 30 1968 ................... | 5th Special Forces |
| Law, Robert D. *............. | Specialist 4.................. | Feb 22 1969 ................... | Co. I 75th Infantry (Ranger) |
| Kerrey, J. Robert............ | Lieutenant.................... | Mar 14 1969 ................... | Seal Team 1 |
| Doane, Stephen H.*....... | 1st Lieutenant.............. | Mar 25 1969 ................... | Co. B 1/5th Infantry |
| Pruden, Robert J.* ......... | Staff Sergeant ............. | Nov 22 1969................... | Co. G 75th Infantry (Ranger) |
| Littrell, Gary L. .............. | Sergeant First Class.... | April 4-8 1970 ................. | Advisory Team 21 (Ranger) |
| Lucas, Andre C.*............ | Lt Colonel .................... | Jul 1-23 1970................. | HHC 2/506 Infantry |
| Gordon, Gary I. *............ | Master Sergeant.......... | Oct 3 1993....................... | Task Force Ranger |
| Shughart, Randall D. * ... | Sergeant First Class.... | Oct 3 1993....................... | Task Force Ranger |

*Awarded posthumously

# SUMMARY OF CHANGES

| | |
|---|---|
| **THROUGHOUT** | UPDATED doctrine, technology, terms; clarified verbiage |
| **CHAPTER 1** | UPDATED duties and responsibilities<br>ADDED FO |
| **CHAPTER 2** | UPDATED WARNO<br>ADDED sample WARNO<br>UPDATED OPORD<br>ADDED terrain model |
| **CHAPTER 3** | ADDED PLOT-CR<br>ADDED risk estimate distances |
| **CHAPTER 4** | IDENTIFIED TL positions during movement |
| **CHAPTER 5** | REORGANIZED combat operations |
| **CHAPTER 7** | ADDED section on antennas |
| **CHAPTER 8** | UPDATED aircraft |
| **CHAPTER 9** | UPDATED rope bridge equipment<br>UPDATED sequence and responsibilities |
| **CHAPTER 10** | UPDATED equipment and descriptions<br>UPDATED climbing commands |
| **CHAPTER 11** | NA |
| **CHAPTER 12** | ADDED first aid techniques |
| **CHAPTER 13** | ADDED figure 'Lighting a time fuse with a match'<br>ADDED expedient explosives<br>ADDED table of US demolitions<br>ADDED breaching and timber cutting charges |
| **CHAPTER 15** | NEW chapter covering vehicle convoy operations |
| **APPENDIX A** | NEW - Resources |
| **APPENDIX B** | NEW - Quick Reference Cards |

# TABLE OF CONTENTS

# Chapter 1
# LEADERSHIP

Leadership, the most essential element of combat power, gives purpose, direction, and motivation in combat. The leader balances and maximizes maneuver, firepower, and protection against the enemy. This chapter discusses how he does this by exploring the principles of leadership (BE, KNOW, DO); the duties, responsibilities, and actions of an effective leader; and the leader's assumption of command.

1-1. **PRINCIPLES**. The principles of leadership are BE, KNOW, and DO.

*BE*

- Technically and tactically proficient
- Able to accomplish to standard all tasks required for the wartime mission.
- Courageous, committed, and candid.
- A leader with integrity.

*KNOW*

- The four major factors of leadership and how they affect each other are--
  - ---Led
  - ---Leader
  - ---Situation
  - ---Communications
- Yourself, and the strengths and weaknesses in your character, knowledge, and skills. Seek continual self-improvement, that is, develop your strengths and work to overcome your weaknesses.
- Your Rangers, and look out for their well-being by training them for the rigors of combat, taking care of their physical and safety needs, and disciplining and rewarding them.

*DO--*

- Seek responsibility and take responsibility for your actions; exercise initiative; demonstrate resourcefulness; and take advantage of opportunities on the battlefield that will lead to you to victory; accept fair criticism, and take corrective actions for your mistakes.
- Assess situations rapidly, make sound and timely decisions, gather essential information, announce decisions in time for Rangers to react, and consider the short- and long-term effects of your decision.
- Set the example by serving as a role model for your Rangers. Set high but attainable standards; be willing do what you require of your Rangers; and share dangers and hardships with them.
- Keep your subordinates informed to help them make decisions and execute plans within your intent, encourage initiative, improve teamwork, and enhance morale.
- Develop a sense of responsibility in subordinates by teaching, challenging, and developing them. Delegate to show you trust them. This makes them want more responsibility.
- Ensure the Rangers understand the task; supervise them, and ensure they accomplish it. Rangers need to know what you expect: when and what you want them to do, and to what standard.
- Build the team by training and cross-training your Rangers until they are confident in their technical and tactical abilities. Develop a team spirit that motivates them to go willingly and confidently into combat.
- Know your unit's capabilities and limitations, and employ them accordingly.

1-2. **DUTIES, RESPONSIBILITIES, AND ACTIONS**. To complete all assigned tasks, every Ranger in the patrol must do his job. Each must accomplish his specific duties and responsibilities and be a part of the team.

*PLATOON LEADER (PL)*

- Is responsible for what the patrol does or fails to do. This includes tactical employment, training, administration, personnel management, and logistics. He does this by planning, making timely decisions, issuing orders, assigning tasks, and supervising patrol activities. He must know his Rangers and how to employ the patrol's weapons. He is responsible for positioning and employing all assigned or attached crew-served weapons and employment of supporting weapons.
- Establishes time schedule using backwards planning. Considers time for execution, movement to the objective, and the planning and preparation phase of the operation.
- Takes the initiative to accomplish the mission in the absence of orders. Keeps higher informed by using periodic situation reports (SITREP).
- Plans with the help of the platoon sergeant (PSG), squad leaders, and other key personnel (team leaders, FO, attachment leaders).
- Stays abreast of the situation through coordination with adjacent patrols and higher HQ; supervises, issues FRAGOs, and accomplishes the mission.
- If needed to perform the mission, requests more support for his patrol from higher headquarters.
- Directs and assists the platoon sergeant in planning and coordinating the patrol's sustainment effort and casualty evacuation (CASEVAC) plan.
- During planning, receives on-hand status reports from the platoon sergeant and squad leaders.
- Reviews patrol requirements based on the tactical plan.
- Ensures that all-round security is maintained at all times.
- Supervises and spot-checks all assigned tasks, and corrects unsatisfactory actions.
- During execution, positions himself where he can influence the most critical task for mission accomplishment; usually with the main effort, to ensure that his platoon achieves its decisive point
- Is responsible for positioning and employing all assigned and attached crew-served weapons.
- Commands through his squad leaders IAW the intent of the two levels higher commanders.
- Conducts rehearsals.

*PLATOON SERGEANT (PSG)*

The PSG is the senior NCO in the patrol and second in succession of command. He helps and advises the patrol leader, and leads the patrol in the leader's absence. He supervises the patrol's administration, logistics, and maintenance, and he prepares and issues paragraph 4 of the patrol OPORD.

*DUTIES*

- Organizes and controls the patrol CP IAW the unit SOP, patrol leader's guidance, and METT-TC factors.
- Receives squad leader's requests for rations, water, and ammunition. Work with the company first sergeant or XO to request resupply. Directs the routing of supplies and mail.
- Supervises and directs the patrol medic and patrol aid-litter teams in moving casualties to the rear.
- Maintains patrol status of personnel, weapons, and equipment; consolidates and forwards the patrol's casualty reports (DA Forms 1155 and 1156); and receives and orients replacements.
- Monitors the morale, discipline, and health of patrol members.
- Supervises task-organized elements of patrol:
  - — Quartering parties.
  - — Security forces during withdrawals.
  - — Support elements during raids or attacks.
  - — Security patrols during night attacks.
- Coordinates and supervises patrol resupply operations.
- Ensures that supplies are distributed IAW the patrol leader's guidance and direction.
- Ensures that ammunition, supplies, and loads are properly and evenly distributed (a critical task during consolidation and reorganization).
- Ensures the casualty evacuation plan is complete and executed properly.

- Ensures that the patrol adheres to the platoon leader's time schedule.
- Assists the platoon leader in supervising and spot-checking all assigned tasks, and corrects unsatisfactory actions.

*ACTIONS DURING MOVEMENT AND HALTS*

- Takes actions necessary to facilitate movement.
- Supervises rear security during movement.
- Establishes, supervises, and maintains security during halts.
- Knows unit location.
- Performs additional tasks as required by the patrol leader and assists in every way possible. Focuses on security and control of patrol.

*ACTIONS AT DANGER AREAS*

- Directs positioning of near-side security (usually conducted by the trail squad or team).
- Maintains accountability of personnel.

*ACTIONS ON THE OBJECTIVE AREA*

- Assists with ORP occupation.
- Supervises, establishes, and maintains security at the ORP.
- Supervises the final preparation of men, weapons, and equipment in the ORP IAW the patrol leader's guidance.
- Assists the patrol leader in control and security.
- Supervises the consolidation and reorganization of ammunition and equipment.
- Establishes, marks, and supervises the planned CCP, and ensures that the personnel status (to include WIA/KIA) is accurately reported to higher.
- Performs additional tasks assigned by the patrol leader and reports status to platoon leader.

*ACTIONS IN THE PATROL BASE*

- Assists in patrol base occupation.
- Assist in establishing and adjusting perimeter.
- Enforces security in the patrol base.
- Keeps movement and noise to a minimum.
- Supervises and enforces camouflage.
- Assigns sectors of fire.
- Ensures designated personnel remain alert and equipment is maintained to a high state of readiness.
- Requisitions supplies, water, and ammunition, and supervises their distribution.
- Supervises the priority of work and ensures its accomplishment.
  - Security plan.
    - Ensures crew-served weapons have interlocking sectors of fire.
    - Ensures Claymores are emplaced to cover dead space.
    - Ensures range cards and sector sketch are complete.
  - Alert plan.
  - Evacuation plan.
  - Withdrawal plan.
  - Alternate patrol base.
  - Maintenance plan.
  - Hygiene plan.
  - Messing plan.
  - Water plan.
  - Rest plan.
- Performs additional tasks assigned by the patrol leader and assists him in every way possible.

*SQUAD LEADER (SL)*

Is responsible for what the squad does or fails to do. He is a tactical leader who leads by example.

*DUTIES*

- Completes casualty feeder reports and reviews the casualty reports completed by squad members.
- Directs the maintenance of the squad's weapons and equipment.

- Inspects the condition of Rangers' weapons, clothing, and equipment.
- Keeps the PL and PSG informed on status of squad.
- Submits ACE report to PSG.

*ACTIONS THROUGHOUT THE MISSION*

- Obtains status report from team leaders and submits reports to the PL and PSG.
- Makes a recommendation to the PL/PSG when problems are observed.
- Delegates priority task to team leaders, and supervises their accomplishment IAW squad leader's guidance.
- Uses initiative in the absence of orders.
- Follows the PL's plan and makes recommendations.

*ACTIONS DURING MOVEMENT AND HALTS*

- Ensures heavy equipment is rotated among members and difficult duties are shared.
- Notifies PL of the status of the squad.
- Maintains proper movement techniques while monitoring route, pace, and azimuth.
- Ensures the squad maintains security throughout the movement and at halts.
- Prevents breaks in contact.
- Ensures subordinate leaders are disseminating information, assigning sectors of fire, and checking personnel.

*ACTION IN THE OBJECTIVE AREA*

- Ensures special equipment has been prepared for actions at the objective.
- Maintains positive control of squad during the execution of the mission.
- Positions key weapons systems during and after assault on the objective.
- Obtains status reports from team leaders and ensures ammunition is redistributed and reports status to PL.

*ACTIONS IN THE PATROL BASE*

- Ensures patrol base is occupied according to the plan.
- Ensures that his sector of the patrol base is covered by interlocking fires; makes final adjustments, if necessary.
- Sends out LP or OPs in front of assigned sector. (METT-TC dependent).
- Ensures priorities of work are being accomplished, and reports accomplished priorities to the PL and PSG.
- Adheres to time schedule.
- Ensures personnel know the alert and evacuation plans, and the locations of key leaders, OPs, and the alternate patrol base.

*WEAPONS SQUAD LEADER*

Is responsible for all that the weapons squad does or fails to do. His duties are the same as those of the squad leader. Also, he controls the machine guns in support of the patrol's mission. He advises the PL on employment of his squad.

*DUTIES*

- Supervises machine gun teams to ensure they follow priorities of work.
- Inspects machine gun teams for correct range cards, fighting positions, and understanding of fire plan.
- Supervises maintenance of machine guns, that is, ensures that maintenance is performed correctly, that deficiencies are corrected and reported, and that the performance of maintenance does not violate the security plan.
- Assists PL in planning.
- Positions at halts and danger areas and according to the patrol SOP any machine guns not attached to squads.
- Rotates loads. Machine gunners normally get tired first.
- Submits ACE report to PSG.
- Designates sectors of fire, principal direction of fire (PDF), and secondary sectors of fire for all guns.
- Gives fire commands to achieve maximum effectiveness of firepower:
  - — Shifts fires.
  - — Corrects windage or elevation to increase accuracy.
  - — Alternates firing guns.
  - — Controls rates of fire and fire distribution.
- Knows locations of assault and security elements, and prevents fratricide.
- Reports to the PL.

*TEAM LEADER (TL)*

Controls the movement of his fire team and the rate and placement of fire. To do this, leads from the front and uses the proper commands and signals. Maintains accountability of his Rangers, weapons, and equipment. Ensures his Rangers maintain unit standards in all areas, and are knowledgeable of their tasks and the operation. The following checklist outlines specific duties and responsibilities of team leaders during mission planning and execution. The TL leads by example:

*ACTIONS DURING PLANNING AND PREPARATION*

- Warning Order.
  - — Assists in control of the squad.
  - — Monitors squad during issuance of the order.
- OPORD Preparation.
  - — Posts changes to schedule.
  - — Posts and updates team duties on warning order board.
  - — Submits ammunition and supply requests.
  - — Picks up ammunition and supplies.
  - — Distributes ammunition and special equipment.
  - — Performs all tasks given in the SL's special instructions paragraph.
- OPORD Issuance and Rehearsal.
  - — Monitors squad during issuance of the order.
  - — Assists SL during rehearsals.
- Takes actions necessary to facilitate movement.
- Enforces rear security.
- Establishes, supervises, and maintains security at all times.
- Performs additional tasks as required by the SL, and assists him in every way possible, particularly in control and security.

*ACTION IN THE ORP*

- Assists in the occupation of the ORP.
- Helps supervise, establish, and maintain security.
- Supervises the final preparation of Rangers, weapons, and equipment in the ORP IAW the SL's guidance.
- Assists in control of personnel departing and entering the ORP.
- Reorganizes perimeter after the reconnaissance party departs.
- Maintains communication with higher headquarters.
- Upon return of reconnaissance party, helps reorganize personnel and redistribute ammunition and equipment; ensures accountability of all personnel and equipment is maintained.
- Disseminates PIR to his team.
- Performs additional tasks assigned by the SL.

*ACTIONS IN THE PATROL BASE*

- Inspects the perimeter to ensure team has interlocking sectors of fire; prepares team sector sketch.
- Enforces the priority of work and ensures it is properly accomplished.
- Performs additional tasks assigned by the SL and assists him in every way possible.

*MEDIC*

Assists the PSG in directing aid and litter teams; monitors the health and hygiene of the platoon.

*DUTIES*

- Treats casualties, conducts triage, and assists in CASEVAC under the control of the PSG.
- Aids the PL or PSG in field hygiene matters. Personally checks the health and physical condition of platoon members.
- Requests Class VIII (medical) supplies through the PSG.
- Provides technical expertise to and supervision of combat lifesavers.
- Ensures casualty feeder reports are correct and attached to each evacuated casualty.
- Carries out other tasks assigned by the PL or PSG.

| |
|---|
| *RADIO OPERATOR*<br>Is responsible for establishing and maintaining communications with higher headquarters and within the patrol.<br>*DUTIES DURING PLANNING*<br>• Enters the net at the specified time.<br>• Ensures that all frequencies, COMSEC fills, and net IDs, are preset in squad/platoon radios.<br>• Informs SL and PL of changes to call signs, frequencies, challenge and password, and number combination based on the appropriate time in the ANCD.<br>• Ensures the proper function of all radios and troubleshoots and reports deficiencies to higher.<br>• Weatherproofs all communications equipment.<br>*DUTIES DURING EXECUTION*<br>• Serves as en route recorder during all phases of the mission.<br>• Records all enemy contact and reports it to higher in a SALUTE format.<br>• Reports all OPSKEDs to higher.<br>• Consolidates and records all PIR. |
| *FORWARD OBSERVER (FO)*<br>Works for the PL. Serves as the eyes of the FA and mortars. Is mainly responsible for locating targets, and for calling for and adjusting indirect fire support. Knows the terrain where the platoon is operating; knows the tactical situation. Knows the mission, the concept, and the unit's scheme of maneuver and priority of fires.<br>*DUTIES DURING PLANNING*<br>• Selects targets to support the platoon's mission based on the company OPORD, platoon leader's guidance, and analysis of METT-TC factors.<br>• Prepares and uses situation maps, overlays, and terrain sketches.<br>*DUTIES DURING EXECUTION*<br>• Informs the FIST headquarters of platoon activities and of the fire support situation<br>• Selects new targets to support the platoon's mission based on the company OPORD, the platoon leader's guidance, and an analysis of METT-TC factors.<br>• Calls for and adjusts fire support.<br>• Operates as a team with the radio operator.<br>• Selects OPs.<br>• Maintains communications as prescribed by the FSO.<br>• Maintains the current 8-digit coordinate of his location at all times. |

1-3. **ASSUMPTION OF COMMAND**. Any platoon/squad member might have to take command of his element in an emergency, so every Ranger must be prepared to do so. During an assumption of command, situation permitting, the Ranger assuming command accomplishes the following tasks (not necessarily in this order) based on METT-TC:

- *INFORMS* the unit's subordinate leaders of the command and notifies higher.
- *CHECKS* security.
- *CHECKS* crew-served weapons.
- *PINPOINTS* location.
- *COORDINATES* and *CHECKS* equipment.
- *CHECKS* personnel status.
- *ISSUES* FRAGO (if required).
- *REORGANIZES* as needed, maintaining unit integrity when possible.
- *MAINTAINS* noise and light discipline.
- *CONTINUES* patrol base activities, especially security, if assuming command in a patrol base.
- *RECONNOITERS* or, at the least, conducts a map reconnaissance.
- *FINALIZES* plan.
- *EXECUTES* the mission.

NOTES

# Chapter 2
# OPERATIONS

This chapter provides techniques and procedures used by Infantry platoons and squads throughout the planning and execution phases of tactical operations. Specifically, it discusses the troop-leading procedures, combat intelligence, combat orders, and planning techniques and tools needed to prepare a platoon to fight. These topics are time sensitive and apply to all combat operations. When they have time, leaders can plan and prepare in depth. If they have less time, they must rely on previously rehearsed actions, battle drills, and standing operating procedures (SOPs).

## 2-1. TROOP-LEADING PROCEDURES

Troop leading procedures comprise the following steps. They are what a leader does to prepare his unit to accomplish a tactical mission. The TLP starts when the leader is alerted for a mission or receives a change or new mission. He can perform Steps 3 through 8 in any order, or at the same time. (He can also use the tools of the tactician shown in Figure 2-1):

| | |
|---|---|
| 1. Receive the mission. | 5. Reconnoiter. |
| 2. Issue a warning order. | 6. Complete the plan. |
| 3. Make a tentative plan. | 7. Issue the complete order. |
| 4. Initiate movement. | 8. Supervise. |

**Figure 2-1. TOOLS OF THE TACTICIAN RELATIONSHIP**

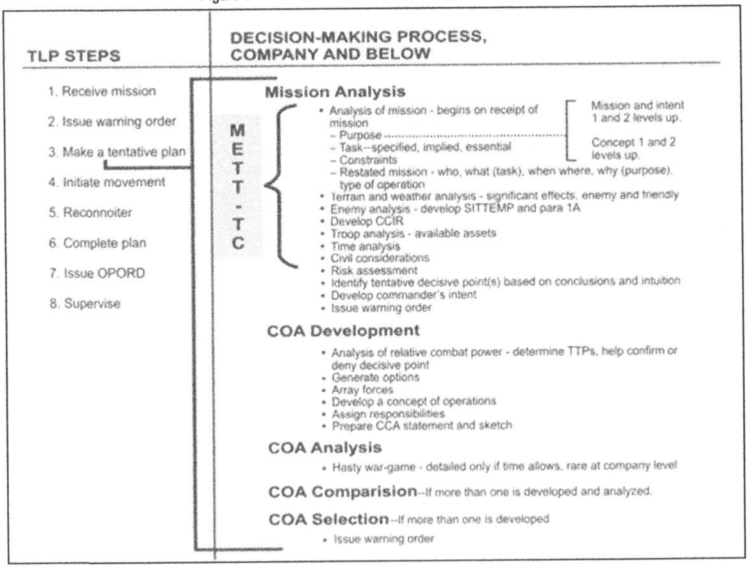

a. **STEP 1--RECEIVE THE MISSION**. The leader may receive the mission in a warning order, an operation order (OPORD), or a fragmentary order (FRAGO). He should use no more than one third of the available time for his own planning and for issuing his OPORD. The remaining two thirds is for subordinates to plan and prepare for the operation. Leaders should also consider other factors such as available daylight and travel time to and from orders and rehearsals.

b. **STEP 2--ISSUE A WARNING ORDER**. The leader provides initial instructions in a warning order. The warning order contains enough information to begin preparation as soon as possible. The warning order mirrors the five-paragraph OPORD format. A warning order may include--

- The mission or nature of the operation (mission statement).
- Time and place for issuance of the operation (coordinating instructions).
- Who is participating in the operation (coordinating instructions).
- Time of the operation (timeline).

c. **STEP 3--MAKE A TENTATIVE PLAN**. The leader uses METT-TC to develop an estimate of the situation, which he will use as the basis for his tentative plan. This set of actions is referred to as the leader's mission analysis:

(1) **Conduct a detailed mission analysis**.

(a) *Concept and Intent.* Higher commanders' concept and intent two levels up. This information is found in paragraph 1b for two levels up and in paragraphs 2 and 3 for one higher.

(b) *Unit Tasks.* Tasks that are clearly stated in the order (Specified Tasks) or tasks that become apparent as the OPORD is analyzed (Implied Tasks).

**EXAMPLE SPECIFIED TASKS**

- Retain hill 545 to prevent envelopment of B Co.
- Provide one squad to the 81-mm platoon to carry ammo.
- Establish an OP vic GA124325 NLT 301500 NOV 89.

**EXAMPLE IMPLIED TASKS**

- Provide security during movement.
- Conduct resupply operations.
- Coordinate with adjacent units.

(c) *Unit Limitations.* The leader next determines all control measures or instructions in the OPORD that restrict his freedom of action; these are called limitations. In every operation, there are some limitations on the unit.

**COMMON LIMITATIONS**

Graphic control measures.
Cross the LD at 100030 OCT 94 (time).
MOPP4 in effect (uniform and environmental considerations).
ADA weapons status, tight; warning status, yellow (rules of engagement).

(d) *Mission-Essential Task(s).* After reviewing all the above factors, the leader identifies his mission-essential task(s). Failure to accomplish a mission-essential task results in the unit's failure to accomplish its primary purpose for that operation. The mission essential task should be found in the maneuver paragraph.

(e) *Restated Mission.* The restated mission statement becomes the focus for the remainder of the estimate process. This is a clear, concise statement of the mission essential task(s) to be accomplished by the unit and the purpose to be achieved. The mission statement will state WHO, WHAT (the task), WHEN (the critical time), WHERE (usually a grid coordinate), and WHY (the purpose the unit must achieve). Some examples of restated missions follow:

| EXAMPLE RESTATED MISSIONS |
|---|
| (WHO) 1st Platoon attacks (WHAT) to seize (WHERE) HILL 482 vic NB 457371 (OBJ BLUE) (WHEN) NLT 09050OZ Dec 92 L 482 (WHY) to enable the company's main effort to destroy enemy command bunker. |
| (WHO) 1st Platoon, C Company defends (WHAT) to destroy from (WHERE) AB163456 to AB163486 to AB123486 to AB123456 (WHEN) NLT 28153OZ Oct 97 (WHY) to prevent enemy forces from enveloping B Company, 1-66 Infantry (L) from the south. |

(2) Analyze the situation and develop a course of action. Each COA must be--suitable, acceptable, feasible, distinguishable, and complete.

(3) Upon developing a COA, the unit leader will assign C2 headquarters, complete generic task organization assigning all organic and attached elements, and prepare COA statement and sketch.

(4) With the restated mission from Step 1 to provide focus, the leader continues the estimate process using the remaining factors of METT-TC.

(a) What is known about the **ENEMY**?

| | |
|---|---|
| ***Composition*** | This is an analysis of the forces and weapons that the enemy can bring to bear. Determine what weapons systems they have available, and what additional weapons and units are supporting him. |
| ***Disposition*** | The enemy's disposition is how he is arrayed on the terrain, such as in defensive positions, in an assembly area, or moving in march formation. |
| ***Strength*** | Percentage strength. |
| ***Recent Activities*** | Identify recent and significant enemy activities that may indicate future intentions. |
| ***Reinforcement Capabilities*** | Determine positions for reserves and estimated time to counterattack or reinforce. |
| ***Possible Courses of Action*** | Determine the enemy's possible COAs. This will help the leader determine how to best array his own forces against the enemy and fight the battle. |

(b) How will **TERRAIN** and **WEATHER** affect the operation? Analyze terrain using OAKOC.

- **Observation and Fields of Fire**. Determine locations that provide the best observation and fields of fire along friendly and enemy avenues of approach, near the objective, and on key terrain. The analysis of fields of fire is mainly concerned with the ability to cover the terrain with direct fire.
- **Avenues of Approach**. Analyze and identify friendly and enemy avenues of approach. Identify avenues of approach en route to the objective, on and around the objective, and for possible enemy counterattack. Also, consider aerial and subterranean avenues.

| | |
|---|---|
| ***Offensive Considerations (Friendly)*** | • How can these avenues support my movement?<br>• What are the advantages/disadvantages of each? (Consider enemy, speed, cover, and concealment.)<br>• What are the likely enemy counterattack routes? |

| | |
|---|---|
| ***Offensive Considerations (Enemy)*** | • How can the enemy use these approaches?<br>• Which avenue is most dangerous? Least? (Prioritize each approach.)<br>• Which avenues would support a counterattack? |

- **Cover and Concealment**. The analysis of cover and concealment is often inseparable from the fields of fires and observation. Weapon positions must have both to be effective and to be survivable. Infantry units are capable of improving poor cover and concealment by digging in and camouflaging their positions. When moving, the terrain is used to provide cover and concealment.
- **Obstacles**. Identify the existing and reinforcing obstacles and hindering terrain that will affect mobility.
- **Key Terrain**. Key terrain is any location or area that the seizure, retention, or control of affords a marked advantage to either combatant. Using the map, aerial photos, and information already gathered, look for key terrain that dominates avenues of approach or the objective area. Next, look for decisive terrain that if held or controlled will have an extraordinary impact on the mission.

(5) Analyze courses of action (war-game). This analysis is conducted by war-gaming friendly courses of action against the enemy's most probable courses of action. The leader can--and should--war-game with his subordinates.

(6) Compare courses of action. The leader compares the COAs and selects the one that is most likely to accomplish the assigned mission. He considers the advantages and disadvantages for each COA. He also considers how the critical events impact on COAs.

(7) Make a decision. The leader selects the COA that he believes has the best chance of accomplishing the mission.

d. **STEP 4--START NECESSARY MOVEMENT**. The unit may need to begin movement while the leader is still planning or forward reconnoitering. This step may occur anytime during the TLP.

e. **STEP 5--RECONNOITER**. If time allows, the leader makes a personal reconnaissance. When time does not allow, the leader must make a map or aerial photo reconnaissance. Sometimes the leader must rely on others, such as scouts, to conduct the reconnaissance.

f. ***STEP 6--COMPLETE THE PLAN***. The leader completes his plan based on the reconnaissance and any changes in the situation.

g. **STEP 7--ISSUE THE COMPLETE ORDER**. Platoon and squad leaders normally issue oral operations orders to aid subordinates in understanding the concept of the mission. Leaders may require subordinates to repeat part of the order, demonstrate it on a terrain model, or sketch their understanding of the operation. Leaders should also quiz their Rangers to ensure that all Rangers understand the mission.

h. **STEP 8--SUPERVISE AND REFINE**. The leader supervises the unit's preparation for combat by conducting rehearsals and inspections.

(1) ***Rehearsals***. Rehearsals include the practice of having squad leaders brief their planned actions in execution sequence to the platoon leader. The leader should conduct rehearsals on terrain that resembles the actual ground, and in similar light conditions. At a minimum, the leader should conduct a backbrief (least preferred), full force rehearsal (most preferred), or reduced force rehearsal (key leaders), and a rock drill.

(a) *Purpose*. The leader uses rehearsals to--

- Practice essential tasks (improve performance).
- Reveal weaknesses or problems in the plan.
- Coordinate the actions of subordinate elements.
- Improve Ranger understanding of the concept of the operation (foster confidence in Rangers).

(b) *Times and Tasks*. The platoon may begin rehearsals of battle drills and other SOP items before the receipt of the operation order. Once the order has been issued, it can rehearse mission specific tasks. Some important tasks to rehearse include--

- Actions on the objective.
- Assaulting a trench, bunker, or building.
- Actions at the assault position.
- Breaching obstacles (mine and wire).
- Using special weapons or demolitions.
- Actions on unexpected enemy contact.

(c) *Types.*

- Backbrief.
  - -- Key leaders sequentially brief the actions required during operation.
  - -- Patrol leader controls.
  - -- Conducted twice: right after FRAGO (confirmation brief), and again after subordinates develop own plan.
- Reduced force.
  - -- Conducted when time is key constraint.
  - -- Conducted when security must be maintained.
  - -- Key leaders normally attend.
  - -- Mock-ups, sand tables, and small scale replicas used.
- Full force.
  - -- Most effective type.
  - -- First executed in daylight and open terrain.
  - -- Secondly conduct in same conditions as operation.
  - -- All Rangers participate.
  - -- May use force on force.
- Techniques.
  - -- Force on force.
  - -- Map (limited value and limited number of attendees).
  - -- Radio (cannot mass leaders; confirms communications).
  - -- Sand table or terrain model (key leaders; includes all control measures).
  - -- Rock drill (similar to sand table/terrain model; subordinates actually move themselves). Very effective for coordinating timing and sequence of events--especially when air assets are used.

(d) *Inspections.* Squad leaders should conduct initial inspections shortly after receipt of the warning order. The platoon sergeant spot-checks throughout the unit's preparation for combat. The platoon leader and platoon sergeant make a final inspection of--

- Weapons and ammunition.
- Uniforms and equipment.
- Mission-essential equipment.
- Soldier's understanding of the mission and their specific responsibilities.
- Communications.
- Rations and water.
- Camouflage.
- Deficiencies noted during earlier inspections.

2-2. **COMBAT INTELLIGENCE**. Gathering information is one of the most important aspects of conducting patrolling operations. The following details what information to collect and how to report it:

a. **Reports**. All information must be quickly, completely, and accurately reported. Use the SALUTE report format for reporting and recording information.

| | |
|---|---|
| SIZE | – Seven enemy personnel |
| ACTIVITY | – Traveling SW |
| LOCATION | – GA123456 |
| UNIT/UNIFORM | – OD uniforms with red six-point star on left shoulder |
| TIME | – 210200JAN99 |
| EQUIPMENT | – Carry one machine gun and one rocket launcher |

b. **Field Sketches**. Try to include a sketch with each report. Include only any aspects of military importance such as targets, objectives, obstacles, sector limits, or troop dispositions and locations (use symbols from FM 1-02). Use notes to explain the drawing, but they should not clutter the sketch.

c. **Captured Documents**. The leader collects documents and turns them in with his reports. He marks each document with the time and place of capture.

d. **Prisoners**. If prisoners are captured during a patrolling operation, they should be treated IAW the Geneva Convention and handled by the 5-S rule:

(1) Search
(2) Silence
(3) Segregate
(4) Safeguard
(5) Speed to rear

e. **Debriefs**. Immediately upon return from a mission, the unit is debriefed. The intelligence officer will generally have a unit-specific format for debriefing a patrol.

2-3. **WARNING ORDER**. Warning orders give subordinates advance notice of upcoming operations. This gives them time to prepare. A warning order must be brief, but complete. *A warning order only authorizes execution when it clearly says so.* (Figure 2-2 shows an example format; Figure 2-3 shows an example warning order.)

**Figure 2-2. WARNING ORDER FORMAT**

## WARNING ORDER ________

*Roll call, pencil/pen/paper, RHB, map, protractor, leader's monitor, hold all questions till the end
References: Refer to higher headquarters' OPORD, and identify map sheet for operation.
Time Zone Used throughout the Order: (Optional)
Task Organization: (Optional; see paragraph 1c.)

1. **SITUATION** (Higher's OPORD para 1a[1]--[3])

a. **Enemy Forces**. Include significant changes in enemy composition, dispositions, and courses of action. Information not available for inclusion in the initial WARNO can be included in subsequent warning orders. (who, what, where)

b. **Friendly Forces**. (Optional) only address if essential to the WARNO.

(1) Higher commander's mission (who, what, where, why).
(2) Higher commander's intent. (Higher's [go to mapboard] OPORD para 1b[2], give task and purpose)

c. **Attachments and Detachments**. Initial task organization, only address major unit changes.

(1) Orient relative to each point on the compass.
(2) Box in the entire AO with grid lines.
(3) Trace each Zone using boundaries.
(4) Familiarize by identifying 3 natural (terrain) and 3 man-made features in each zone.
(5) Point out the enemy and friendly locations on the map board.

2. **MISSION**. State mission twice (who, what, when, where, and why) and concisely state task and purpose.

3. **EXECUTION**.

a. **Concept of Operation**. Provide as much information as available. The concept should describe the form of maneuver or defensive technique, critical events, decisive point of the operation and why it is decisive, task and purpose for the main and supporting efforts, purposes of the warfighting systems, such as engineers, fire support, intelligence etc), acceptable risk, and the desired endstate. After the desired endstate is stated, leaders should talk through the concept of the operation like a story. Use a simple sketch to assist in relaying the information.

b. **Tasks to Maneuver Units**. Provide specified tasks to subordinate units. These are tactical instructions on how to execute the mission for each element in task organization. Planning guidance consists of tasks assigned to subordinate units, special teams, and key individuals.

c. **Tasks to Combat Support Units**. See paragraph 3b.

d. **Coordinating Instructions**. Include any information available at that time. If you know it, then at least cover--

- Uniform and equipment common to all. Consider the factors of METT-TC and tailor the load for each Ranger.
- Time line. (State 4 W's and specified and implied times. Reverse plan. Use 1/3-2/3 rule.)
- CCIR.
- Risk guidance.
- Deception guidance.
- Specific priorities, in order of completion.
- Guidance on orders and rehearsals.
- Orders group meeting (attendees, location, and time).
- Earliest movement time and degree of notice.

4. **SERVICE SUPPORT**. Include any known logistics preparation for the operation.

a. Special equipment. Identify requirements and any coordination measures the unit needs to take to transfer the special equipment. (State the equipment you will use, need, or want for the mission.)

b. Transportation. Identify requirements and any coordination needed to pre-position assets. (State unit's means of infil/exfil.)

5. **COMMAND AND SIGNAL**.

a. **Command**. State the chain of command if different from unit SOP.

b. **Signal**. Identify current SOI edition, and pre-position signal assets to support operation.

- Give subordinates guidance on tasks to complete for preparation of the OPORD and the mission.
- Give time, place, and uniform of the OPORD.
- Give a time hack and ask for questions.

NOTES

Figure 2-3. EXAMPLE WARNING ORDER

## WARNING ORDER (SQUAD)

ROLL CALL, CAMP DARBY SPECIAL (CDS) 1:50,000 MAP, PEN, PAPER, PENCIL, PROTRACTOR, RANGER HANDBOOK (RHB), HOLD ALL QUESTIONS, AND TLs MONITOR AND TASK-ORGANIZE

1. **SITUATION**.
   a. **Enemy**.
      WHO: The Aragon Liberation Front (ALF).
      WHAT: Has been conducting aggressive offensive operations.
      WHERE: In Zones A, B, C.
   b. **Friendly**:
      (1) **Higher's Mission**.
         WHO: 2nd PLT, C CO.
         WHAT: Conducts area ambushes to destroy enemy personnel and equipment.
         WHERE: Zone D.
         WHY: To prevent enemy logistical resupply.
      (2) **Higher Commander's Intent**.
         - Rapid infiltration into Zone D.
         - Find, fix, and finish enemy forces in Zone D.
         - Enemy personnel and equipment located and targeted.
         - Squads ready to conduct future combat operations.
   c. **Attachments/Detachments**. MG Tm #1/2 PLT 160630-190630.
2. **MISSION**.
   WHO: 1st SQD, 2nd PLT
   WHAT: Conducts a point ambush to destroy enemy personnel and equipment.
   WHERE: GA 1805 7878.
   WHEN: NLT 25 1830 DEC 2007.
   WHY: To prevent enemy logistical resupply and gather PIR.
3. **EXECUTION**.

a. **Concept of Operation**. (Orient Rangers to sketch or terrain model). We are currently located at Camp Darby, GA 1962 7902, which is on a ridge. Our ground tactical plan will begin when we will move generally southwest to our ORP at Grid GA 187778. Terrain feature there is a draw for 1,000 meters, by foot and it will take 1 hour. Here will we finalize the preparing of M, W, and E. We will then move generally southwest to our objective at Grid GA 184774. Terrain feature there is a ridge for 200 meters, by foot, and it will take us 30 minutes to an hour. The reason for this is stealth on the objective while we occupy our positions.

b. **Tasks to Maneuver Units**.

(1) HQs: HQ is responsible for command and control and communications. Commo/HQ is 2nd in the OOM. M240 will be placed in a position to provide supporting fires in the kill zone. RTO will be the recorder en route and during actions on the objective. SL will be located where he can best control his squad and observe the kill zone. SL is responsible for dissemination of information.

(2) ATM: ATM will be 1st OOM/responsible for navigation, and for clearing and securing the far side at all linear danger areas. ATM is flank security for AOO, will provide early warning and seal off objective/1-2 man EPW, and search TM/1-2 man aid and litter TM/1-2 man demo TM/1 man (P), 1 man (A) ORP clearing TM/ 2-2 man flank security. TM for AOO/ 1 compassman, 1 pace man, or ATL will be the sec. TM LDR along the most likely avenue of approach.

(3) BTM: BTM 3RD OOM/responsible for securing the near side of all linear danger areas/BTM. Is assault for AOO. Must provide suppressive fire and assault thru kill zone/1-2 man EPW and search team/1-2 man aid and litter team/1-2 man demolitions team; 1 man (P); 1 man (A) ORP clearing team; 1-2 man S/O team; 1-2 man security team for reentry; 1 compassman; 1 pace man; and BTL. You will be the assault element leader for AOO.

c. **Tasks to Combat Support Units**. None.

d. **Coordinating Instructions**.

(1) Uniform and equipment common to all. At a minimum, each Ranger wears or carries--

(a) A ballistic helmet w/camouflaged band and luminous tape; clean and serviceable ACUs, camelback, black belt w/buckle, blackened combat boots, serviceable OD or black socks, serviceable OD/brown T-shirt, ID tags w/breakaway lanyard around neck (silenced). Right cargo pocket; (RHB, CDS [map], protractor, pencil, pen, and paper) waterproofed, left cargo pocket; patrol cap w/ luminous tape, ID card in right breast pocket, SL will carry sterile fire support overlay in left breast pocket.

(b) LCE, consisting of pistol belt, suspenders, 2 ammo pouches, seven 30-round magazines, one first aid dressing with pouch, one compass and pouch, two 1-quart canteens w/covers and cups, one whistle, and four M67 fragmentation grenades. All are tied down IAW 4th Bn SOP, except that the pace cord is tied to suspenders.

(c) Rucksack, complete; one set serviceable ACUs, two 2-quart canteens w straps; E-tool with cover; two pair socks; two brown T-shirts; weapons cleaning kit; one waterproof bag; a sewing kit; a patrol cap; MREs; poncho and liner; wet weather top; personal hygiene kit; one brown towel; and camouflage stick. RTO also packs and ties off the ANCD.

(d) Optional bug juice, 550 cord, foot powder, black tape, knife, all other items will be approved by SL 15 minutes prior to completion time of OPORD.

(2) Time schedule.

| WHEN | WHAT | WHERE | WHO | REF |
|---|---|---|---|---|
| *0630 | WARNING ORDER | BAY AREA | All | 3D1 |
| 0700 | INITIAL INSPECTION | BAY AREA | All | |
| *0730 | Required Ammo/Supplies | CO TOC | BTL | 4B1 |
| *0745 | P/U-Ammo/Supplies | CO TOC | BTL/Detail | 4B1 |
| 0800 | Test Fire | T/F AREA | All | |
| *0830 | S-2/S-3 Coord | CO BAY AO | SL/RTO | 3D1 |
| *0830 | Fire Spt Coord | BN TOC | SL/ RTO | 3A2 |
| 0850 | Adj Unit Coord | BN TOC | ATL/compass man | 3D1 |
| *0900 | Enter Net | BAY AREA | RTO | 5B1 |
| 0930 | Complete Plan | BAY AREA | All | |
| *1000 | Issue OPORD | BAY AREA | All | 3D1 |
| 1300 | Rehearsal | BAY AREA | All | |
| 1400 | Final Inspection | BAY AREA | All | |
| 1430 | Conduct Move | BAY AREA | All | |
| 1800 | In ORP | GA | All | |
| 2000 | In Position | GA 184774 | All | |
| *2300 | Mission Complete | GA 186874 | All | 3D1 |
| *0200 | Reenter FFU | GA | All | |
| 0500 | S-2 Debrief | BN TOC | All | |
| * Specified time | | | | |

4. **SERVICE AND SUPPORT**.

a. **Special Equipment**.

(1) HQs: One hand-held radio and extra batteries; one AN/PRC-119 radio and extra batteries, one ANCD, one AN/PSN-11 and extra batteries, one M-22 binos, one AN/PVS-14 and extra batteries; one sheets overlay material, three sheets carbon paper; two HE smoke, two red smoke, one M240-B MG complete w/tripod, flex mount, T&E, and Pintle, one AN/PVS-4 and extra batteries, one AN/PVS-7D, and extra batteries.

(2) ATM: One hand-held radio and extra batteries, one M-24 binos, two AN/PVS-14 and extra batteries, two HE smoke, two yellow smoke, one thermite grenade, two gags, two flex cuffs, two sandbags, one CLS bag, one pole-less litter, two M18A1 claymore mines, one AT-4, one LAW, one M-203 vest, two M-249 SAW pouches, one small metal flashlight, and extra batteries secured to rifle.

(3) BTM: One hand-held radio and extra batteries, one M-24 binos, two AN/PVS 14 and extra batteries, two HE smoke, two yellow smoke, one thermite grenade; two each gags, flex cuffs, and sandbags; one CLS bag, one pole-less Litter, two M18A1 claymore mines, one AT-4, one LAW, one M-203 vest, two M249 SAW pouches, one small metal flashlight, and extra batteries secured to rifle.

b. **Ammo**. (per weapon system).

| **M16/M4** | **M249** | **M203** | **M240** | **SL AND TLs** |
|---|---|---|---|---|
| 210 Rounds<br>5.56-MM Ball | 625 Rounds<br>5.56-MM (4:1 Mix) | 16 Rounds 40-MM HEDP<br>5 Rounds 40-MM Shot<br>4 Rounds 40-MM Illumination<br>2 Rounds 40-MM TP-T | 825 Rounds<br>7.62-MM (4:1 Mix) | 42 Rounds<br>5.56-MM<br>Tracer |

c. **Transportation**. NA

5. **COMMAND AND SIGNAL**.

a. **Command**.

| **PERSONNEL** | **POS** | **SOC** |
|---|---|---|
| **HQ/SUPPORT** | | |
| BARNETT | SL | 1 |
| MARENGO | RTO | 4 |
| CASELY | MG | 11 |
| WADE | AG | 12 |
| **ATM/SECURITY** | | |
| TRUITT | ATL | 3 |
| SILCOX | GRN | 5 |
| WISE | R | 6 |
| EDMUNDS | R | 7 |
| **BTM/ASSAULT** | | |
| SPENCER | BTL | 2 |
| CHRISTENSEN | GRN | 8 |
| RUFUS | R | 9 |
| ROBINSON | R | 10 |

| SUBORDINATE GUIDANCE: | |
|---|---|
| ATL: | Write paragraph 1 of squad OPORD, and for drawing all sketches, terrain models, routes, and nonsterile fire support overlays. |
| BTL: | You are second in the chain of command and in charge at all times during my absence.<br>Write paragraph 4 of squad OPORD Write the supply, DX, and ammo lists, draw and issue all items, ensure that everyone conducts a test fire, and all equipment is tied down IAW 4th RTBn SOP, and update the squad status card and hand receipt. |
| RTO: | Write paragraph 5 of squad OPORD, ensure all radios are operational with proper freqs loaded, and ensure we enter the net on time. |
| TL: | TLs: You will update the Warning Order board with all the correct information. As a task is accomplished, you will line it out. Also, post your succession of command (SOC), duty positions, and job descriptions (specialty teams and key individuals). Come and see me for further guidance at the conclusion of this WARNO. |

b. **Signal**.

| | |
|---|---|
| TACTICAL FREQUENCY | 62,000 |
| MEDEVAC FREQUENCY | 35,000 |
| CHALLENGE | BAYONET |
| PASSWORD | MILLETT |
| RUNNING PASSWORD | RANGER |
| NUMBER COMBO | 7 |
| TIME, PLACE, AND UNIFORM FOR OPORD. TIME HACK. ANY QUESTIONS? | |

NOTES

2-4. **OPERATIONS ORDER**. An operations order (OPORD) is a directive issued by a leader to his subordinates in order to effect the coordinated execution of a specific operation. A five-paragraph format (shown below) is used to organize the briefing, to ensure completeness, and to help subordinate leaders understand and follow the order. Use a terrain model or sketch along with a map to explain the order. When possible, such as in the defense, give the order while observing the objective. The platoon/squad leader briefs his OPORD orally off notes that follow the five-paragraph format. Before the issuance of the OPORD, the leader ensures that the following resources are in place: pencil, pen, paper; RHB; map; protractor; leader's monitor. Then he calls roll and says "*Please hold all questions till the end.*" Figure 2-4 shows an example OPORD format. Figure 2-5 shows an OPORD shell.

**Figure 2-4. EXAMPLE OPORD FORMAT**

## OPERATIONS ORDER

[Plans and orders normally contain a code name and are numbered consecutively within a calendar year.]

*References:* [The heading of the plan or order includes a list of maps, charts, datum, or other related documents the unit will need to understand the plan or order. The user need not reference the SOP, but may refer to it in the body of the plan or order. He refers to a map by map series number (and country or geographic area, if required), sheet number and name, edition, and scale, if required. "*Datum*" refers to the mathematical model of the earth that applies to the coordinates on a particular map. It is used to determine coordinates. Different nations use different datum for printing coordinates on their maps. The datum is usually referenced in the marginal information of each map.]

*Time zone used throughout the order:* The time zone used throughout the order (including annexes and appendixes) is the time zone applicable to the operation. Operations across several time zones use Zulu time.

*Task organization:* Describe the allocation of forces to support the commander's concept. Task organization may be shown in one of two places: preceding paragraph one, or in an annex, if the task organization is long and complicated.

**Weather and light data, and general forecast (Only discuss what will affect your patrol. Note effects on friendly and enemy personnel and equipment.):*

High
Moonrise
Sunrise
Low
Moonset
Sunset
Wind Speed
Moon Phase
BMNT
Wind Direction
Percent Illumination
EENT

**This is the information the leader received from higher's OPORD.*

*Terrain:* Make a statement concerning the effects terrain will have on both friendly and enemy forces in the area of operation using the OAKOC format.

1. **SITUATION**.

a. **Enemy Forces**. The enemy situation in higher headquarters' OPORD (paragraph 1a) is the basis for this, but the leader refines this to provide the detail required by his subordinates.

(1) Enemy's composition, disposition, strength.
(2) Recent activities.
(3) Known/suspected locations and capabilities.
(4) Describe the enemy's most likely and most dangerous course of action on a map.

b. **Friendly Forces**. This information is in paragraphs 1b, 2, and 3 of higher headquarters' OPORD.

(1) State the mission of the Higher Unit (2 levels up).

(2) State the mission of the Higher Unit (1 level up).

(3) State intent 2 levels up.

(4) State locations of units to the left, right, front, and rear. State those units' tasks and purposes; and say how those units will influence your unit's patrol.

(a) Show other units locations on map board.

(b) Include statements about the influence each of the above patrols will have on your mission, if any.

- Obtain this information from higher's OPORD. It gives each leader an idea of what other units are doing and where they are going. This information is in paragraph 3a (1) (execution, concept of the operation, maneuver).
- Also include any information obtained when the leader conducts adjacent unit coordination.

c. **Attachments and Detachments**. Do not repeat information already listed under Task Organization. Try to put all information in the Task Organization. However, when not in the Task Organization, list units that are attached or detached to the headquarters that issues the order. State when attachment or detachment is to be effective if different from when the OPORD is effective, such as on order, on commitment of the reserve). Use the term "*remains attached*" when units will be or have been attached for some time.

2. **MISSION**. State the mission derived during the planning process. There are no subparagraphs in a mission statement. Include the 5 W's: Who, What (task), Where, When, and Why (purpose).

- Make it a clear and concise statement and read it twice.
- Go to map and point out the exact location of the OBJ and point out the unit's present location.

3. **EXECUTION**. State the commander's intent.

a. **Concept of Operation**. The concept statement should be concise and clear. It describes how the unit will accomplish its mission from start to finish. The number of subparagraphs, if any, is based on what the leader considers appropriate for the level of leadership and complexity of the operation. Generally, the concept should describe the form of maneuver or defensive technique and critical events; identify what the decisive point of the operation is and why; describe the task and purpose for the main effort and supporting efforts; state the purpose of the warfighting functions, such as engineers, fire support, or intelligence; define acceptable risk; and identify the desired endstate. After stating the desired endstate, leaders should talk through the concept of the operation like a story, using a simple sketch to help them convey the information. Figure 1 shows the six warfighting functions (WFF):

**Figure 1. Warfighting functions.**

- Fire support
- Movement and Maneuver
- Protection
- Command and Control
- Intelligence
- Sustainment (formerly called "CSS")

(1) ***Maneuver***. The maneuver paragraph details the mechanics of the operations. This subparagraph specifically addresses all subordinate units and attachments by name, giving each its mission in the form of a task and purpose. The main effort's mission must be supported by the subordinate unit's missions. Actions on the objective will comprise the majority of this paragraph and should include a detailed plan for engagement/disengagement criteria, an alternate plan in the event of compromise or unplanned movement of enemy forces, and a withdrawal plan.

(2) ***Fires***. State scheme of fires to support the overall concept. This paragraph should state which maneuver unit has priority of fires. Fires should be planned using the PLOT-CR format (purpose, location, observer, trigger, communication method, resources). A target list worksheet and overlay are referenced here, if applicable. Specific targets are discussed and pointed out on the terrain model (see Chapter 3, Fire Support).

(3) ***Casualty Evacuation***.

(a) *Plan*. State a detailed CASEVAC plan during each phase of the operation. Include CCP locations, tentative extraction points, and methods of extraction.

b. **Tasks to Maneuver Units**. Clearly state the missions or tasks for each maneuver unit that reports directly to the headquarters issuing the order. List the units in the task organization, including reserves. Use a separate subparagraph for each maneuver unit. Only state the tasks that are necessary for comprehension, clarity, and emphasis. Place tactical tasks that affect two or more units in Coordinating Instructions (subparagraph 3d). Platoon leaders task their subordinate squads. Those squads may be tasked to provide any of the following special teams: reconnaissance and security, assault, support, aid and litter, EPW and search, clearing, and demolitions. Detailed instructions may also be given to the platoon sergeant, RTO, compassman, and paceman.

(1) ***Tasks to Combat Support Units***. Add subparagraphs here only as necessary. List CS units in subparagraphs from the task organization. Use CS subparagraphs to list only those specific tasks that CS units must accomplish and that are not specified or implied elsewhere. Include organization for combat if not clear from task organization.

c. **Coordinating Instructions**. This is always the last subparagraph under paragraph 3. List only the instructions that apply to two or more units, and which are seldom covered in unit SOPs. Refer the user to an annex for more complex instructions. The following information is required:

*1. Time schedule (state time, place, uniform, and priority of rehearsals, backbriefs, inspections, and movement).*

*2. Give the commander's critical information requirements (CCIR), which includes priority intelligence requirements (PIR), essential elements of friendly information (EEFI), and friendly force information requirements (FFIR):*

*a. PIR is intelligence that the commander must have for planning and decision making.*

*b. EEFI are critical aspects of friendly operations that, if known by the enemy, would compromise, lead to failure, or limit success of the operation.*

*c. FFIR include what the commander needs to know about friendly forces available for the operation. It can include personnel status, ammunition status, and leadership capabilities.*

*3. Risk reduction control measures. These are measures unique to the operation. They supplement the unit SOP and can include mission-oriented protective posture, operational exposure guidance, vehicle recognition signals, and fratricide prevention measures.*

*4. Rules of engagement (ROE).*

*5. Environmental considerations.*

*6. Force protection*

*7. Movement plan. Use terrain model, sketch, or both. State azimuths, directions, and grid coordinates. Use names of real Rangers on the sketches.*

*a. Sketch out the order of movement, the formation, and the movement technique.*

*b. Sketch actions at halts (long and short).*

*c. Routes (use terrain model):*

*1) Orient the terrain model (North, South, East, and West).*

*2) Identify the grid lines, both North-South and East-West.*

*3) Brief the legend.*

*4) Box in the AO with distinguishable natural or man-made terrain features such as roads, ridgelines, rivers, and streams.*

*5) Brief all the routes depicted on the terrain model from start to finish using specific azimuths and distance.*

*a) The primary route from the start point to the ORP, from the ORP to the objective, and from the objective to the patrol base while describing the terrain over which the squad/platoon will travel. (Include near/far side rally points for danger areas.)*

*b) The alternate route from the start point to the ORP, from the ORP to the objective, and from the objective to the patrol base while describing the terrain over which the squad/platoon will travel. (Include near/far side rally points for danger areas.)*

*6) Brief the fire support plan (if not given in fires paragraph). When covering fires the leader should cover the following (PLOT-CR):*

- *Purpose.*
- *Location and terrain feature to the targets.*
- *Observer.*
- *Trigger.*
- *Communication method.*
- *Resources.*

*7) State where the sterile fire support overlay is being carried.*

*a) Rally points and actions at all rally points (include grid location and terrain reference, and use sketch).*

*b) Actions at danger areas. This is a general plan for unknown linear, small open areas and large open areas; specific plan for all known danger areas that unit will encounter along the route (use sketch).*

*c) Actions on enemy contact; battle drills (use sketch).*

4. **SERVICE SUPPORT**. Address service support in the areas shown below as needed to clarify the service support concept. Subparagraphs can include:

a. **General**. Refer to the SOPs that govern the sustainment operations of the unit. Provide current and proposed company trains locations, casualty, and damaged equipment collection points and routes.

b. **Materiel and Services**.

(1) ***Supply***.

Class I--Rations plan.
Class III--Petroleum.
Class V--Ammunition.
Class VII--Major end items.
Class VIII--Medical.
Class IX--Repair parts.
Distribution Methods.

(2) ***Transportation***.

(3) ***Services***. (Laundry and showers.)

(4) ***Maintenance***. (Weapons and equipment.)

(a) *Medical Evacuation and Hospitalization*. Method of evacuating dead and wounded, friendly and enemy personnel. Include priorities and location of CCP.

(b) *Personnel Support*. Method of handling EPWs and designation of the EPW collection point.

5. **COMMAND AND SIGNAL**.

This paragraph states where command and control facilities and key leaders are located during the operation. Use the OPORD shell to save time and organize your thoughts while preparing an OPORD.

a. **Command**.

- Location of the higher unit commander and CP.
- Location of key personnel and CP during each phase of the operation.

- Succession of command.
- Adjustments to the patrol SOP.

b. **Signal**.
- SOI index in effect.
- Methods of communication in priority.
- Pyrotechnics and signals, to include arm and hand signals (demonstrate).
- Code words such as OPSKEDs.
- Challenge and password (use behind friendly lines).
- Number combination (use forward of friendly lines).
- Running password.
- Recognition signals (near/far and day/night).

c. **Special Instructions to RTOs**.

d. **Actions after Issuance of OPORD--**
(1) Issue annexes.
(2) Give time hack.
(3) ASK for questions.

**Figure 2-5. OPORD SHELL**

## OPORD

DTG:

REFERENCE:

TASK ORGANIZATION:

1ST PLT 2ND PLT TM CONTROL:

1. **SITUATION**.

BMNT: SR: SS: EENT: MR: MS: HI: LO: ILLUM:

a. **Enemy**.

Terrain | End state

(1) Higher mission:

Cdr's intent:

(2) Units to N, S, E, W.

Terrain | End state

2. **MISSION**:

3. **EXECUTION**:

a. **Concept of Operations**.

**Cdr's intent**:

| | | | | | |
|---|---|---|---|---|---|
| (ME):<br>T:<br>P: | (SE1):<br>T:<br>P: | (SE2):<br>T:<br>P: | (SE3):<br>T:<br>P: | (SE4):<br>T:<br>P: | (SE5):<br>T:<br>P: |

b. **Maneuver**

PHASE 1:
ME:
SE1:
SE2:
SE3:
SE4:
SE5:
FIRES:
MTRS:
C2:

PHASE 2:
ME:
SE1:
SE2:
SE3:
SE4:
SE5:
FIRES:
MTRS:
C2:

PHASE 3:
ME:
SE1:
SE2:
SE3:
SE4:
SE5:
FIRES:
MTRS:
C2:

PHASE 4:
ME:
SE1:
SE2:
SE3:
SE4:
SE5:
FIRES:
MTRS:
C2:

c. **Tasks to Maneuver Units.**

(1) Risk-reduction control measures.
Rules of engagement.
(2) Environmental considerations.
(3) Non-Airborne personnel.

(ME): (SE1): (SE2):

(SE3): (SE4): (SE5):

d. **Tasks to Subordinate Units**

Mortars

Coordinating instructions:

PIR:

EEFI:

FFIR:

Timeline.

Coordinating Instructions.

| WHEN | WHO | WHAT | WHERE |
|---|---|---|---|
| | | | |
| | | | |
| | | | |
| | | | |
| | | | |
| | | | |
| | | | |
| | | | |
| | | | |
| | | | |
| | | | |
| | | | |
| | | | |
| | | | |
| | | | |
| | | | |

4. **SERVICE SUPPORT.**

a. **General.**

b. **Material and Services.**

(1) Supply:

Class I.

Class III:

Class V:

Class VIII:

Class IX:

(2) LOGSTAT.

(3) The chain of command is responsible to ensure that ammo lasts between resupplies.

c. **Medical Evacuation.**

(1) Air MEDEVAC is on call through [this] frequency. JP capable; use standard 9-line, 24-hr coverage.

(2) Establish friendly CCPs in AAs. Adjust CCPs as mission progresses.

(3) Troop medical clinic located vic [this location].

(4) Precedence for evacuation of WIA is US military, host nation military, host nation civilian, and enemy military.

(5) Precedence for evacuation of KIA is same. Evacuation of KIA will be as per mission constraints.

d. **Personnel.**

(1) Method of handling EPWs:

**ANNEXES:**

5. **COMMAND AND SIGNAL.**

a. **Command.**

(1) TF CP:

(2) Succession of Command:

b. **Signal.**

(1) Challenge and Password:

(2) Number Combination:

(3) Running Password:

(4) Recognition:

(a) Far:

(b) Near:

(7) Code word:

(8) OPSKED:

(9) Reports. SALUTE to higher on enemy contact.

2-5. **FRAGMENTARY ORDER.** A FRAGO is an abbreviated form of an operations order, usually issued on a day-to-day basis that eliminates the need for restating information contained in a basic operations order. It is issued after an OPORD to change or modify that order or to execute a branch or sequel to that order. Figure 2-6 shows an annotated FRAGO format:

**Figure 2-6. ANNOTATED FRAGO FORMAT**

**FRAGMENTARY ORDER**__________

*Time Zone referenced throughout order:*

*Task Organization:*

*Weather and Light Data:*

| | | |
|---|---|---|
| High: | BMNT: | Moonrise: |
| Low: | Sunrise: | Moonset: |
| Wind Speed: | Sunset: | % Illum: |
| Wind Direction: | EENT: | |
| Forecast: | | |

*Terrain* (changes that will effect operation in new area of operations):

OAKOC: Observation and fields of fire, avenues of approach, key terrain, observation, and cover and concealment.

1. **SITUATION.** (Brief changes from base OPORD specific to this day's operation.)

a. **Enemy Situation.**

(1) Composition, disposition and strength.

(2) Capabilities.

(3) Recent activities.
(4) Most likely COA.
b. **Friendly Situation**.
(1) Higher mission.
(2) Adjacent patrols task/purpose.
(3) Adjacent patrol objective/route (if known).

2. **MISSION** (Who, what [task], when, where, why [purpose]—from higher HQ's maneuver paragraph).

3. **EXECUTION**

a. **Concept of Operation**. (Explain how unit will accomplish mission in general terms. Identify mission essential task, designate the main effort and how the supporting efforts support the main effort.)

(1) ***Maneuver***. (Assign task/purpose and discuss actions on the objective in detail. Use a sketch or terrain model to brief.)

(2) ***Fires***. (In support of today's mission, portions that do not change are not briefed.) PLOT-CR:
(a) Purpose.
(b) Location.
(c) Observer.
(d) Trigger.
(e) Communication method.
(f) Resource.

(3) ***CASEVAC***.

b. **Tasks to Maneuver Units**. (List unique tasks that apply to each squad.)

c. **Coordinating Instructions**.

(1) ***Timeline***.
(a) Hit time:
(b) ORP time:
(c) Movement time from PB:
(d) Final inspection:
(e) Rehearsals:
(f) FRAGO complete:

(2) ***Movement Plan***. (Use a sketch to brief.)
(a) Routes (primary and alternate):
(b) OOM:
(c) Formations:
(d) Movement technique:

(3) ***PIR***. (Specific to this mission):

(4) ***Rehearsal Plan***.

(5) ***Patrol Base Plan***. (If not IAW SOP):
(a) Teams:
(b) Occupation plan:
(c) Operations plan (security plan, alert plan, Black and Gold):
(d) Priorities of work:

(6) ***Air Assault Plan***. (Provided with higher's FRAGO, if applicable):
(a) Number, type, and ACL aircraft:
(b) PZ grid, DOL, and PZ posture time:
(c) Load time/lift time/flight time:
(d) Number of lifts and composition:
(e) Air checkpoints en route:
(f) LZ grid/DOL/LZ time:
(g) Actions after getting off aircraft:
(h) Actions on contact on LZ:

(7) ***Linkup Plan***. (If applicable):
(a) Time of linkup:
(b) Location of linkup site:
(c) Stationary element:
(d) Moving element:
(e) Rally points:
(f) Actions at linkup point:
(g) Near/far recognition signals (day and night):

4. **SUPPORT**. (Only cover changes from base order that apply for today's mission).
a. **Material and Services**.
(1) Any changes in classes of supply:
(2) Resupply plan:
(3) Water resupply plan:
(4) Aerial resupply plan (if applicable) :
(5) Truck plan:
(6) Maintenance issues specific to plan:
b. **Medical Evacuation Plan**. (Specific to mission):
(1) CCP point and markings:
(2) Aid and litter duties (if not SOP):

5. **COMMAND AND SIGNAL**.
a. **Command**.
(1) Location of Company CP:
(2) Location of PL:
(3) Location of key leaders:
(4) Succession of command:
b. **Signal**.
(1) Location of radios:
(a) During movement:
(b) During actions on the objective:
(2) SOI in effect:

| | |
|---|---|
| (a) Bn command freq: | Co CP call sign: |
| (b) Bn MEDEVAC freq: | PL call sign: |
| (c) Retrans freq: | PSG call sign: |
| (d) Company freq: | 1 SL call sign: |
| (e) Platoon freq: | 2 SL call sign: |
| | 3 SL call sign: |
| | WSL call sign: |

(3) Pyro signals in use:
(4) Running password/number combination:
(5) OPSKEDs in effect:
(a) The field FRAGO should take no more than 40 minutes to issue, with 30 minutes for the target. The proposed planning guide is as follows:

| | |
|---|---|
| Paragraphs 1 and 2: | 5 minutes |
| Paragraph 3: | 20 to 30 minutes |
| Paragraphs 4 and 5: | 5 minutes |

(b) The FRAGO should focus on actions on the objective. The PL may use subordinates to prepare para 1, 4, 5 and routes and fires for the FRAGO. It is acceptable for subordinates to brief the portions of the FRAGO they prepare.
(c) Use of sketches and a terrain model are critical to allow rapid understanding of the operation/FRAGO.
(d) Rehearsals are critical as elements of the constrained planning model. The FRAGO used in conjunction with effective rehearsals reduces preparation time and allows the PL more time for movement and recon.
(e) Planning in a field environment will necessarily reduce the amount of time leaders have for in-depth mission planning. The TLP gives leaders a framework to plan missions and produce orders when time is short.

2-6. **ANNEXES**. Operation order annexes are necessary to complete the plan and to provide greater clarity and understanding during complex or critical aspects of the operation. Information issued in annex form include the: aerial re-supply, truck movement, air assault, patrol base, small boat, linkup, and stream crossing annexes. Annexes are prepared only if the subject is not addressed thoroughly enough in the OPORD; brevity remains the standard. Annexes are always issued after the operation order.

## AIR MOVEMENT ANNEX

1. **SITUATION**.
   a. **Enemy Situation**.
      (1) Enemy air capability.
      (2) Enemy ADA capability.
      (3) Include in weather: % Illum, illum angle, NVG window, ceiling, and visibility.
   b. **Friendly Situation**.
      (1) Unit(s) supporting operation.
      (2) Friendly ADA status.

2. **MISSION**.

3. **EXECUTION**.
   a. **Concept of Operation**.
   b. **Subunit Missions**.
   c. **Coordinating instructions**.
      (1) ***Pickup Zone***.
         (a) Name/Number/
         (b) Coordinates/
         (c) Load Time/
         (d) Takeoff Time/
         (e) Markings/
         (f) Control/
         (g) Landing Formation/
         (h) Approach/Departure Direction.
         (i) Alternate PZ Name/Number.
         (j) Penetration Points.
         (k) Extraction Points.
      (2) ***Landing Zone***.
         (a) Name/Number.
         (b) Coordinates.
         (c) H-Hour.
         (d) Markings.
         (e) Control.
         (f) Landing Formation/Direction.
         (g) Alt LZ Name/Number.
         (h) Deception Plan.
         (i) Extraction LZ.
      (3) ***Laager Site***.
         (a) Communications.
         (b) Security Force.
      (4) ***Flight Routes and Alternates***.
      (5) ***Abort Criteria***.
      (6) ***Down Aircraft/Crew*** (designated area of recovery [DAR]).
      (7) ***Special Instructions***.
      (8) ***Cross-FLOT Considerations***.
      (9) ***Aircraft Speed***.

(10) ***Aircraft Altitude***.
(11) ***Aircraft Crank Time***.
(12) ***Rehearsal Schedule/Plan***.
(13) ***Actions on Enemy Contact*** (en route and on the ground).

4. **SERVICE SUPPORT**.
 a. **Forward Area Refuel/Rearm Points**.
 b. **Classes I, III, and V** (specific).

5. **COMMAND AND SIGNAL**.
 a. **Command**.
 b. **Signal**.
  (1) Air/ground call signs and frequencies.
  (2) Air/ground emergency code.
  (3) Passwords/number combinations.
  (4) Fire net/Quickfire net.
  (5) Time zone.
  (6) Time hack.

## AERIAL RESUPPLY ANNEX

1. **SITUATION**.
 a. **Enemy Forces** (include weather).
 b. **Friendly Forces**.
 c. **Attachments and Detachments**.

2. **MISSION**.

3. **EXECUTION**.
 a. **Concept of Operation**.
  (1) ***Maneuver***.
  (2) ***Fires***.
 b. **Tasks to Combat Units**.
  (1) ***Command and Control***.
  (2) ***Security***.
  (3) ***Marking***.
  (4) ***Recovery/Transport***.
 c. **Tasks to Combat Support Units**.
 d. **Coordinating Instructions**.
  (1) Flight Route.
   (a) General.
   (b) Checkpoints.
   (c) Communication checkpoint (CCP).
    - Marking of CCP.
    - Report time.
   (d) Heading from CCP.
  (2) Landing/Drop Zone.
   (a) Location.
    - Primary.
    - Alternate.
   (b) Marking.
    - Near.
    - Far.

(3) Drop Information.
   (a) Date/time of resupply (and alternates).
   (b) Code letter at DZ/LZ.
   (c) Length of DZ in seconds or dimensions of LZ.
   (d) Procedures for turning off DZ/LZ.
   (e) Formation, altitude, and air speed.
      - En route.
      - At DZ/LZ.
(4) Actions on enemy contact during resupply.
(5) Abort Criteria: En route and at DZ/LZ.
(6) Actions at DZ/LZ.
   - Rehearsals.

4. **SERVICE SUPPORT**.

5. **COMMAND AND SIGNAL**.
   a. **Command**.
      (1) Location of patrol leader.
      (2) Location of assistant patrol leader.
      (3) Location of members not involved in resupply.
   b. **Signal**.
      (1) Air to ground call-signs and frequencies (primary and alternate).
      (2) Long range visual signals.
      (3) Short range visual signals.
      (4) Emergency procedures and signals.
      (5) Air drop communication procedures.
      (6) Code words.

## PATROL BASE ANNEX

1. **SITUATION**
   a. Enemy Forces
   b. Friendly Forces
   c. Attachments and Detachments

2. **MISSION**.

3. **EXECUTION**.
   a. **Concept of Operation**.
      (1) ***Maneuver***.
      (2) ***Fires***.
   b. **Tasks to Combat Units**.
      (1) ***Teams***.
         - Security
         - Recon
         - Surveillance
         - LP/OPs
      (2) ***Individuals***.
   c. **Tasks to Combat Support Units**.
   d. **Coordinating instructions**.
      (1) Occupation plan.
      (2) Operations plan.
         - Security Plan.
         - Alert Plan.

- Priority of work.
- Evacuation plan.
- Alternate patrol base (used when primary is unsuitable or compromised).

4. **SERVICE SUPPORT**.
   a. Water plan.
   b. Maintenance plan.
   c. Hygiene plan.
   d. Messing plan.
   e. Rest plan.

5. **Command and Signal**.
   a. **Command**.
      (1) Location of patrol leader.
      (2) Location of assistant patrol leader.
      (3) Location of patrol CP.
   b. **Signal**.
      (1) Call signs and frequencies.
      (2) Code words.
      (3) Emergency signals.

## SMALL BOAT ANNEX

1. **SITUATION**.
   a. **Enemy Forces**.
      (1) **Weather**.
         (a) Tide.
         (b) Surf.
         (c) Wind.
      (2) **Terrain**.
         (a) River width.
         (b) River depth and water temperature.
         (c) Current.
         (d) Vegetation.
      (3) **Identification, Location, Activity, and Strength**.
   b. **Friendly Forces** (unit furnishing support).
   c. **Attachments and Detachments**.
   d. **Organization for Movement**.

2. **MISSION**.

3. **EXECUTION**.
   a. **Concept of Operation**.
      (1) ***Maneuver***.
      (2) ***Fires***.
   b. **Tasks to Combat Units**.
      (1) Security.
      (2) Tie-down teams.
         (a) Load equipment.
         (b) Secure equipment.
      (3) Designation of coxswains and boat commanders.
      (4) Selection of navigator(s) and observer(s).

c. **Coordinating Instructions**.
(1) Formations and order of movement.
(2) Route and alternate route.
(3) Method of navigation.
(4) Actions on enemy contact.
(5) Rally points.
(6) Embarkation plan.
(7) Debarkation plan.
(8) Rehearsals.
(9) Time schedule.

4. **SERVICE SUPPORT**.
a. **Ration Plan**.
b. **Weapons and Ammunition**.
c. **Uniform and Equipment**.
(1) Method of distribution of paddles and life jackets.
(2) Disposition of boat, paddles and life jackets upon debarkation.

5. **COMMAND AND SIGNAL**.
a. **Command**.
(1) Location of patrol leader.
(2) Location of assistant patrol leader.
b. **Signal**.
(1) Signals used between boats and in boats.
(2) Code words.

## STREAM CROSSING ANNEX

1. **SITUATION**.
a. **Enemy Forces**.
(1) ***Weather***.
(2) ***Terrain***.
(a) River width.
(b) River depth and water temperature.
(c) Current.
(d) Vegetation.
(e) Obstacles.
(3) ***Enemy*** (location, identification, activity).
b. **Friendly Forces**.
c. **Attachments and Detachments**.

2. **MISSION**.

3. **EXECUTION**.
a. **Concept of Operation**.
(1) ***Maneuver***.
(2) ***Fires***.
b. **Tasks to Combat Units**.
(1) ***Elements***.
(2) ***Teams***.
(3) ***Individuals***.

c. **Tasks to Combat Support Units.**
d. **Coordinating Instructions.**
(1) Crossing procedure/techniques.
(2) Security.
(3) Order of crossing.
(4) Actions on enemy contact.
(5) Alternate plan.
(6) Rallying points.
(7) Rehearsal plan.
(8) Time schedule.

4. **SERVICE SUPPORT.**

5. **COMMAND AND SIGNAL.**
a. **Command.**
(1) Location of patrol leader.
(2) Location of assistant patrol leader.
(3) Location of CP.
b. **Signal.**
(1) Emergency signals.
(2) Signals.

## TRUCK ANNEX

1. **SITUATION.**
a. **Enemy Forces.**
b. **Friendly Forces.**
c. **Attachments and Detachments.**

2. **MISSION.**

3. **EXECUTION.**
a. **Concept of Operation.**
(1) ***Maneuver.***
(2) ***Fires.***
b. **Tasks to Combat Units.**
c. **Tasks to Combat Support Units.**
d. **Coordinating Instructions.**
(1) Time of departure and return.
(2) Loading plan and order of movement.
(3) Route (primary and alternate).
(4) Air Guards.
(5) Actions on enemy contact (vehicle ambush) during movement, loading, and downloading.
(6) Actions at the detrucking point.
(7) Rehearsals.
(8) Vehicle speed, separation, and recovery plan.
(9) Broken vehicle instructions.

4. **SERVICE SUPPORT**.

5. **COMMAND AND SIGNAL**.
   a. **Command**. Location of PL and PSG
   b. **Signal**.
      (1) Radio call signs and frequencies
      (2) Code words

2-7. **COORDINATION CHECKLISTS**. The following checklists are items that a platoon/squad leader must check when planning for a combat operation. In some cases, he will coordinate directly with the appropriate staff section, in most cases this information will be provided by the company commander or platoon leader. The platoon/squad leader, to keep him from overlooking anything that may be vital to his mission, may carry copies of these checklists:



**Figure 2-7. INTELLIGENCE COORDINATION CHECKLIST**

## INTELLIGENCE COORDINATION CHECKLIST

In this coordination, the leader is informed of any changes in the situation as given in the operation order of mission briefing. He must keep himself constantly updated to ensure the plan is sound:

1. Identification of enemy unit.
2. Weather and light data.
3. Terrain update.
   a. Aerial photos.
   b. Trails and obstacles not on map.
4. Known or suspected enemy locations.
5. Weapons.
6. Probable course of action.
7. Recent enemy activities.
8. Reaction time of reaction forces.
9. Civilians on the battlefield.
10. Update to CCIR.

**Figure 2-8. OPERATIONS COORDINATION CHECKLIST**

## OPERATIONS COORDINATION CHECKLIST

This coordination occurs with the platoon leader/company commander so that the platoon/squad leader can confirm his mission and operational plan, receive any last-minute changes to his mission or plan, and to update his subordinates or issue a FRAGO, if required:

1. Mission backbrief.
2. Identification of friendly units.
3. Changes in the friendly situation.
4. Route selection, LZ/PZ/DZ selection.
5. Linkup procedures.
6. Transportation/movement plan.

7. Resupply (in conjunction with S4).
8. Signal plan.
9. Departure and re-entry of forward units.
10. Special equipment requirements.
11. Adjacent units operating in the area of operations.
12. Rehearsal areas.
13. Method of insertion/extraction.

**Figure 2-9. FIRE SUPPORT COORDINATION CHECKLIST**

## FIRE SUPPORT COORDINATION CHECKLIST

The platoon/squad leader will normally coordinate the following with the platoon forward observer (FO):

1. Mission backbrief.
2. Identification of supporting unit.
3. Mission and objective.
4. Route to and from the objective (include alternate routes).
5. Time of departure and expected time of return.
6. Unit target list (from fire plan).
7. Type of available support (artillery, mortar, naval gunfire, and aerial support to include Army, Navy, and Air Force) and their locations.
8. Ammunition available (to include different fuses).
9. Priority of fires.
10. Control measures.
    a. Checkpoints.
    b. Boundaries.
    c. Phase lines.
    d. Fire support coordination measures.
    e. Priority targets (target list).
    f. RFA (restrictive fire area).
    g. RFL (restrictive fire line).
    h. NFA (no-fire areas).
    i. Precoordinated authentication.
11. Communication (include primary and alternate means, emergency signals and code words).

**Figure 2-10. COORDINATION WITH FORWARD UNIT CHECKLIST**

## COORDINATION WITH FORWARD UNIT CHECKLIST

A platoon/squad that requires foot movement through a friendly forward unit must coordinate with that unit's commander for a safe and orderly passage. If no time and place has been designated for coordination with the forward unit, the platoon/squad leader should set a time and place when he coordinates with the S-3. He must talk with someone at the forward unit who has the authority to commit that unit in assisting the platoon/squad during departure. Coordination entails a two-way exchange of information:

1. Identification (yourself and your unit).
2. Size of platoon/squad.
3. Time(s) and place(s) of departure and return, location(s) of departure point(s), ERRP, and detrucking points.
4. General area of operations.
5. Information on terrain and vegetation.
6. Know or suspected enemy positions or obstacles.
7. Possible enemy ambush sites.
8. Latest enemy activity.
9. Detailed information on friendly positions such as crew-served weapons, FPF.

10. Fire and barrier plan.
   a. Support the unit can furnish. How long and what can they do?
      (1) Fire support.
      (2) Litter teams.
      (3) Navigational signals and aids.
      (4) Guides.
      (5) Communications.
      (6) Reaction units.
      (7) Other.
   b. Call signs and frequencies.
   c. Pyrotechnic plan.
   d. Challenge and password, running password, number combination.
   e. Emergency signals and code words.
   f. If the unit is relieved, pass the information to the relieving unit.
   g. Recognition signals.

**Figure 2-11. ADJACENT UNIT COORDINATION CHECKLIST**

## ADJACENT UNIT COORDINATION CHECKLIST

Immediately after the operation order of mission briefing, the platoon/squad leader should check with other platoon/squad leaders who will be operating in the same areas. If the leader is not aware of any other units operating is his area, he should check with the S3 during the operations coordination. The S3 can help arrange this coordination if necessary. The platoon/squad leaders should exchange the following information with other units operating in the same area:

1. Identification of the unit.
2. Mission and size of unit.
3. Planned times and points of departure and reentry.
4. Route(s).
5. Fire support and control measures.
6. Frequencies and call signs.
7. Challenge and password, running password, number combination.
8. Pyrotechnic plan.
9. Any information that the unit may have about the enemy.
10. Recognition signals.

**Figure 2-12. REHEARSAL AREA COORDINATION CHECKLIST**

## REHEARSAL AREA COORDINATION CHECKLIST

This coordination is conducted with the platoon leader/company commander to facilitate the unit's safe, efficient and effective use of rehearsal area prior to its mission:

1. Identification of your unit.
2. Mission.
3. Terrain similar to objective site.
4. Security of the area.
5. Availability of aggressors.
6. Use of blanks, pyrotechnics, and ammunition.
7. Mock-ups available.
8. Time the area is available (preferably when light conditions approximate light conditions of patrol).
9. Transportation.
10. Coordination with other units using area.

**Figure 2-13. ARMY AVIATION COORDINATION CHECKLIST**

## ARMY AVIATION COORDINATION CHECKLIST

This coordination is conducted with the platoon leader/company commander and/or S3 Air to facilitate the time, detailed and effective use of aviation assets as they apply to your tactical mission:

1. **SITUATION**.
   a. **Enemy Situation**.
      (1) Enemy air capability.
      (2) Enemy ADA capability.
      (3) Include in Weather: Percent illum, illum angle, NVG window, ceiling, and visibility.
   b. **Friendly Situation**.
      (1) Unit(s) supporting operation, Axis of movement/corridor/routes.
      (2) Friendly ADA status.

2. **MISSION**.

3. **EXECUTION**.
   a. **Concept of the Operation**. Overview of what requesting unit wants to accomplish with the air assault/air movement.
   b. **Tasks to Combat Units**.
      (1) Infantry.
      (2) Attack aviation.
   c. **Tasks to Combat Support Units**.
      (1) Artillery.
      (2) Aviation (lift).
   d. **Coordinating Instructions**.
      (1) ***Pickup Zone***.
      - Direction of landing.
      - Time of landing/flight direction.
      - Location of PZ/alternate PZ.
      - Loading procedures.
      - Marking of PZ (panel, smoke, SM, lights).
      - Flight route planned (SP, ACP, RP).
      - Formations: PZ, en route, LZ.
      - Code words:
        -- PZ secure (prior to landing), PZ clear (lead bird and last bird).
        -- Alternate PZ (at PZ, en route, LZ), names of PZ/alt PZ.
      - TAC air/artillery.
      - Number of pax per bird and for entire lift.
      - Equipment carried by individuals.
      - Marking of key leaders.
      - Abort criteria (PZ, en route, LZ).

      (2) ***Landing Zone***.
      - Direction of landing.
      - False insertion plans.
      - Time of landing (LZ time).
      - Location of LZ and Alternate LZ.
      - Marking of LZ (panel, smoke, SM, lights).
      - Formation of landing.
      - Code words, LZ name, alternate LZ name.
      - TAC air/artillery preparation, fire support coordination.
      - Secure LZ or not?

4. **SERVICE AND SUPPORT**.
   a. Number of aircraft per lift and number of lifts.
   b. Refuel/rearm during mission or not?
   c. Special equipment/aircraft configuration for weapons carried by unit personnel.
   d. Bump plan.
5. **COMMAND AND SIGNAL**.
   a. Frequencies, call signs and code words.
   b. Locations of air missions commander ground tactical commander, and air assault task force commander.

**Figure 2-14. VEHICULAR MOVEMENT COORDINATION CHECKLIST**

## VEHICULAR MOVEMENT COORDINATION CHECKLIST

This is coordinated with the supporting unit through the platoon sergeant/first sergeant to facilitate the effective, detailed, and efficient use of vehicular support and/or assets:

1. Identification of the unit.
2. Supporting unit identification.
3. Number and type of vehicles and tactical preparation.
4. Entrucking point.
5. Departure/loading time.
6. Preparation of vehicles for movement.
   a. Driver responsibilities.
   b. Platoon/squad responsibilities.
   c. Special supplies/equipment required.
7. Availability of vehicles for preparation/rehearsals/inspection (time and location).
8. Routes.
   a. Primary.
   b. Alternate.
   c. Checkpoints.
9. Detrucking points.
   a. Primary.
   b. Alternate.
10. March internal/speed.
11. Communications (frequencies, call signs, codes).
12. Emergency procedures and signals.

2-8. **DOCTRINAL TERMS**

**TACTICAL TASK**--A clearly defined, measurable activity accomplished by individuals and organizations. Tasks are specific activities that contribute to the accomplishment of encompassing missions or other requirements. A task should be definable, measurable, and decisive (achieve the purpose).

| Actions by Friendly Force | | Effects on Enemy Force |
|---|---|---|
| Assault<br>Attack-by-Fire<br>Breach<br>Bypass<br>Clear<br>Combat Search and Rescue<br>Consolidation and Reorganization<br>Control<br>Counterreconnaissance<br>Disengagement<br>Exfiltrate | Follow and Assume<br>Follow and Support<br>Linkup<br>Occupy<br>Reconstitution<br>Reduce<br>Retain<br>Secure<br>Seize<br>Support-by-Fire<br>Suppress | Block<br>Canalize<br>Contain<br>Defeat<br>Destroy<br>Disrupt<br>Fix<br>Interdict<br>Isolate<br>Neutralize<br>Penetrate<br>Turn |

**PURPOSE** (in order to)--The desired or intended result of the tactical operation stated in terms related to the enemy or the desired situation. The why of the mission statement. The most important component of the mission statement.

| | | |
|---|---|---|
| Allow<br>Cause<br>Create<br>Deceive<br>Deny | Divert<br>Enable<br>Envelop<br>Influence<br>Open | Prevent<br>Protect<br>Support<br>Surprise |

**OPERATION**--A military action or the carrying out of a military action to gain the objectives of any battle or campaign.

| | | | |
|---|---|---|---|
| Types of operations include:<br>Movement to Contact<br>  Search and Attack<br>Attack<br>  Ambush<br>  Demonstration<br>  Feint<br>  Raid<br>  Spoiling Attack<br>Exploitation<br>Pursuit | Forms of Offensive Maneuver:<br>  Envelopment<br>  Frontal Attack<br>  Infiltration<br>  Penetration<br>  Turning Movement | Types of Defensive Techniques:<br>Area Defense<br>Mobile Defense<br>Retrograde Ops<br>  Delay<br>  Withdrawal<br>  Retirement | Other Operations:<br>Reconnaissance Ops<br>Security Ops<br>Information Ops<br>Combined Arms<br>  Breach<br>Passage of Lines<br>Relief in Place<br>River Crossing Ops<br>Troop Movement<br>  Admin<br>  Approach March<br>  Road March |

2-9. **TERRAIN MODEL**. The terrain model (Figure 2-15) is an outstanding visual means used during the planning process to communicate the patrol routes and also detailed actions on the objective. At a minimum, the model is required to during planning to display routes to the objective and to highlight prominent terrain features the patrol will encounter during movement. A second terrain model is usually constructed of the objective area, enlarged to a sufficient size and detail to brief the patrol's actions on the objective.

a. **Checklist**.

(1) North-seeking arrow.
(2) Scale.
(3) Grid lines.
(4) Objective location.
(5) Exaggerated terrain relief, water obstacles.
(6) Friendly patrol locations.
(7) Targets (indirect fires, including grid and type of round).
(8) Routes, primary and alternate.
(9) Planned RPs (ORP, L/URP, RP).
(10) Danger areas (roads, trails, open areas).
(11) Legend.
(12) Blowup of objective area.

b. **Construction**. The following are some field-expedient techniques to aid in terrain model construction:

(1) Use a 3 x 5 card, MRE box, or piece of paper to label the objective or key sites.

(2) Grid lines can be made using string from the guts of 550 cord or colored tape. (Grid lines are identified by writing numbers on small pieces of paper.)

(3) Trees and vegetation are replicated by using moss, green or brown spray paint, pine needles, crushed leaves, or cut up grass.

(4) Water is designated by blue chalk, blue spray paint, blue yarn, tin foil, or MRE creamer.

(5) North seeking arrows are made from sharpened twigs, pencils, or colored yarn.

(6) Enemy positions are designated using red yarn, M-16 rounds, toy Rangers, or poker chips.

(7) Friendly positions such as security elements, support by fire, and assault elements are made using M-16 rounds, toy Rangers, poker chips, small MRE packets of sugar and coffee, or pre-printed acetate cards.

(8) Small pieces of cardboard or paper can identify target reference points and indirect fire targets. Ensure grids are shown for each point.

(9) Breach, support by fire, and assault positions are made using colored yarn or string so that these positions can be easily identified.

(10) Bunkers and buildings are constructed using MRE boxes or tongue depressors/sticks.

(11) Perimeter wire is constructed from a spiral notebook.

(12) Key phase lines are constructed with colored string or yarn.

(13) Trench lines are replicated by colored tape or yarn, by digging a furrow and coloring it with colored chalk or spray paint.

***Note:*** All symbols used on the terrain model must be clearly identified in the legend.

Figure 2-15. TERRAIN MODEL

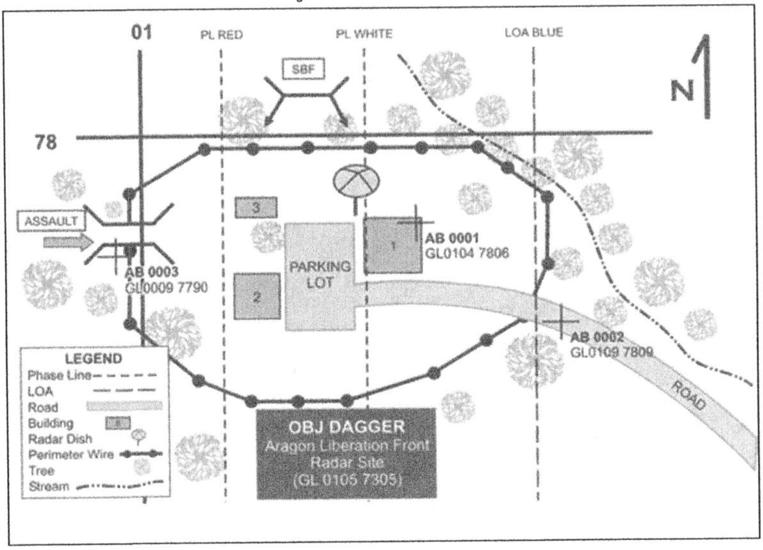


NOTES

## Chapter 3

# FIRE SUPPORT

Indirect fire support can greatly increase the combat effectiveness and survivability of any Infantry unit. The ability to plan for and effectively use this asset is a task that every Ranger and small unit leader should master. Fire support assets can assist a unit suppress, fix, destroy, or neutralize the enemy. Employment of indirect fire support should be considered throughout every offensive and defensive operation. This chapter discusses planning, tasks, capabilities, risk estimate distances, target overlays, close air support, elements and sequence of call for fire, and example call-for-fire transmissions.

3-1. **PLANNING**. Planning is the continual process of selecting targets on which fires are prearranged to support a phase of the commander's plan. The principles of planning follow:

- Consider what the commander wants to do.
- Plan early and continuously.
- Exploit all available targeting assets.
- Use all available lethal and nonlethal fire support means.
- Use the lowest echelon able to furnish effective support.
- Observe all fires.
- Use the most effective fire support asset available.
- Provide adequate fire support.
- Avoid unnecessary duplication.
- Provide for safety of friendly forces and installations.
- Provide for flexibility.
- Furnish the type of fire support requested.
- Consider the airspace.
- Provide rapid and effective coordination.
- Keep all fire support informed.

3-2. **TASKS**.

a. **All Operations**.

(1) Locate targets.
(2) Integrate all available assets.
(3) Destroy, neutralize, or suppress all enemy direct and indirect fire systems.
(4) Provide illumination and smoke.
(5) Provide fires in support of JAAT and SEAD missions.
(6) Deliver scatterable mines.
(7) Prepare for future operations.
(8) Provide positive clearance of fires.

b. **Offensive Operations**.

(1) Support the movement to contact, chance contact.
(2) Soften enemy defenses before the attack by arranging short, violent preparations, where required.
(3) Provide support during the attack by attacking high payoff targets.
(4) Plan for deep and flanking fires.
(5) Plan fires during consolidation.
(6) Provide counterfires.

c. **Defensive Operations**.

(1) Execute all plans as the commander intends.
(2) Disorganize, delay, and weaken the enemy before the attack.
(3) Provide counterfire.
(4) Provide fires in support of planned engagement areas (EAs).
(5) Attack high payoff targets.
(6) Plan fires in support of barriers and obstacle plans.
(7) Plan for deep, flanking, and rear fires.

(8) Provide fires in support of counterattacks.
(9) Plan final protective fire (FPFs).

3-3. **CAPABILITIES**. Tables 3-1 and 3-2 show capabilities of field artillery and mortars.

**Table 3-1. CAPABILITIES OF FIELD ARTILLERY**

| WEAPON | MAX RANGE (meters) | MIN RANGE (meters) | MAX RATE Rds per Min | Burst Radius | SUSTAINED RATE Rds per Min |
|---|---|---|---|---|---|
| 105-mm Howitzer M119, Towed | 14,000 m | 0 | 6 for 2 min | 35 m | 3 Rds for 30 min, then 1 rd per min |
| 155-mm Howitzer M198, Towed | 18 m,100 m<br>30,000 m (RAP) | 0 | 4 for 3 min<br>2 for 30 min | 50 m | 1 rd per min Temp Dependent |
| 155-mm Howitzer M109A6 SP | 18 m, 100 m<br>30,000 m (RAP) | 0 | 4 for 3 min | 50 m | 1 for 60 min<br>0.5 |

**Table 3-2. CAPABILITIES OF MORTARS**

| WPN | MUNITIONS AVAILABLE | MAX HE RANGE | MIN RANGE (meters) | MAX RATE (Rds per Min) | BURST RADIUS | SUSTAINED RATE (Rds per Min) |
|---|---|---|---|---|---|---|
| 60mm | HE, WP, Illum | 3,500 m | 70 m (HE) | 30 for 4 min | 30 m | 20 |
| 81mm | HE, WP, Illum | 5,600 m | 70 m (HE) | 25 for 2 min | 38 m | 8 |
| 120mm | HE, Smoke, Illum | 7,200 m | 180 m (HE) | 15 for 1 min | 60 m | 5 |

## DANGER

**DANGER CLOSE**

1. **When the target is within 600 meters of any friendly troops (for mortars and field artillery), announce DANGER CLOSE in the Method of Engagement portion of the call for fire.**
2. **When adjusting 5-inch or smaller naval guns on targets within 750 meters, announce DANGER CLOSE. For larger naval guns, announce DANGER CLOSE for targets within 1,000 meters. Failure to adhere to this guidance can result in fratricide.**
3. **Avoid making corrections using the bracketing method of adjustment, because doing so can cause serious injury or death. Use only the creeping method of adjustment during DANGER CLOSE missions. Make corrections of no more than 100 meters by creeping the rounds to the target.**

3-4. **RISK ESTIMATE DISTANCES**. RED applies to combat only. Minimum safe distances (Table 3-3) apply to training IAW AR 350-1. RED takes into account the bursting radius of particular munitions and the characteristics of the delivery system and associates this combination with a percentage representing the likelihood of becoming a casualty, that is, the percentage of risk. RED is defined as the minimum distance friendly troops can approach the effects of friendly fires without suffering appreciable casualties of 0.1% PI or higher.

> **WARNING**
> **[Commanders] use risk-estimate distance formulas to determine acceptable risk levels in combat only. Specifically, use them to identify the risk to your Rangers at various distances from their targets. Risk estimate distances apply only in combat. In training, use minimum safe distances (MSD).**

a. **Casualty Criterion**. The casualty criterion is the 5-minute assault criterion for a prone soldier in winter clothing and helmet. Physical incapacitation means that a soldier is physically unable to function in an assault within a 5-minute period after an attack. A PI value of less than 0.1 percent can be interpreted as being less than or equal to one chance in one thousand.

**Table 3-3. RISK ESTIMATE DISTANCES FOR MORTARS AND CANNON ARTILLERY**

| | Risk Estimate Distances (Meters) | | | | | |
|---|---|---|---|---|---|---|
| | 10 % PI | | | 0.1 % PI | | |
| **System** | 1/3 range | 2/3 range | Max range | 1/3 range | 2/3 range | Max range |
| 60-mm mortar | 60 | 65 | 65 | 100 | 150 | 175 |
| 81-mm mortar | 75 | 80 | 80 | 165 | 185 | 230 |
| 120-mm mortar | 100 | 100 | 100 | 150 | 300 | 400 |
| 105-mm howitzer | 85 | 85 | 90 | 175 | 200 | 275 |
| 155-mm howitzer | 100 | 100 | 125 | 200 | 280 | 450 |
| 155-mm DPICM | 150 | 180 | 200 | 280 | 300 | 475 |

b. **Risk**. Using echelonment of fires within the specified RED for a delivery system requires the unit to assume some risks. The maneuver commander determines by delivery system how close he will allow fires to fall in proximity to his forces. The maneuver commander makes the decision for this risk level, but he relies heavily on the FSO's expertise.

3-5. **TARGET OVERLAYS**.

a. Figure 3-1 shows contents of fire support overlay.

**Figure 3-1. CONTENTS OF FIRE SUPPORT OVERLAY**

- Unit and official capacity of person making overlay.
- Date the overlay was prepared.
- Map Sheet Number.
- Effective Period of Overlay (DTG).
- Priority target.
- ORP Location.
- Call signs and frequencies. (PRI/ALT)
- Routes--Primary/Alternate.
- Phase Lines/Checkpoints used by the patrol.
- Spares.
- Index marks to position overlay on map.
- Objective.
- Target Symbols.
- Description, location and remarks column, complete.

b. Using a PLOT-CR checklist helps ensure the leader's fire support plan is complete:
   (1) Purpose--Purpose of the planned indirect fires.
   (2) Location--Planned targets with an 8-digit grid (minimum).
   (3) Observer--Planned observer to see the impact of the rounds and adjust.
   (4) Trigger--Method to initiate fires.
   (5) Communication--Method to communicate between the observer and the supporting unit.
   (6) Resources--Planned allocated resource for each target.

c. Sterile overlay includes---
   (1) Index marks to position overlay on map.
   (2) Target symbols.

d. Target overlay symbols (Figure 3-2) include--
   (1) Point target:
   (2) Linear target:
   (3) Circular target:

**Figure 3-2. TARGET OVERLAY SYMBOLS--POINT (TOP), LINEAR (CENTER), AND CIRCULAR (BOTTOM)**

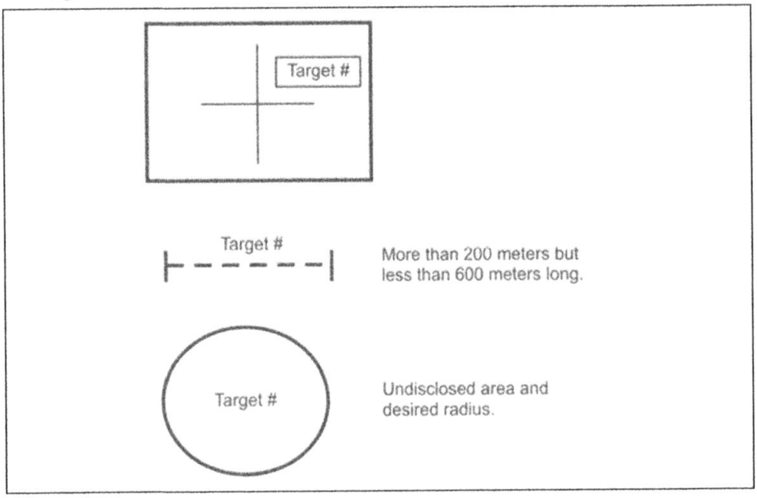


3-6. **CLOSE AIR SUPPORT**. The two types of close air support requests are planned and immediate. Planned requests are processed by the Army chain to corps for approval. Immediate requests are initiated at any level and processed by the battalion S-3 FSO, and air liaison officer.

a. Format for requesting immediate CAS:
   (1) Observer identification.
   (2) Warning order (request close air).
   (3) Target location (grid).

(4) Target description. (Description must include, as a minimum: type and number of targets, activity or movement; point or area targets, include desired results on target and time on target.

(5) Location of friendly forces.

(6) NAV details (elevation).

(7) Threats--ADA, small arms, and so on.

(8) Hazards--friendly aircraft in area.

(9) Wind direction.

b. Close air support capabilities are shown in Table 3-4.

**Table 3-4. CLOSE AIR SUPPORT**

| AIRCRAFT | SERVICE | CHARACTERISTICS |
|---|---|---|
| A-10 * | AF | Specialized CAS aircraft; subsonic; typical load 6,000 lbs, max load 16,000 lbs; 30-mm gun. |
| F-16 * | AF | A multirole aircraft; complements the F-4 and F-15 in an air-to-air role. Most accurate ground delivery system in the inventory; supersonic; typical load 6,000 lbs, max load 10,600 lbs. |
| F-18 * | N/MC | A multirole fighter scheduled to replace the F-4. A wide variety of air-to-surface weapons. Typical load 7,000 lbs, max load 17,000 lbs, 20-mm gun mounted in the nose; air-to-air missiles. |
| AC-130 | AF/R | A specialized CAS/RACO aircraft, propeller driven. Two models: The "A" model has two 40-mm guns, two 20-mm guns and two 7.62-mm miniguns. The similar "H" model has no 7.62-mm miniguns, and has a 105-mm howitzer in place of one of the 40-mm guns. Both models have advanced sensors and a target-acquisition system, including FLIR and low light TV. Very accurate. Vulnerable to enemy air defense systems, so must operate in a low-threat environment. |
| *FM capability. | | |

3-7. **CALL FOR FIRE**.

a. **Observer's Identification - Call Signs**.

b. **Warning Order**.

(1) ***Type of mission***.

(a) Adjust fire.

(b) Fire for effect.

(c) Suppress.

(d) Immediate suppression/immediate smoke.

(2) ***Size of element to Fire for Effect***. When the observer does not specify what size element to fire, the battalion FDC will decide.

(3) ***Method of Target Location***.

(a) Polar plot.

(b) Shift from a known point.

(c) Grid.

(4) **Location of Target**

(a) *Grid Coordinate - 6 Digit*. 8 digit if greater accuracy is required.

(b) *Shift from a Known Point*. Send observer-target (OT) direction:

- Mils (nearest 10).
- Degrees.
- Cardinal direction.
- Send lateral shift (right/left, nearest 10 mils).
- Send range shift (add/drop, nearest 100 mils).
- Send vertical shift (up/down). Use only if it exceeds 35 mils (nearest 5 mils).

(c) *Polar Plot*. Send direction. (Nearest 10 mils)

- Send distance (nearest 100 mils).
- Send vertical shift (nearest 5 mils).

(5) ***Description of Target***.

(a) Type.

(b) Activity.

(c) Number.

(d) Degree of protection.

(e) Size and shape (length/width or radius).

(6) ***Method of Engagement, Type of Adjustment***. When the observer does not request a specific type of fire control adjustment, area fire is issued.

(a) Area fire--moving target.

(b) Precision fire--point target.

(7) ***Danger Close***. Use when friendly troops are within--

(a) 600 mils for mortars.

(b) 600 mils for artillery.

(c) 750 mils for naval guns 5 inches or smaller.

(8) ***Mark***. Use to orient observer or to indicate targets.

(9) ***Trajectory***.

(a) Low angle (standard).

(b) High angle (mortars or if requested).

(10) ***Ammunition***. Use HE quick used unless observer specifies otherwise.

(a) Projectile (HE, illum, ICM, smoke, and so on).

(b) Fuse (quick, time, and so on).

(c) Volume of fire (observer may request the number of rounds to be fired).

(11) ***Distribution***.

(a) 100 mil sheaf (standard).

(b) Converged sheaf (used for small hard targets).

(c) Special sheaf (any length, width, and attitude).

(d) Open sheaf (separate bursts).

(e) Parallel sheaf (linear target).

(12) ***Method of Fire and Control***.

(a) *Method of Fire*. Specific guns and a specific interval between rounds. Normally adjust fire, one gun is used with a 5-second interval between rounds.

(b) *Method of Control*.

- AT MY COMMAND - FIRE--Remains in effect until observer announces, "*Cancel at my command*."
- CANNOT OBSERVE--Observer can't see the target.
- TIME ON TARGET--Observer tells FDC when he wants the rounds to impact.
- CONTINUOUS ILLUMINATION. Calculated by the FDC, otherwise observer indicates interval between rounds in seconds.
- COORDINATED ILLUMINATION. Observer may order the interval between illum and HE shells.
- CEASE LOADING--Stop loading rounds.
- CHECK FIRING--Stop firing at once.
- CONTINUOUS FIRE--Load and fire as fast as possible.
- REPEAT--Fire another round(s) with or without adjustments.

(13) ***Authentication***. Challenge and reply.

(14) ***Message to Observer***.

(a) Battery(ies) to fire for effect.

(b) Adjusting battery.

(c) Changes to the initial call for fire.

(d) Number of rounds (per tube) to be fired for effect.

(e) Target numbers.

(f) Additional information:

- Time of flight. Moving target mission.
- Probable error in range. 38 meters or greater (normal mission).
- Angle T. 500 mils or greater.

c. **Correction of Errors**. When FDC has made an error when reading back he fire support data, the observer announces "*CORRECTION*" and transmits the correct data in its entirety.

d. **Call-for-Fire Transmissions**. Example call-for-fire transmissions are shown in Table 3-5.

**Table 3-5. EXAMPLE CALL-FOR-FIRE TRANSMISSIONS.**

| ***GRID MISSION*** | |
|---|---|
| **OBSERVER** | **FIRING UNIT** |
| F24, this is J42, ADJUST FIRE, OVER. | J42, this is F24, AJUST FIRE, OUT. |
| GRID WM180513, DIRECTION 0530, OVER. | GRID WM180513, DIRECTION 0530, OUT. |
| Infantry platoon dug in, OVER. | Infantry platoon dug in, OUT. |
| | SHOT OVER. |
| SHOT OUT. | |
| | SPLASH, OVER. |
| SPLASH OUT. | |
| End of mission, 15 casualties, platoon dispersed, OVER. | End of mission, 15 casualties, platoon dispersed, OUT. |
| ***SHIFT FROM KNOWN POINT*** | |
| **OBSERVER** | **FIRING UNIT** |
| J42, THIS IS F24, ADJUST FIRE, SHIFT AB1001, OVER. | F24, THIS IS J42, ADJUST FIRE, SHIFT AB1001, OUT. |
| DIRECTION 2420, RIGHT 400, ADD 400, OVER. | DIRECTION 2420, RIGHT 400, ADD 400, OUT. |
| FIVE T-72 TANKS AT POL SITE, OVER. | FIVE T-72 TANKS AT POL SITE, AUTHENTICATE JULIET NOVEMBER, OVER. |
| I AUTHENTICATE TANGO, OVER. | |
| | SHOT, OVER. |
| SHOT OUT. | SPLASH, OVER. |
| SPLASH OUT. | |
| END OF MISSION, 2 TANKS DESTROYED, 3 IN WOODLINE, OVER | END OF MISSION, 2 TANKS DESTROYED, 3 IN WOODLINE, OUT |

| POLAR | |
|---|---|
| **OBSERVER** | **FIRING UNIT** |
| *J42, this is F24, adjust fire, polar OVER.* | F24, this is J42, adjust fire, polar, OUT |
| *Direction 2300, distance 4,000, OVER.* | Direction 2300, distance 4,000, OUT. |
| *Infantry platoon dug in, OVER* | Infantry platoon dug in, OUT |
| | Shot OVER. |
| *Shot OUT.* | Splash, OVER. |
| *Splash OUT.* | |
| *End of mission, 15 casualties, platoon dispersed, OVER.* | End of mission, 15 casualties, platoon dispersed, OUT. |

NOTES

# Chapter 4
# MOVEMENT

To survive on the battlefield, stealth, dispersion, and security must be enforced in all tactical movements. The leader must be skilled in all movement techniques.

4-1. **FORMATIONS**. Movement formations are comprised of elements and Soldiers in relation to each other. Fire teams, squads and platoons use several movement formations. Formations provide the leader control based on a METT-TC analysis. Leaders position themselves where they can best command and control formations. The formations below allow the fire team leader to lead by example, "*Follow me, and do as I do.*" All Soldiers in the team must be able to see their leader. The formations in Figure 4-1 reflect fire team formations. Squad formations are very similar with more Soldiers. Squads can operate in lines and files similar to fire teams. When squads operate in wedges or in echelon, the fire teams use those formations and simply arrange themselves in column or one team behind the other. Squads may also use the vee, with one team forming the lines of the vee, and the SL at the apex for command and control. Platoons work on the same basic formations as the Squads. When operating as a platoon, the platoon leader must carefully select the location for his machine guns in the movement formation.

**Figure 4-1. FORMATIONS**

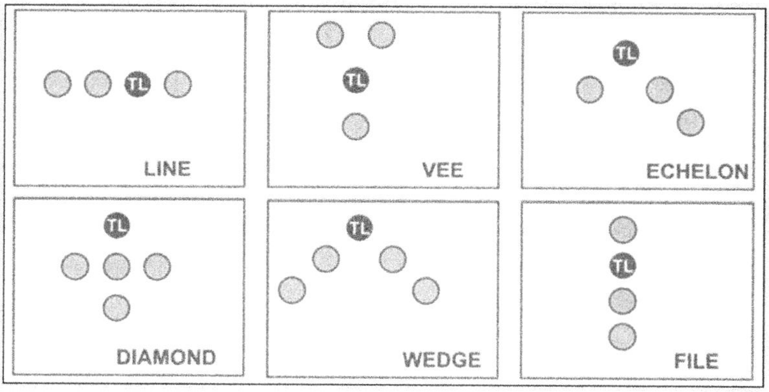


a. **Techniques**. A movement technique is the method a unit uses to traverse terrain. There are three movement techniques: traveling, traveling overwatch, and bounding overwatch. The selection of a movement technique is based on the likelihood of enemy contact and the need for speed. Factors to consider for each technique are control, dispersion, speed, and security. Movement techniques are not fixed formations. They refer to the distances between Soldiers, teams, and squads that vary based on METT-TC. Soldiers must be able to see their fire team leaders. The platoon leader should be able to see his lead squad leader. Leaders control movement with hand-and-arm signals and use radios only when needed.

b. **Standards**.

(1) Unit moves on designated route or arrives at specified location IAW OPORD maintaining accountability of all assigned/attached personnel.

(2) Unit uses movement formation and technique ordered by the leader based on METT-TC.

(3) Leaders remain oriented (within 200 meters) and follow planned route unless METT-TC dictates otherwise.

(4) Unit maintains 360 degree security and remains 100% alert during movement.

(5) Unit maintains 360 degree security and a minimum of 75% security during halts.

(6) If contact with the enemy is made, it is made with the smallest element possible.

(7) Control measures are used during movement such as head counts, rally points, or phase lines.

c. **Fundamentals**.

(1) ***Land Navigation***. Mission accomplishment depends on successful land navigation. The patrol should use stealth and vigilance to avoid chance contact. Designate a primary and alternate compass and pace man per patrol.

***Note:*** The point man will not be tasked to perform compass or pace duties. The point man's sole responsibility is forward security for the element.

(2) ***Avoidance of Detection***. Patrols must use stealth, and use the cover and concealment of the terrain to its maximum advantage. Whenever possible, move during limited visibility in order to maximize technological advantages gained by night vision devices and to hinder the enemy's ability to detect the patrol. Exploit the enemy's weaknesses, and attempt to time movements to coincide with other operations that distract the enemy. The enemy threat and terrain determines which of the three movement techniques will be used:

(a) Fire teams maintain visual contact, but the distance between them is such that the entire patrol does not become engaged if contact is made. Fire teams can spread their formations as necessary to gain better observation to the flanks. Although widely spaced, men retain their relative position in their wedge and follow their team leader. Only in extreme situations should the file be used.

(b) The lead squad must secure the front along with assuming responsibility for navigation. For a long movement, the PL may rotate the lead squad responsibilities. The fire team/squad in the rear is charged with rear security.

(c) Vary movement techniques to meet the changing situation.

(d) With the exception of fire team leaders, leaders move inside their formations where they can maintain the best control.

(3) ***Security***. The patrol must use both active and passive security measures constantly. Assign subunits responsibility for security at danger areas, patrol bases, and most importantly in the objective area.

(4) ***Fire Support***. Plan fire support (mortars, artillery, tactical air, attack helicopter, naval gunfire).

(5) ***Three-Dimensional Battlefield***. 360 Degree Security is achieved through high and low security. Within a fire team, squad, and so on, the leader must assign appropriate sectors of fire to their subordinate in order to ensure all aspects of the battlefield are covered. This includes trees, multiple storied structures, tunnels, sewers, ditches.

d. **Movement Techniques**.

(1) The traveling technique is used when enemy contact is not likely but speed is necessary.

(2) The traveling overwatch technique is used when enemy contact is possible.

(3) The bounding overwatch technique is used when enemy contact is likely, or when crossing a danger area.

e. **Traveling**. In the traveling technique, the distance between individuals is about 10 meters with 20 meters between squads. It has the following characteristics:

(1) More control than traveling overwatch but less than bounding overwatch.

(2) Minimum dispersion.

(3) Maximum speed.

(4) Minimum security.

f. **Traveling Overwatch**. The traveling overwatch technique is the basic movement technique. The distance between individuals is about 20 meters, between teams about 50 meters.

(1) In platoon traveling overwatch, the lead squad must be far enough ahead of the rest of the platoon to detect or engage any enemy before the enemy observes or fires on the main body. However, it must be close enough to be supported by the platoon's small arm's fires. This is normally between 50 to 100 meters, depending on terrain, vegetation, and light and weather conditions.

(2) In a column formation, only the lead squad should use the traveling overwatch; however, if greater dispersion is desired, all squads may use it.

(3) In other formations, all squads use traveling overwatch unless the platoon leader specifies otherwise.

(4) Traveling overwatch has the following characteristics:

- Good control.
- Good dispersion.
- Good speed.
- Good security forward.

g. **Bounding Overwatch**. In the bounding overwatch technique (Figure 4-2), the distance between men remains about 20 meters. The distance between teams and squads varies.

(1) The squad or platoon has a bounding element and an overwatch element. The bounding element moves while the overwatch element occupies an overwatch position that can cover the route of the bounding element by fire. Each bound is within supporting range of the overwatch element.

(2) There are two types of bounding, successive and alternating (Figure 4-2). Successive is nothing more than one squad moving to a position, then the overwatching squad moving to a position generally online with the first squad. Alternating bounding is when one squad moves into position, then the overwatching squad moves to a position in front of the first squad.

**Figure 4-2. SQUAD BOUNDING OVERWATCH**

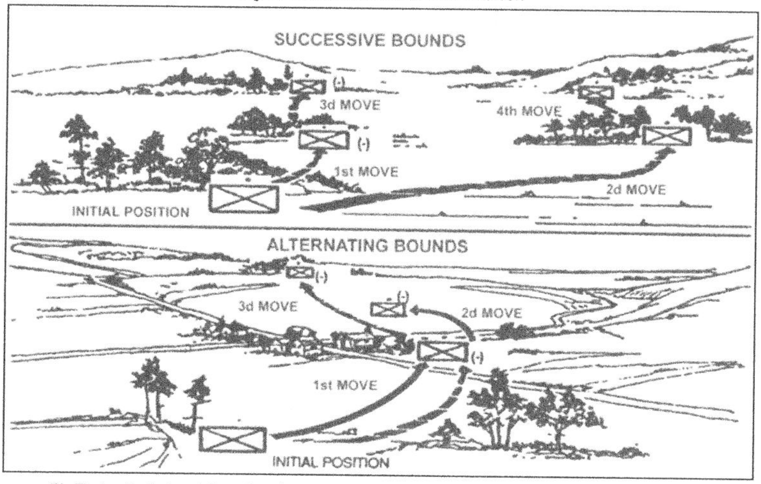


(3) The length of a bound depends on the terrain, visibility, and control.

(4) Before a bound, the leader gives the following instructions to his subordinates:

- Direction of the enemy if known
- Position of overwatch elements
- Next overwatch position
- Route of the bounding element
- What to do after the bounding element reaches the next position
- How the elements receive follow-on orders

(5) The characteristics of bounding overwatch are:

- Maximum control
- Maximum dispersion
- Minimum speed
- Maximum security

h. **Platoon Bounding Overwatch** (Figure 4-3).

(1) ***Method One***. When platoons use bounding overwatch, one squad bounds and one squad overwatches; the third squad awaits orders. Forward observers stay with the overwatching squad to call for fire. Platoon leaders normally stay with the overwatching squad who use machine guns and attached weapons to support the bounding squad.

(2) ***Method Two***. Another way is to have one squad use bounding overwatch and have the other two squads use traveling or traveling overwatch technique.

**Figure 4-3. PLATOON BOUNDING OVERWATCH**

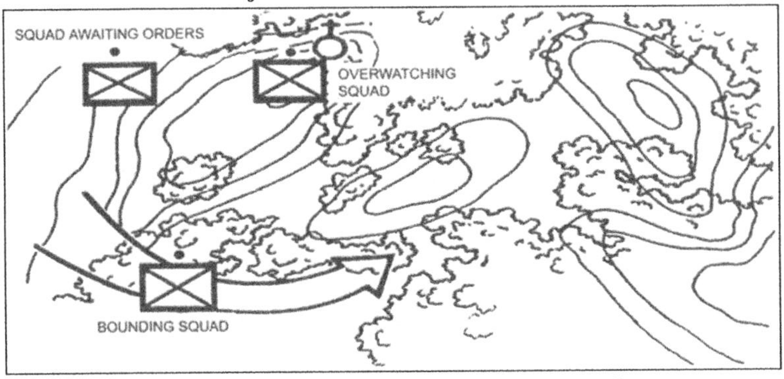


(3) ***Movement Considerations***. When deciding where to move the bounding element, consider--

- Where the enemy is likely to be.
- The mission.
- The routes to the next overwatch position.
- The weapons ranges of the overwatching unit.
- The responsiveness of the rest of the unit.
- The fields of fire at the next overwatch position.

4-2. **TACTICAL MARCHES**. Platoons conduct two types of marches with the company: foot marches and motor (road) marches.

a. **Purpose/General**. A successful foot march is when troops arrive at their destination at the prescribed time, physically able to execute their tactical mission.

b. **Standard**.

(1) The unit crosses the start point and release point at the time specified in the order.

(2) The unit follows the prescribed route, rate of march, and interval without deviation unless required otherwise by enemy action or higher headquarters action.

c. **Fundamentals**.

(1) Effective control.

(2) Detailed planning.

(3) Rehearsals.

d. **Considerations**.

(1) ***METT-TC***.

- Mission ....................................Task and purpose.
- Enemy .....................................Intentions, capabilities, and course of action.
- Terrain and weather.................Road condition/trafficability, and visibility.

- Troops/equipment....................Condition of Soldiers and their loads, numbers and types of weapons and radios.
- Time........................................Start time, release time, rate of march, time available.
- Civilians ..................................Movement through populated areas, refugees, OPSEC.

(2) ***Task Organization***:
- Security...................................Advance and trail teams.
- Main body ...............................Two remaining line squads and weapons squad.
- Headquarters...........................Command and control.
- Control measures.

(3) ***Start Point and Release Point*** (given by higher):
(a) Checkpoints............................At checkpoints report to higher and use checkpoint to remain oriented.
(b) Rally or rendezvous points ......Used when elements become separated.
(c) Location of leaders .................Where they can best control their elements.
(d) Commo plan ...........................Locations of radios, frequencies, call signs, and OPSKEDs.
(e) Dispersion between Soldiers--
- 3 to 5 meters / day.
- 1 to 3 meters / night.

(4) ***March Order***. May be issued as an OPORD, FRAGO, or annex to either (must use operational overlay or strip map)"
(a) Formations and order of movement
(b) Route of march.......................Assembly area, start point, release point, rally points, checkpoints, break/halt points.
(c) Start point time, release point time, and rate of march.
(d) March interval for squads, teams, and individuals.
(e) Actions on enemy contact........Air and ground.
(f) Actions at halts.
(g) Fires .......................................Detailed plan of fire support for the march.
(h) Water supply plan.
(i) MEDEVAC plan.

e. **Duties and Responsibilities**.

(1) ***Platoon Leader***:
- Before .........................................Issues warning order and FRAGO; inspects and supervises.
- During .........................................Ensures unit makes movement time; ensures interval is maintained and that unit remains oriented; maintains security; checks condition of men; and enforces water discipline and field sanitation.
- After ............................................Ensures men are prepared to accomplish their mission, supervises SLs, and ensures medical coverage is provided to men as needed.

(2) ***Platoon Sergeant***:
- Before .....................................Assists PL, makes recommendations, and enforces uniform and packing lists.
- During .....................................Controls stragglers, helps platoon leader maintain proper interval and security.
- At Halts ...................................Enforces security, ensures welfare of men, enforces field sanitation and litter discipline, and enforces use of preventive medicine.
- After .........................................Coordinates for water, rations, and medical supplies. Recovers casualties.

(3) ***Squad Leaders***:
- Before .....................................Provides detailed instruction to TLs, inspects boots and socks for serviceability and proper fit, inspects adjustment of equipment, inspects to ensure canteens are full, and ensures equal distribution of loads.
- During .....................................Controls squad, maintains proper interval between men and equipment, enforces security, remains oriented.

- At Halts ................................... Ensures security is maintained, provides men for water resupply as detailed. Physically checks the men in his squad, ensures they drink water, and that they change socks as necessary. Rotates heavy equipment. (Units should plan this in detail to avoid confusion before, during, and after halts.)
- After ......................................... Occupies squad sector of assembly area, conducts foot inspection, reports condition of men to PL, and prepares men to accomplish the mission.

(4) ***Security Squad***:

- Lead Team.
  - -- Serves as point element for platoon.
  - -- Recons route to SP.
  - -- Calls in checkpoints.
  - -- Provides early warning.
  - -- Maintains rate of march.
  - -- Moves 10 to 20 meters in front of main body.
- Trail Team.
  - -- Provides rear security.
  - -- Moves 10 to 20 meters behind main body.

(5) ***Medic***:

- Assesses and treats march casualties.
- Advise chain of command on evacuation and transportation requirements of casualties.

(6) ***Individual***:

- Maintains interval.
- Follows TL's examples.
- Relays hand and arm signals; remains alert during movement and at halts.
- Remains alert during movement and at halts.

4-3. **MOVEMENT IN LIMITED VISIBILITY CONDITIONS**. During hours of limited visibility, a platoon will use surveillance, target acquisition, and night observation (STANO) devices to enhance their effectiveness. Leaders must be able to control, navigate, maintain security, and move during limited visibility.

a. **Control**. When visibility is poor, the following methods aid in control.

(1) Use of night vision devices.
(2) Leaders move closer to the front.
(3) Platoon reduces speed.
(4) Use of luminescent tape on equipment.
(5) Reduce intervals between men and elements.
(6) Conducts headcount regularly.

b. **Navigation**. While navigating during limited visibility the same techniques are used as during the day; however, leaders exercise more care to keep the patrol oriented.

c. **Security**.

(1) Enforce strict noise and light discipline.
(2) Use radio-listening silence.
(3) Use camouflage.
(4) Use terrain to avoid detection by enemy surveillance or night vision devices.
(5) Make frequent listening halts; Conduct SLLS (Stop, Look, Listen, Smell).
(6) Mask the sounds of movement when possible. (Rain, wind, and flowing water will mask the sounds of movement).

d. **Rally Points**. Plan actions to be taken at rally points in detail. All elements must maintain communications at all time. There are two techniques for actions at rally points:

(1) ***Minimum Force***: Patrol members assemble at the rally point, and the senior leader assumes command. When the minimum force (designated in the OPORD) is assembled and organized the patrol will continue the mission.

(2) ***Time Available***: The senior leader determines if the patrol has enough time remaining to accomplish the mission.

e. **Actions at Halts**. During halts, post security and cover all approaches into the sector with key weapons.

(1) ***Short Halt***. Typically 1-2 minutes long. Soldiers seek immediate cover and concealment and take a knee. Leaders assign sectors of fire.

(2) ***Long Halt***. More than 2 minutes. Soldiers assume the prone position behind cover and concealment. (Ensure soldiers have clear fields of fire.) Leaders assign sectors of fire.

4-4. **DANGER AREAS**. A danger area is any place on a unit's route where the leader determines his unit may be exposed to enemy observation or fire. Some examples of danger areas are open areas, roads and trails, built-up areas, enemy positions, and natural and man-made obstacles. Bypass danger areas whenever possible.

a. **Standards**.

(1) The unit prevents the enemy from surprising the main body.
(2) The unit moves all personnel and equipment across the danger area.
(3) The unit prevents decisive engagement by the enemy.

b. **Fundamentals**.

(1) Designate near and far side rally points.
(2) Secure near side, left and right flank, and rear security.
(3) Recon and secure the far side.
(4) Cross the danger area.
(5) Plan for fires on all known danger areas.

c. **Technique for Crossing Danger Areas**.

(1) ***Linear Danger Area*** (LDA, Figure 4-4) ***Actions for a Squad***:

**STEP 1:** The alpha team leader (ATL) observes the linear danger area and sends the hand and arm signal to the SL who determines to bound across.

**STEP 2:** SL directs the ATL to move his team across the LDA far enough to fit the remainder of the squad on the far side of the LDA. Bravo team moves to the LDA to the right or left to provide an overwatch position prior to A team crossing.

**STEP 3:** SL receives the hand and arm signal that it is safe to move the rest of the squad across (B team is still providing overwatch).

**STEP 4:** SL moves himself, radio operator, and B team across the LDA. (A team provides overwatch for squad missions.)

**STEP 5:** A team assumes original azimuth at SL's command or hand and arm signal.

Figure 4-4. LINEAR DANGER AREA

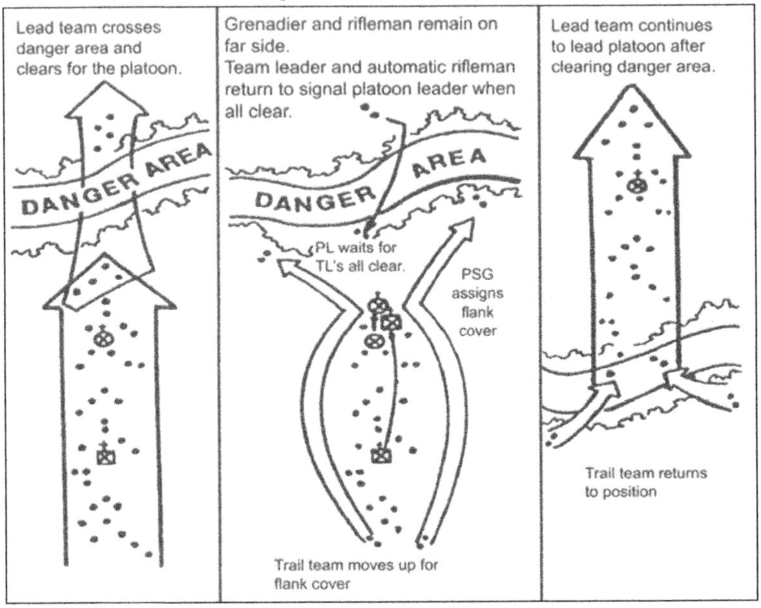


(2) ***LDA Crossing for a Platoon***:

(a) The lead squad halts the platoon, and signals danger area.

(b) The platoon leader moves forward to the lead squad to confirm the danger area and decides if current location is a suitable crossing site.

(c) The platoon leader confirms danger area/crossing site and establishes near and far side rally points.

(d) On the platoon leader's signal, the A team of the lead squad establishes an overwatch position to the left of the crossing site. Prior to crossing, the compassman with the lead two squads confirm azimuth and pace data.

(e) B team of the lead squad establishes an overwatch position to the right of the crossing site.

(f) Once overwatch positions are established, the platoon leader gives the second squad in movement the signal to bound across by fire team.

(g) Once across, the squad is now lead in movement and continues on azimuth

(h) Once Stop, Look, Listen, and Smell (SLLS) is conducted, squad leader signals platoon leader all clear.

- Day time—hand and arm signal such as thumbs up.
- Night time—Clandestine signal such as IR, red lens.

(i) Platoon leader receives all clear and crosses with radio operator, FO, WSL, and 2 gun teams.

(j) Once across, PL signals the 3rd squad in movement to cross.

(k) PSG with medic and one gun team crosses after 2nd squad is across (sterilizing central crossing site).

(l) PSG signals security squad to cross at their location.

***Notes:*** Platoon leader will plan for fires at all known LDA crossing sites.
Nearside security in overwatch will sterilize signs of the patrol .

(3) ***Danger Area (Small/Open)***

(a) The lead squad halts the platoon and signals danger area.

(b) The PL moves forward to the lead squad to confirm the danger area.

(c) The platoon leader confirms danger area and establishes near and far side rally points.

(d) The PL designates lead squad to bypass danger area using the detour-bypass method.

(e) Paceman suspends current pace count and initiates an interim pace count. Alternate pace/compass man moves forward and offsets compass 90 degrees left or right as designated and moves in that direction until clear of danger area.

(f) After moving set distance (x-meters as instructed by PL). Lead squad assumes original azimuth, and primary pace man resumes original pace.

(g) After the open area, the alternate pace/compass man offsets his compass 90 degrees left or right and leads the platoon/squad the same distance (x-meters) back to the original azimuth.

(4) ***Danger Areas (Series)***: A series of danger areas is two or more danger areas within an area that can be either observed or covered by fire.

- Double linear danger area (use linear danger area technique and cross as one LDA).
- Linear/small open danger area (use by-pass/contour technique. Figure 4-5).
- Linear/large open danger area (use platoon wedge when crossing).

***Note:*** A series of danger areas is crossed using the technique which provides the most security.

(5) ***Danger Area (Large)***:

(a) Lead squad halts the platoon, and signals danger area.

(b) The platoon leader moves forward with RTO and FO and confirms danger area.

(c) The platoon leader confirms danger area and establishes near and far side rally points.

(d) PL designates direction of movement.

(e) PL designates change of formation as necessary to ensure security.

***Notes:*** Before point man steps into danger area, PL and FO adjust targets to cover movement.
If far side of danger area is within 250 meters, PL establishes overwatch, and designates lead squad to clear wood line on far side.

Figure 4-5. SMALL OPEN AREA

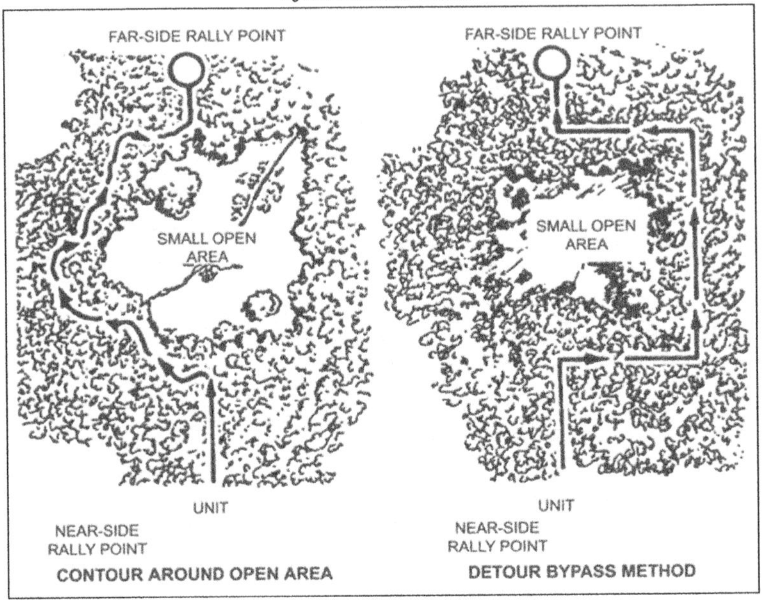


NOTES

# Chapter 5
# PATROLS

Infantry platoons and squads primarily conduct two types of patrols: reconnaissance and combat. This chapter describes the principles of patrolling, planning considerations, types of patrols, supporting tasks, patrol base, and movement to contact (FM 7-8, FM 3-0, and FM 101-5-1). In this chapter, the terms "*element*" and "*team*" refer to the squads, fire teams, or buddy teams that perform the tasks as described.

5-1. **PRINCIPLES**. All patrols are governed by five principles:

a. **Planning**. Quickly make a simple plan and effectively communicate it to the lowest level. A great plan that takes forever to complete and is poorly disseminated isn't a great plan. Plan and prepare to a realistic standard and rehearse everything.

b. **Reconnaissance**. Your responsibility as a Ranger leader is to confirm what you think you know, and to learn that which you do not already know.

c. **Security**. Preserve your force as a whole. Every Ranger and every rifle counts; anyone could be the difference between victory and defeat.

d. **Control**. Clear understanding of the concept of the operation and commander's intent, coupled with disciplined communications, to bring every man and weapon available to overwhelm the enemy at the decisive point.

e. **Common Sense**. Use all available information and good judgment to make sound, timely decisions.

5-2. **PLANNING**. This paragraph provides the planning considerations common to most patrols. It discusses task organization, initial planning and coordination, completion of the plan, and contingency planning.

a. **Task Organization**. A patrol is a mission, not an organization. To accomplish the patrolling mission, a platoon or squad must perform specific tasks, for example, secure itself, cross danger areas, recon the patrol objective, breach, support, or assault. As with other missions, the leader tasks elements of his unit in accordance with his estimate of the situation, identifying those tasks his unit must perform and designating which elements of his unit will perform which tasks. Where possible, in assigning tasks, the leader should maintain squad and fire team integrity. The chain of command continues to lead its elements during a patrol. Squads and fire teams may perform more than one task in an assigned sequence; others may perform only one task. The leader must plan carefully to ensure that he has identified and assigned all required tasks in the most efficient way. Elements and teams for platoons conducting patrols include--

(1) Elements common to all patrols:

(a) *Headquarters Element*. The headquarters consists of the platoon leader (PL), RTO, platoon sergeant (PSG), FO, RTO, and medic. It may include any attachments that the PL decides that he or the PSG must control directly.

(b) *Aid and Litter Team*. Aid and litter teams are responsible for buddy aid and evacuation of casualties.

(c) *Enemy Prisoner of War Team*. EPW teams control enemy prisoners using the five S's and the leader's guidance.

(d) *Surveillance Team*. The surveillance team keeps watch on the objective from the time that the leader's reconnaissance ends until the unit deploys for actions on the objective. They then rejoin their parent element.

(e) *En Route Recorder*. Part of the HQ element, maintains communications with higher and acts as the recorder for all CCIR collected during the mission.

(f) *Compass Man*. The compass man assists in navigation by ensuring the patrol remains on course at all times. Instructions to the compass man must include initial and subsequent azimuths. As a technique, the compass man should preset his compass on the initial azimuth before the unit moves out, especially if the move will be during limited visibility conditions. The platoon or squad leader should also designate an alternate compass man.

(g) *Point/Pace Man*. As required, the PL designates a primary and alternate point man and a pace man for the patrol. The pace man aids in navigation by keeping an accurate count of distance traveled. The point man selects the actual route through the terrain, guided by the compass man or team leader. In addition, the point man also provides frontal security.

(2) Elements common to all combat patrols:

(a) *Assault Element*. The assault element seizes and secures the objective and protects special teams as they complete their assigned actions on the objective.

(b) *Security Element*. The security element provides security at danger areas, secures the ORP, isolates the objective, and supports the withdrawal of the rest of the patrol once actions on the objective are complete. The security element may have separate security teams, each with an assigned task or sequence of tasks.

(c) *Support Element*. The support element provides direct and indirect fire support for the unit. Direct fires include machine guns, medium and light antiarmor weapons, small recoilless rifles. Indirect fires available may include mortars, artillery, CAS, and organic M203 weapon systems.

(d) *Demolition Team*. Demolition teams are responsible for preparing and detonating the charges to destroy designated equipment, vehicles, or facilities on the objective.

(e) *EPW and Search Teams*. The assault element may provide two-man (buddy teams) or four-man (fire team) search teams to search bunkers, buildings, or tunnels on the objective. These teams will search the objective or kill zone for any PIR that may give the PL an idea of the enemy concept for future operations. Primary and alternate teams may be assigned to ensure enough prepared personnel are available on the objective.

(f) *Breach Element*. The breach team conducts initial penetration of enemy obstacles to seize a foothold and allow the patrol to enter an objective. This is typically done IAW METT-TC and the steps outlined in the "*Conduct an Initial Breach of a Mined Wire Obstacle*" battle drill in Chapter 6 of this Handbook.

(3) Elements common to all reconnaissance patrols:

(a) *Reconnaissance Team*. Reconnaissance teams reconnoiter the objective area once the security teams are in position. Normally these are two-man teams (buddy teams) to reduce the possibility of detection.

(b) *Reconnaissance and Security Teams*. R&S teams are normally used in a zone reconnaissance, but may be useful in any situation when it is impractical to separate the responsibilities for reconnaissance and security.

(c) *Security Element*. When the responsibilities of reconnaissance and security are separate, the security element provides security at danger areas, secures the ORP, isolates the objective, and supports the withdrawal of the rest of the platoon once the recon is complete. The security element may have separate security teams, each with an assigned task or sequence of tasks.

b. **Initial Planning and Coordination**. Leaders plan and prepare for patrols using the troop-leading procedures and the estimate of the situation, as described in Chapter 2. Through an estimate of the situation, leaders identify required actions on the objective (mission analysis) and plan backward to departure from friendly lines and forward to reentry of friendly lines. Because patrolling units act independently, move beyond the direct-fire support of the parent unit, and operate forward of friendly units, coordination must be thorough and detailed. Coordination is continuous throughout planning and preparation. PLs use checklists to preclude omitting any items vital to the accomplishment of the mission.

(1) ***Coordination with Higher Headquarters***. This coordination includes Intelligence, Operations, and Fire Support IAW Chapter 2, *Coordination Checklists* (**page 2-34**). This initial coordination is an integral part of Step 3 of Troop-Leading Procedures, *Make a Tentative Plan*.

(2) ***Coordination with Adjacent Units***. The leader also coordinates his unit's patrol activities with the leaders of other units that will be patrolling in adjacent areas at the same time, IAW Chapter 2-7, *Coordination Checklists* (**page 2-34**).

c. **Completion of Plan**. As the PL completes his plan, he considers--

(1) ***Specified and Implied Tasks***. The PL ensures that he has assigned all specified tasks to be performed on the objective, at rally points, at danger areas, at security or surveillance locations, along the route(s), and at passage lanes. These make up the maneuver and tasks to maneuver units subparagraphs of the Execution paragraph.

(2) ***Key travel and Execution Times***. The leader estimates time requirements for movement to the objective, leader's reconnaissance of the objective, establishment of security and surveillance, completion of all assigned tasks on the objective, and passage through friendly lines. Some planning factors are--

- Movement: Average of 1 Kmph during daylight hours in Woodland Terrain; Average limited visibility ½ Kmph. Add additional time for restrictive, or severely restrictive terrain such as mountains, swamps, or thick vegetation.
- Leader's recon: NLT 1.5 hr.
- Establishment of security and surveillance: 0.5 hr.

(3) ***Primary and Alternate Routes***. The leader selects primary and alternate routes to and from the objective. The return routes should differ from the routes to the objective. The PL may delegate route selection to a subordinate, but is ultimately responsible for the routes selected.

(4) ***Signals***. The leader should consider the use of special signals. These include hand-and-arm signals, flares, voice, whistles, radios, and infrared equipment. Primary and alternate signals must be identified and rehearsed so that all Soldiers know their meaning.

(5) ***Challenge and Password Forward of Friendly Lines***. The challenge and password from the unit's ANCD must not be used beyond the FLOT.

(a) *Odd-Number System*. The leader specifies an odd number. The challenge can be any number less than the specified number. The password will be the number that must be added to it to equal the specified number, for example, the number is 7, the challenge is 3, and the password is 4.

(b) *Running Password*. ANCDs may also designate a running password. This code word alerts a unit that friendly Soldiers are approaching in a less than organized manner and possibly under pressure. The number of Soldiers approaching follows the running password. For example, if the running password is "*Ranger*," and five friendly Soldiers are approaching, they would say "*Ranger five*."

(6) ***Location of Leaders***. The PL considers where he and the PSG and other key leaders are located during each phase of the mission. The PL positions himself where he can best control the actions of the patrol. The PSG is normally located with the assault element during a raid or attack to help the PL control the use of additional assaulting squads, and will assist with securing the OBJ. The PSG will locate himself at the CCP to facilitate casualty treatment and evacuation. During a reconnaissance mission, the PSG will stay behind in the ORP to facilitate the transfer of Intel to the higher headquarters, and control the recon elements movement into and out of the ORP.

(7) ***Actions on Enemy Contact***. Unless required by the mission, the unit avoids enemy contact. The leader's plan must address actions on chance contact at each phase of the patrol mission. The unit's ability to continue will depend on how early contact is made, whether the platoon is able to break contact successfully (so that its subsequent direction of movement is undetected), and whether the unit receives any casualties because of the contact. The plan must address the handling of seriously wounded Soldiers and KIAs. The plan must also address the handling of prisoners who are captured because of chance contact and are not part of the planned mission.

(8) ***Contingency Plans***. The leader leaves his unit for many reasons throughout the planning, coordination, preparation, and execution of his patrol mission. Each time the leader departs the patrol main body, he must issue a five-point contingency plan to the leader left in charge of the unit. The contingency plan is described by the acronym GOTWA, as follows. The patrol leader will additionally issue specific guidance stating what tasks are to be accomplished in the ORP in his absence:

- G: Where the leader is GOING.
- O: OTHERS he is taking with him.
- T: TIME he plans to be gone.
- W: WHAT to do if the leader does not return in time.
- A: The unit's and the leader's ACTIONS on chance contact while the leader is gone.

(9) ***Rally Points***. The leader considers the use and location of rally points. A rally point is a place designated by the leader where the unit moves to reassemble and reorganize if it becomes dispersed. Soldiers must know which rally point to move to at each phase of the patrol mission should they become separated from the unit. They must also know what actions are required there and how long they are to wait at each rally point before moving to another.

(a) *Criteria*. Rally points must be--

- Easily identifiable in daylight and limited visibility.
- Show no signs of recent enemy activity.
- Covered and concealed.
- Away from natural lines of drift and high-speed avenues of approach.
- Defendable for short periods of time.

(b) *Types*. The most common types of rally points are initial, en route, objective, and near-and-far-side rally points.

(10) ***Objective Rally Point***. The ORP is typically 200 to 400m from the objective, or at a minimum, one major terrain feature away. Actions at the ORP include--

- Conduct SLLS and pinpoint location.
- Leaders Recon of the Objective.
- Issuing a FRAGO, if needed.
- Making final preparations before continuing operations, for example, recamouflaging, preparing demolitions, lining up rucksacks for quick recovery. Preparing EPW bindings, first aid kits, litters, and inspecting weapons.
- Accounting for Soldiers and equipment after actions at the objective are complete.
- Reestablishing the chain of command after actions at the objective are complete.
- Disseminating information from reconnaissance, if contact was not made.

(11) ***Leader's reconnaissance of the objective***. The plan must include a leader's reconnaissance of the objective once the platoon or squad establishes the ORP. Before departing, the leader must issue a 5-point contingency plan. During his

reconnaissance, the leader pinpoints the objective, selects reconnaissance, security, support, and assault positions for his elements, and adjusts his plan based on his observation of the objective. Each type of patrol requires different tasks during the leader's reconnaissance. The platoon leader will bring different elements with him. (These are discussed separately under each type of patrol). The leader must plan time to return to the ORP, complete his plan, disseminate information, issue orders and instructions, and allow his squads to make any additional preparations. During the Leader's Reconnaissance for a Raid or Ambush, the PL will leave surveillance on the OBJ.

(12) ***Actions on the objective***. Each type of patrol requires different actions on the objective. Actions on the objective are discussed under each type of patrol.

5-3. **RECONNAISSANCE PATROLS**. Recon patrols are one of the two types of patrols. provide timely and accurate information on the enemy and terrain. They confirm the leader's plan before it is executed. Units on reconnaissance operations collect specific information (*priority intelligence requirements* [PIR]) or general information (*information requirements* [IR]) based on the instructions from their higher commander. The two types of recon patrols discussed here are area and zone. This section discusses the fundamentals of reconnaissance, task standards for the two most common types of recon, and actions on the objective for those types of recon.

a. **Fundamentals of Reconnaissance**. In order to have a successful area reconnaissance, the platoon leader must apply the fundamentals of the reconnaissance to his plan during the conduct of the operation.

(1) ***Gain all required information***. The parent unit tells the patrol leader (PL) what information is required. This is in the form of the IR and PIR. The platoon's mission is then tailored to what information is required. During the entire patrol, members must continuously gain and exchange all information gathered, but cannot consider the mission accomplished unless all PIR has been gathered.

(2) ***Avoid detection by the enemy***. A patrol must not let the enemy know that it is in the objective area. If the enemy knows he is being observed, he may move, change his plans, or increase his security measures. Methods of avoiding detection are--

(a) Minimize movement in the objective area (area reconnaissance).

(b) Move no closer to the enemy than necessary.

(c) If possible, use long-range surveillance or night vision devices.

(d) Camouflage, stealth, noise, and light discipline.

(e) Minimize radio traffic.

(3) ***Employ security measures***. A patrol must be able to break contact and return to the friendly unit with what information is gathered. If necessary, break contact and continue the mission. Security elements are emplaced so that they can overwatch the reconnaissance elements and suppress the enemy so the reconnaissance element can break contact.

(4) ***Task organization***. When the platoon leader receives the order, he analyzes his mission to ensure he understands what must be done. Then he task organizes his platoon to best accomplish the mission IAW METT-TC. Recons are typically squad-sized missions.

b. **Task Standards**.

(1) ***Area Reconnaissance***. The area recon patrol collects all available information on PIR and other intelligence not specified in the order for the area. The patrol completes the recon and reports all information by the time specified in the order. The patrol is not compromised.

(2) ***Zone Reconnaissance***. The zone recon patrol determines all PIR and other intelligence not specified in the order for its assigned zone. The patrol reconnoiters without detection by the enemy. The patrol completes the recon and reports all information by the time specified in the order.

c. **Actions on the Objective, Area Reconnaissance** (Figure 5-1).

(1) The element occupies the ORP as discussed in the section on occupation of the ORP. The RTO calls in spare for occupation of ORP. The leader confirms his location on map while subordinate leaders make necessary perimeter adjustments.

(2) The PL organizes the platoon in one of two ways: separate recon and security elements, or combined recon and security elements.

(3) The PL takes subordinates leaders and key personnel on a leader's recon to confirm the objective and plan.

(a) Issues a 5-point contingency plan before departure.

(b) Establishes a suitable release point that is beyond sight and sound of the objective if possible, but that is definitely out of sight. The RP should also have good rally point characteristics.

(c) Allow all personnel to become familiar with the release point and surrounding area.

(d) Identifies the objective and emplaces surveillance. Designates a surveillance team to keep the objective under surveillance. Issues a contingency plan to the senior man remaining with the surveillance team. The surveillance team is positioned with one man facing the objective, and one facing back in the direction of the release point.

(e) Takes subordinate leaders forward to pinpoint the objective, emplace surveillance, establish a limit of advance, and choose vantage points.

(f) Maintains commo with the platoon throughout the leader's recon.

(4) The PSG maintains security and supervises priorities of work in the ORP.

(a) Reestablishes security at the ORP.

(b) Disseminates the PLs contingency plan.

(c) Oversees preparation of recon personnel (personnel recamouflaged, NVDs and binos prepared, weapons on safe with a round in the chamber).

(5) The PL and his recon party return to the ORP.

(a) Confirms the plan or issues a FRAGO.

(b) Allows subordinate leaders time to disseminate the plan.

(6) The patrol conducts the recon by long-range observation and surveillance if possible.

(a) R&S elements move to observation points that offer cover and concealment and that are outside of small-arms range.

(b) Establishes a series of observation posts (OP) if information cannot be gathered from one location.

(c) Gathers all PIR using the SALUTE format.

(7) If necessary, the patrol conducts its recon by short-range observation and surveillance.

(a) Moves to an OP near the objective.

(b) Passes close enough to the objective to gain information.

(c) Gathers all PIR using the SALUTE format.

(8) R&S teams move using a technique such as the cloverleaf method to move to successive OP's. In this method, R&S teams avoid paralleling the objective site, maintain extreme stealth, do not cross the limit of advance, and maximize the use of available cover and concealment.

(9) During the conduct of the recon, each R&S team will return to the release point when any of the following occurs:

- They have gathered all their PIR.
- They have reached the limit of advance.
- The allocated time to conduct the recon has elapsed.
- Contact has been made.

(10) At the release point, the leader will analyze what information has been gathered and determine if he has met the PIR requirements.

(11) If the leader determines that he has not gathered sufficient information to meet the PIR requirements, or if the information he and the subordinate leader gathered differs drastically, he may have to send R&S teams back to the objective site. In this case, R&S teams will alternate areas of responsibilities. For example, if one team reconnoitered from the 6 – 3 – 12, then that team will now recon from the 6 – 9 – 12.

(12) The R&S element returns undetected to the ORP by the specified time.

(a) Disseminates information to all patrol members through key leaders at the ORP, or moves to a position at least one terrain feature or one kilometer away to disseminate. To disseminate, the leader has the RTO prepare three sketches of the objective site based on the leader's sketch and provides the copies to the subordinate leaders to assist in dissemination.

(b) Reports any information requirements and/or any information requiring immediate attention to higher headquarters, and departs for the designated area.

(13) If contact is made, move to the release point. The recon element tries to break contact and return to the ORP, secure rucksacks, and quickly move out of the area. Once they have moved a safe distance away, the leader will inform higher HQ of the situation and take further instructions from them.

(a) While emplacing surveillance, the recon element withdraws through the release point to the ORP, and follows the same procedures as above.

(b) While conducting the reconnaissance, the compromised element returns a sufficient volume of fire to allow them to break contact. Surveillance can fire an AT-4 at the largest weapon on the objective. All elements will pull off the objective and move to the release point. The senior man will quickly account for all personnel and return to the ORP. Once in the ORP, follow the procedures previously described.

Figure 5-1. ACTIONS ON THE OBJECTIVE, AREA RECONNAISSANCE

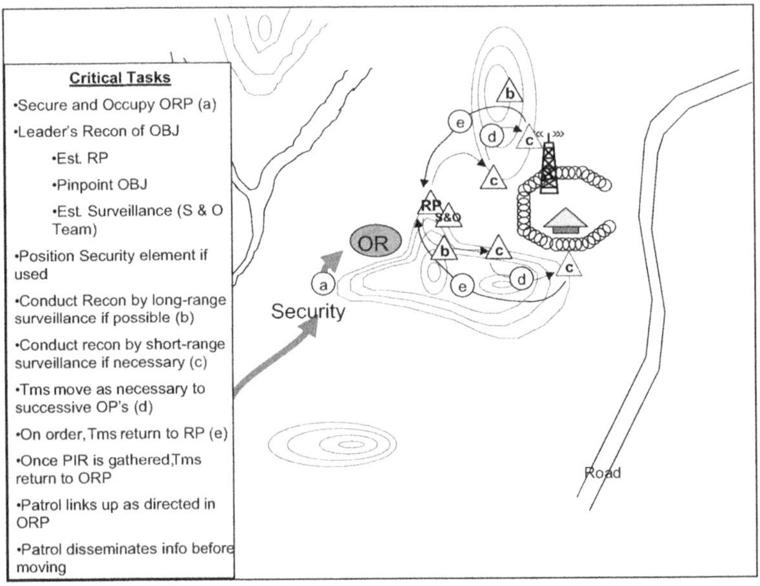


d. **Actions on the Objective, Zone Reconnaissance.**

(1) The element occupies the initial ORP as discussed in the section occupation of the ORP. The radio operator calls in spare for occupation of ORP. The leader confirms his location on map while subordinate leaders make necessary perimeter adjustments.

(2) The recon team leaders organize their recon elements.

(a) Designate security and recon elements.

(b) Assign responsibilities (point man, pace man, en route recorder, and rear security), if not already assigned.

(c) Designates easily recognizable rally points.

(d) Ensure local security at all halts.

(3) The patrol recons the zone.

(a) Moves tactically to the ORPs.

(b) Occupies designated ORPs.

(c) Follows the method designated by the PL (fan, converging routes, or box method, Figure 5-2).

Figure 5-2. COMPARISON OF ZONE RECONNAISSANCE METHODS

| *FAN METHOD* | *CONVERGING ROUTES METHOD* | *BOX METHOD* |
|---|---|---|
| • Uses a series (fan) of ORPs.<br>• Patrol establishes security at first ORP.<br>• Each recon element moves from ORP along a different fan-shaped route. Route overlaps with that of other recon elements. This ensures recon of entire area.<br>• Leader maintains reserve at ORP.<br>• When all recon elements return to ORP, PL collects and disseminates all info before moving to next ORP. | • PL selects routes from ORP thru zone to a linkup point at the far side of the zone from the ORP.<br>Each recon element moves and recons along a specified route. They converge (link up) at one time and place. | • PL sends recon elements from the first ORP along routes that form a box.<br>He sends other elements along routes throughout the box. All teams link up at the far side of the box from the ORP. |

(d) The recon teams reconnoiter.

- During movement, the squad will gather all PIR specified by the order.
- Recon team leaders will ensure sketches are drawn or digital photos are taken of all enemy hard sites, roads, and trails.
- Return to the ORP, or link up at the rendezvous point on time.
- When the squad arrives at new rendezvous point or ORP, the recon team leaders report to the PL with all information gathered.

(e) The PL continues to control the recon elements.

- PL moves with the recon element that establishes the linkup point.
- PL changes recon methods as required.
- PL designates times for the elements to return to the ORP or to linkup.
- PL collects all information and disseminates it to the entire patrol. PL will brief all key subordinate leaders on information gathered by other squads, establishing one consolidated sketch if possible, and allow team leaders time to brief their teams.
- PL and PSG account for all personnel.

(f) The patrol continues the recon until all designated areas have been reconned, and returns undetected to friendly lines.

5-4. **COMBAT PATROLS**. Combat patrols are the second type of patrol. Combat patrols are further divided into raids and ambushes Units conduct combat patrols to destroy or capture enemy Soldiers or equipment; destroy installations, facilities, or key points; or harass enemy forces. Combat patrols also provide security for larger units. This section describes overall combat patrol planning considerations, task considerations for each type of combat patrol, and finally actions on the objective for each type.

a. **Planning Considerations (General)**. In planning a combat patrol, the PL considers the following:

(1) ***Tasks to Maneuver Units***. Normally the platoon headquarters element controls the patrol on a combat patrol mission. The PL makes every try to maintain squad and fire team integrity as he assigns tasks to subordinates units.

(a) The PL must consider the requirements for assaulting the objective, supporting the assault by fire, and security of the entire unit throughout the mission.

- For the assault on the objective, the PL considers the required actions on the objective, the size of the objective, and the known or presumed strength and disposition of the enemy on and near the objective.
- The PL considers the weapons available, and the type and volume of fires required to provide fire support for the assault on the objective.
- The PL considers the requirement to secure the platoon at points along the route, at danger areas, at the ORP, along enemy avenues of approach into the objective, and elsewhere during the mission.
- The PL will also designate engagement/disengagement criteria.

(b) The PL assigns additional tasks to his squads for demolition, search of EPWs, guarding of EPWs, treatment and evacuation (litter teams) of friendly casualties, and other tasks required for successful completion of patrol mission (if not already in the SOP).

(c) The PL determines who will control any attachments of skilled personnel or special equipment.

(2) ***Leader's Reconnaissance of the Objective***. In a combat patrol, the PL has additional considerations for the conduct of his reconnaissance of the objective from the ORP.

(a) *Composition of the leader's reconnaissance party.* The platoon leader will normally bring the following personnel.

- Squad leaders to include the weapons squad leader.
- Surveillance team.
- Forward observer.
- Security element (dependent on time available).

(b) *Conduct of the leader's reconnaissance.* In a combat patrol, the PL considers the following additional actions in the conduct of the leader's reconnaissance of the objective.

- The PL designates a release point approximately half way between the ORP and this objective. The PL posts the surveillance team. Squads and fire teams separate at the release point, and then they move to their assigned positions.
- The PL confirms the location of the objective or kill zone. He notes the terrain and identifies where he can emplace claymores to cover dead space. Any change to his plan is issued to the squad leaders (while overlooking the objective if possible).
- If the objective is the kill zone for an ambush, the leader's reconnaissance party should not cross the objective; to do so will leave tracks that may compromise the mission.
- The PL confirms the suitability of the assault and support positions and routes from them back to the ORP.
- The PL issues a five-point contingency plan before returning to the ORP.

b. **Ambush**.

(1) ***Planning Considerations***. An ambush is a surprise attack from a concealed position on a moving or temporarily halted target. Ambushes are classified by category--hasty or deliberate; type--point or area; and formation--linear or L-shaped. The leader uses a combination of category, type, and formation in developing his ambush plan. The key planning considerations include--

(a) Coverage of entire kill zone by fire.

(b) METT-TC.

(c) Use of existing or reinforcing obstacles, including Claymores, to keep the enemy in the kill zone.

(d) Security teams (typically equipped with hand-held antitank weapons such as AT-4 or LAW; Claymores; and various means of communication.

(e) Protect the assault and support elements with claymores or explosives.

(f) Use security elements or teams to isolate the kill zone.

(g) Assault through the kill zone to the limit of advance (LOA). (The assault element must be able to move quickly through its own protective obstacles.)

(h) Time the actions of all elements of the platoon to preclude loss of surprise. In the event any member of the ambush is compromised, he may immediately initiate the ambush.

(i) When the ambush must be manned for a long time, use only one squad to conduct the entire ambush and determining movement time of rotating squads from the ORP to the ambush site.

(2) ***Categories***.

(a) *Hasty Ambush*. A unit conducts a hasty ambush when it makes visual contact with an enemy force and has time to establish an ambush without being detected. The actions for a hasty ambush must be well rehearsed so that Soldiers know what to do on the leader's signal. They must also know what action to take if the unit is detected before it is ready to initiate the ambush.

(b) *Deliberate Ambush*. A deliberate ambush is conducted at a predetermined location against any enemy element that meets the commander's engagement criteria. The leader requires the following detailed information in planning a deliberate ambush: size and composition of the targeted enemy, and weapons and equipment available to the enemy.

(3) ***Types***.

(a) Point ambush. In a point ambush, Soldiers deploy to attack an enemy in a single kill zone.

(b) Area ambush. In an area, Soldiers deploy in two or more related point ambushes.

(4) ***Formations***. (Figure 5-3).

(a) *Linear Ambush*. In an ambush using a linear formation, the assault and support elements deploy parallel to the enemy's route. This positions both elements on the long axis of the kill zone and subjects the enemy to flanking fire. This formation can be used in close terrain that restricts the enemy's ability to maneuver against the platoon, or in open terrain provided a means of keeping the enemy in the kill zone can be effected.

(b) *L-Shaped Ambush*. In an L-shaped ambush, the assault element forms the long leg parallel to the enemy's direction of movement along the kill zone. The support element forms the short leg at one end of and at right angles to the assault element. This provides both flanking (long leg) and enfilading fires (short leg) against the enemy. The L-shaped ambush can be used at a sharp bend in a trail, road, or stream. It should not be used where the short leg would have to cross a straight road or trail.

**Figure 5-3. AMBUSH FORMATIONS**

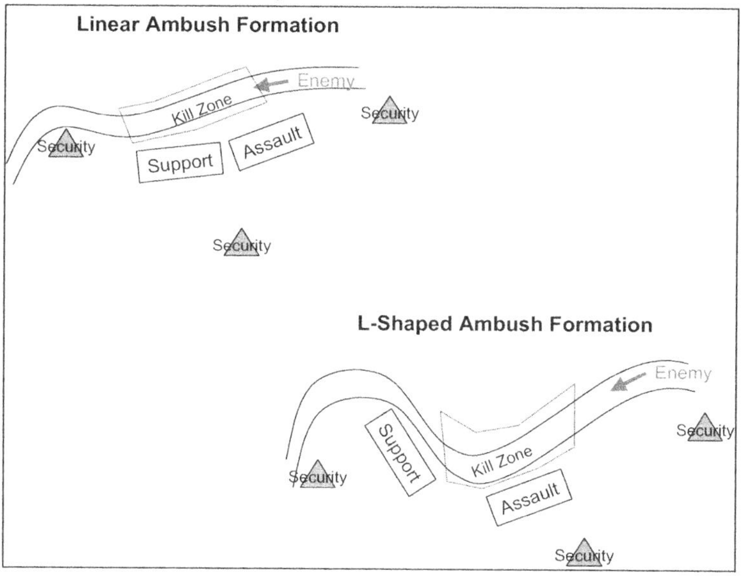

c. **Hasty Ambush**.

(1) ***Task Standards***. The platoon moves quickly to concealed positions. The ambush is not initiated until the majority of the enemy is in the kill zone. The unit does not become decisively engaged. The platoon surprises the enemy. The patrol captures, kills, or forces the withdrawal of all of the enemy within the kill zone. On order, the patrol withdraws all personnel and equipment in the kill zone from observation and direct fire. The unit does not become decisively engaged by follow-on elements. The platoon continues follow-on operations.

(2) ***Actions on the Objective (Hasty Ambush)***. See Figure 5-4.

(a) Using visual signals, any Soldier alerts the unit that an enemy force is in sight. The Soldier continues to monitor the location and activities of the enemy force until his team or squad leader relieves him, and gives the enemy location and direction of movement.

(b) The platoon or squad halts and remains motionless.

- The PL gives the signal to conduct a hasty ambush, taking care not to alert the enemy of the patrol's presence.
- The leader determines the best nearby location for a hasty ambush. He uses arm-and-hand signals to direct the unit members to covered and concealed positions.

(c) The leader designates the location and extent of the kill zone.

d) Teams and squads move silently to covered and concealed positions, ensuring positions are undetected and have good observation and fields of fire into the kill zone.

(e) Security elements move out to cover each flank and the rear of the unit. The leader directs the security elements to move a given distance, set up, and then rejoin the unit on order or, after the ambush (the sound of firing ceases). At squad level, the two outside buddy teams normally provide flank security as well as fires into the kill zone. At platoon level, fire teams make up the security elements.

(f) The PL assigns sectors of fire and issues any other commands necessary such as control measures.

(g) The PL initiates the ambush, using the greatest casualty-producing weapon available, when the largest percentage of enemy is in the kill zone. The PL--

- Controls the rate and distribution of fire.
- Employs indirect fire to support the ambush.
- Orders cease fire.
- (If the situation dictates) Orders the patrol to assault through the kill zone.

(h) The PL designates personnel to conduct a hasty search of enemy personnel and process enemy prisoners and equipment.

(i) The PL orders the platoon to withdraw from the ambush site along a covered and concealed route.

(j) The PL gains accountability, reorganizes as necessary, disseminates information, reports the situation, and continues the mission as directed.

d. **Deliberate (Point/Area) Ambush**.

(1) ***Task Standards***. The ambush is emplaced NLT the time specified in the order. The patrol surprises the enemy and engages the enemy main body. The patrol kills or captures all enemy in the kill zone and destroys equipment based on the commander's intent. The patrol withdraws all personnel and equipment from the objective, on order, within the time specified in the order. The patrol obtains all available PIR from the ambush and continues follow-on operations.

(2) ***Actions on the Objective (Deliberate Ambush).*** See Figure 5-5.

(a) The PL prepares the patrol for the ambush in the ORP.

(b) The PL prepares to conduct a leader's reconnaissance. He--

- Designates the members of the leader's recon party (typically includes squad leaders, surveillance team, FO, and possibly the security element.
- Issues a contingency plan to the PSG.

Figure 5-4. ACTIONS ON THE OBJECTIVE -- HASTY AMBUSH

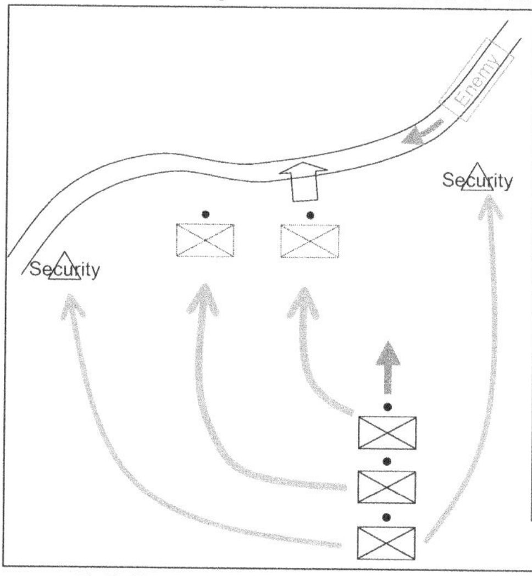


**Critical Tasks**

- Patrol detects an enemy unit; PL is notified
- Patrol halts and remains motionless
- PL gives signal for Hasty Ambush
- PL directs elements to covered and concealed positions
- Security Elements move to flanks of patrol
- PL establishes control measures
- PL initiates and controls ambush
- PL directs a hasty search
- Patrol consolidates, reorganizes, withdraws, reports, and continues mission

(c) The PL conducts his leader's reconnaissance. He--
- Ensures the leader's recon party moves undetected.
- Confirms the objective location and suitability for the ambush.
- Selects a kill zone.
- Posts the surveillance team at the site and issues a contingency plan.
- Confirms suitability of assault and support positions, and routes from them to the ORP.
- Selects the position of each weapon system in the support-by-fire position, and then designates sectors of fire.
- Identifies all offensive control measures to be used. Identifies the PLD, the assault position, LOA, any boundaries or other control measures. If available, the PL can use infrared aiming devices to identify these positions on the ground.

(d) The PL adjusts his plan based on info from the reconnaissance. He--
- Assigns positions.
- Designates withdrawal routes.

(e) The PL confirms the ambush formation.

(f) The security team(s) occupy first, securing the flanks of the ambush site, and providing early warning. The security element must be in position before the support and assault elements move forward of the release point. A security team remains in the ORP if the patrol plans to return to the ORP after actions on the objective. If the ORP is abandoned, a rear security team should be emplaced.

(g) Support element leader assigns sectors of fire. He--
- Emplaces claymores and obstacles as designated.
- Identifies sectors of fire and emplaces limiting stakes to prevent friendly fires from hitting other elements.
- Overwatches the movement of the assault element into position.

(h) Once the support element is in position, or on the PLs order, the assault element--

- Departs the ORP and moves into position.
- Upon reaching the PLD, the assault element transitions from the movement formation to the battle formation.
- Identifies individual sectors of fire as assigned by the PL. Emplaces aiming stakes or uses metal-to-metal contact with the machine gun tripods to prevent fratricide on the objective.
- Emplaces claymores to help destroy the enemy in the kill zone.
- Camouflages positions.

(i) The security element spots the enemy and notifies the PL, and reports the direction of movement, size of the target, and any special weapons or equipment carried. The security element also keeps the platoon leader informed if any enemy forces are following the lead force.

(j) The PL alerts other elements, and determines if the enemy force is too large, or if the ambush can engage the enemy successfully.

(k) The PL initiates the ambush using the highest casualty-producing device. He may use a command-detonated claymore. He must also plan a backup method for initiating the ambush, in case his primary means fails. This should also be a casualty-producing device such as his individual weapon. He passes this information to all Rangers, and practices it during rehearsals.

(l) The PL ensures that the assault and support elements deliver fire with the heaviest, most accurate volume possible on the enemy in the kill zone. In limited visibility, the PL may use infrared lasers to further define specific targets in the kill zone.

(m) Before assaulting the target, the PL gives the signal to lift or shift fires.

(n) The assault element--

- Assaults before the remaining enemy can react.
- Kills or captures enemy in the kill zone.
- Uses individual movement techniques or bounds by fire teams to move.
- Upon reaching the limit of advance, halts and establishes security. If needed, it reestablishes the chain of command and remains key weapon systems. All Soldiers will load a fresh magazine or drum of ammunition using the buddy system. ACE reports will be submitted through the chain of command. The PL will submit an initial contact report to higher.

(o) The PL directs special teams (EPW search, aid and litter, demo) to accomplish their assigned task once the assault element has established its LOA.

- Once the kill zone had been cleared, collect and secure all EPWs and move them out of the kill zone before searching bodies. Coordinate for an EPW exchange point to link up with higher to extract all EPWs and treat them IAW the five S's.
- Search from one side to the other and mark bodies that have been searched to ensure the area is thoroughly covered. Units should use the "*clear out, search in*" technique, clear from the center of the objective out ensuring the area is clear of all enemy combatants; then search all enemy personnel towards the center of the objective. Search all dead enemy personnel using two-man search techniques.
  - -- As the search team approaches a dead enemy Soldier, one-man guards while the other man searches. First, he kicks the enemy weapon away.
  - -- Second, he rolls the body over (if on the stomach) by lying on top and when given the go ahead by the guard (who is positioned at the enemy's head), the searcher rolls the body over on him. This is done for protection in case the enemy Soldier has a grenade with the pin pulled underneath him.
  - -- The searchers then conduct a systematic search of the dead Soldier from head to toe removing all papers and anything new (different type rank, shoulder boards, different unit patch, pistol, weapon, or NVD). They note if the enemy has a fresh or shabby haircut and the condition of his uniform and boots. They note the radio frequency, and then they secure the SOI, maps, documents, and overlays.
  - -- Once the body has been thoroughly searched, the search team will continue in this manner until all enemy personnel in and near the kill zone have been searched.
- Identify, collect, and prepare all equipment to be carried back or destroyed.
- Evacuate and treat friendly wounded first, then enemy wounded, time permitting.

- The demolition team prepares dual-primed explosives or incendiary grenades and awaits the signal to initiate. This is normally the last action performed before the unit departs the objective and may signal the security elements to return to the ORP.
- Actions on the objective with stationary assault line; all actions are the same with the exception of the search teams. To provide security within the teams to the far side of the kill zone during the search, they work in three-Ranger teams. Before the search begins, the Rangers move all KIAs to the near side of the kill zone.

(p) If enemy reinforcements try to penetrate the kill zone, the flank security will engage to prevent the assault element from being compromised.

(q) The platoon leader directs the unit's withdrawal from the ambush site:

- Elements normally withdraw in the reverse order that they established their positions.
- The elements may return to the RP or directly to the ORP, depending on the distance between elements.
- The security element of the ORP must be alert to assist the platoon's return to the ORP. It maintains security for the ORP while the rest of the platoon prepares to leave.
- If possible, all elements should return to the location at which they separated from the main body. This location should usually be the RP.

(r) The PL and PSG direct actions at the ORP, to include accountability of personnel and equipment and recovery of rucksacks and other equipment left at the ORP during the ambush.

(s) The platoon leader disseminates information, or moves the platoon to a safe location (no less than one kilometer or one terrain feature away from the objective) and disseminates information.

(t) As required, the PL and FO execute indirect fires to cover the platoon's withdrawal.

**Figure 5-5. ACTIONS ON THE OBJECTIVE -- DELIBERATE AMBUSH**

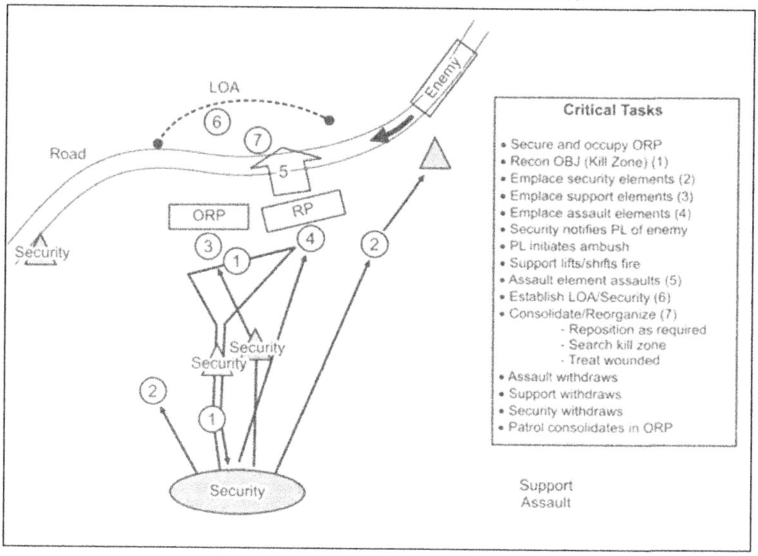

e. **Perform Raid**. The patrol initiates the raid NLT the time specified in the order, surprises the enemy, assaults the objective, and accomplishes its assigned mission within the commander's intent. The patrol does not become decisively engaged en route to the objective. The patrol obtains all available PIR from the raid objective and continues follow-on operations.

(1) ***Planning Considerations***. A Raid is a form of attack, usually small scale, involving a swift entry into hostile territory to secure information, confuse the enemy, or destroy installations followed by a planned withdrawal. Squads do not conduct raids. The sequence of platoon actions for a raid is similar to those for an ambush. Additionally, the assault element of the platoon may have to conduct a breach of an obstacle. It may have additional tasks to perform on the objective such as demolition of fixed facilities. Fundamentals of the raid include--

- Surprise and speed. Infiltrate and surprise the enemy without being detected.
- Coordinated fires. Seal off the objective with well-synchronized direct and indirect fires.
- Violence of action. Overwhelm the enemy with fire and maneuver.
- Planned withdrawal. Withdraw from the objective in an organized manner, maintaining security.

(2) ***Actions on the Objective (Raid)***. See Figure 5-6.

(a) The patrol moves to and occupies the ORP IAW the patrol SOP. The patrol prepares for the leader's recon.

(b) The PL, squad leaders, and selected personnel conduct a leader's recon.

- PL leaves a five point contingency plan with the PSG.
- PL establishes the RP, pinpoints the objective, contacts the PSG to prep men, weapons, and equipment, emplaces the surveillance team to observe the objective, and verifies and updates intelligence information.
- Leader's recon verifies location of and routes to security, support, and assault positions.
- Security teams are brought forward on the leader's reconnaissance and emplaced before the leader's recon leaves the RP
- Leaders conduct the recon without compromising the patrol.
- Leaders normally recon support by fire position first, then the assault position.

(c) The PL confirms, denies, or modifies his plan and issues instructions to his squad leaders.

- Assigns positions and withdrawal routes to all elements.
- Designates control measures on the objective (element objectives, lanes, limits of advance, target reference points, and assault line).
- Allows SLs time to disseminate information, and confirm that their elements are ready.

(d) Security elements occupy designated positions, moving undetected into positions that provide early warning and can seal off the objective from outside support or reinforcement.

(e) The support element leader moves the support element to designated positions. The support element leader ensures his element can place well-aimed fire on the objective.

(f) The PL moves with the assault element into the assault position. The assault position is normally the last covered and concealed position before reaching the objective. As it passes through the assault position the platoon deploys into its assault formation; that is, its squads and fire teams deploy to place the bulk of their firepower to the front as they assault the objective.

- Makes contact with the surveillance team to confirm any enemy activity on the objective.
- Ensures that the assault position is close enough for immediate assault if the assault element is detected early.
- Moves into position undetected, and establish local security and fire control measures.

(g) Element leaders inform the PL when their elements are in position and ready.

(h) The PL directs the support element to fire.

(i) Upon gaining fire superiority, the PL directs the assault element to move towards the objective.

- Assault element holds fire until engaged, or until ready to penetrate the objective.
- PL signals the support element to lift or shift fires. The support element lifts or shifts fires as directed, shifting fire to the flanks of targets or areas as directed in the FRAGO.

(j) The assault element attacks and secures the objective. The assault element may be required to breech a wire obstacle. As the platoon, or its assault element, moves onto the objective, it must increase the volume and accuracy of fires. Squad leaders assign specific targets or objectives for their fire teams. Only when these direct fires keep the enemy suppressed can the rest of the unit maneuver. As the assault element gets closer to the enemy, there is more emphasis on suppression and less on maneuver. Ultimately, all but one fire team may be suppressing to allow that one fire team to break into the enemy position. Throughout the assault, Soldiers use proper individual movement techniques, and fire teams retain their basic shallow wedge formation. The platoon does not get "*on-line*" to sweep across the objective.

- Assault element assaults through the objective to the designated LOA.
- Assault element leaders establish local security along the LOA, and consolidate and reorganize as necessary. They provide ACE reports to the PL and PSG. The platoon establishes security, operates key weapons, provides first aid, and prepares wounded Soldiers for MEDEVAC. They redistribute ammunition and supplies, and they relocate selected weapons to alternate positions if leaders believe that the enemy may have pinpointed them during the attack. They adjust other positions for mutual support. The squad and team leader provide ammunition, casualty, and equipment (ACE) reports to the platoon leader. The PL/PSG reorganizes the patrol based on the contact.
  -- On order, special teams accomplish all assigned tasks under the supervision of the PL, who positions himself where he can control the patrol.
  -- Special team leaders report to PL when assigned tasks are complete.

(k) On order or signal of the PL, the assault element withdraws from the objective. Using prearranged signals, the assault line begins an organized withdrawal from the objective site, maintaining control and security throughout the withdrawal. The assault element bounds back near the original assault line, and begins a single file withdrawal through the APL's choke point. All Rangers must move through the choke point for an accurate count. Once the assault element is a safe distance from the objective and the headcount is confirmed, the platoon can withdraw the support element. If the support elements were a part of the assault line, they withdraw together, and security is signaled to withdraw. Once the support is a safe distance off the objective, they notify the platoon leader, who contacts the security element and signals them to withdraw. All security teams link up at the release point and notify the platoon leader before moving to the ORP. Personnel returning to the ORP immediately secure their equipment and establish all-round security. Once the security element returns, the platoon moves out of the objective area as soon as possible, normally in two to three minutes.

- Before withdrawing, the demo team activates demo devices and charges.
- Support element or designated personnel in the assault element maintain local security during the withdrawal.
- Leaders report updated accountability and status (ACE report) to the PL and PSG.

(l) Squads withdraw from the objective in the order designated in the FRAGO to the ORP.

- Account for personnel and equipment.
- Disseminate information.
- Redistribute ammunition and equipment as required.

(m) The PL reports mission accomplishment to higher and continues the mission.

- Reports raid assessment to higher.
- Informs higher of any IR/PIR gathered.

NOTES

Figure 5-6. ACTIONS ON THE OBJECTIVE -- RAID

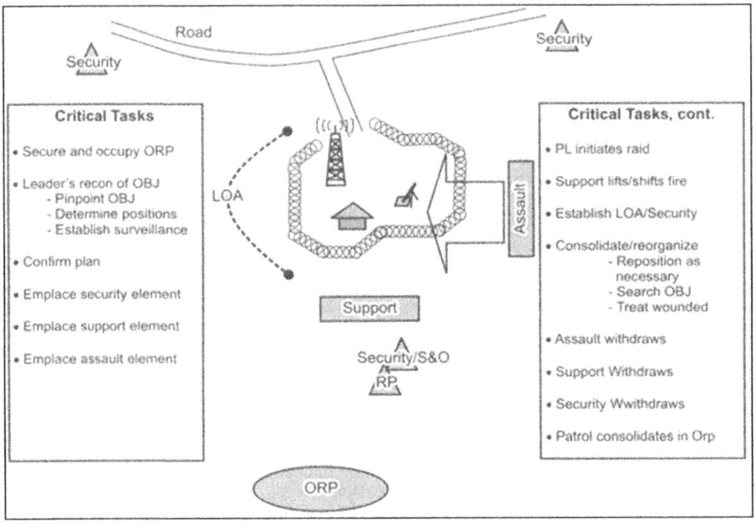


5-5. **SUPPORTING TASKS**. This section covers Linkup, Patrol Debriefing, and Occupation of an ORP.

a. **Linkup**. A linkup is a meeting of friendly ground forces. Linkups depend on control, detailed planning, communications, and stealth.

(1) ***Task Standard***. The units link up at the time and place specified in the order. The enemy does not surprise the main bodies. The linkup units establish a consolidated chain of command.

(2) ***Linkup Site Selection***. The leader identifies a tentative linkup site by map reconnaissance, other imagery, or higher headquarters designates a linkup site. The linkup site should have the following characteristics:

- Ease of recognition.
- Cover and concealment.
- No tactical value to the enemy.
- Location away from natural lines of drift.
- Defendable for a short period of time.
- N=Multiple access and escape routes.

(3) ***Execution***. Linkup procedure begins as the unit moves to the linkup point. The steps of this procedure are:

(a) The stationary unit performs linkup actions.

- Occupies the linkup rally point (LRP) NLT the time specified in the order.
- Establishes all-around security, establishes commo, and prepares to accept the moving unit.
- The security team clears the immediate area around the linkup point. It then marks the linkup point with the coordinated recognition signal. The security team moves to a covered and concealed position and observes the linkup point and immediate area around it.

(b) The moving unit--

- Performs linkup actions.
- The unit reports its location using phase lines, checkpoints, or other control measures.
- Halts at a safe distance from the linkup point in a covered and concealed position (the linkup rally point).

(c) The PL and a contact team--

- Prepare to make physical contact with the stationary unit.
- Issue a contingency plan to the PSG.
- Maintain commo with the platoon; verify near and far recognition signals for linkup (good visibility and limited visibility).
- Exchange far and near recognition signals with the linkup unit; conduct final coordination with the linkup unit.

(d) The stationary unit--

- Guides the patrol from its linkup rally point to the stationary unit linkup rally point.
- Linkup is complete by the time specified in the order.
- The main body of the stationary unit is alerted before the moving unit is brought forward.

(e) The patrol continues its mission IAW the order.

(4) ***Coordination Checklist***. The PL coordinates or obtains the following information from the unit that his patrol will link up with:

- Exchange frequencies, call signs, codes, and other communication information.
- Verify near and far recognition signals.
- Exchange fire coordination measures.
- Determine command relationship with the linkup unit; plan for consolidation of chain of command.
- Plan actions following linkup.
- Exchange control measures such as contact points, phase lines, contact points, as appropriate.

b. **Debrief**. Immediately after the platoon or squad returns, personnel from higher headquarters conduct a thorough debrief. This may include all members of the platoon or the leaders, RTOs, and any attached personnel. Normally the debriefing is oral. Sometimes a written report is required. Information on the written report should include--

- Size and composition of the unit conducting the patrol.
- Mission of the platoon such as type of patrol, location, and purpose.
- Departure and return times.
- Routes. Use checkpoints, grid coordinates for each leg or include an overlay.
- Detailed description of terrain and enemy positions that were identified.
- Results of any contact with the enemy.
- Unit status at the conclusion of the patrol mission, including the disposition of dead or wounded Soldiers.
- Conclusions or recommendations.

c. **Objective Rally Point**. The ORP is a point out of sight, sound, and small arms range of the objective area. It is normally located in the direction that the platoon plans to move after completion of actions on the objective. The ORP is tentative until the objective is pinpointed.

(1) ***Occupation of the ORP***. See Figure 5-7.

(a) The patrol halts beyond sight and sound of the tentative ORP (200-400m in good visibility, 100-200m in limited visibility).

(b) The patrol establishes a security halt IAW the unit SOP.

(c) After issuing a five-point contingency plan to the PSG, the PL moves forward with a recon element to conduct a leader's recon of the ORP.

(d) For a squad-sized patrol, the PL moves forward with a compass man and one member of each fire team to confirm the ORP.

- After physically clearing the ORP location, the PL leaves two Rangers at the 6 o'clock position facing in opposite directions.
- The PL issues a contingency plan and returns with the compass man to guide the patrol forward.
- The PL guides the patrol forward into the ORP, with one team occupying from 3 o'clock through 12 o'clock to 9 o'clock, and the other occupying from 9 o'clock through 6 o'clock to 3 o'clock.

(e) For a platoon-size patrol, the PL, RTO, WSL, three ammo bearers, a team leader, a SAW gunner, and riflemen go on the leaders recon for the ORP and position themselves at 10, 2, and 6 o'clock.

- The first squad in the order of march is the base squad, occupying from 10 to 2 o'clock.
- The trail squads occupy from 2 to 6 o'clock and 6 to 10 o'clock, respectively.
- The patrol headquarters element occupies the center of the triangle.

(2) ***Actions in the ORP***. The unit prepares for the mission in the ORP. Once the leader's recon pinpoints the objective, the PSG generally lines up rucks IAW unit SOP in the center of the ORP.

**Figure 5-7. OCCUPATION OF THE ORP**

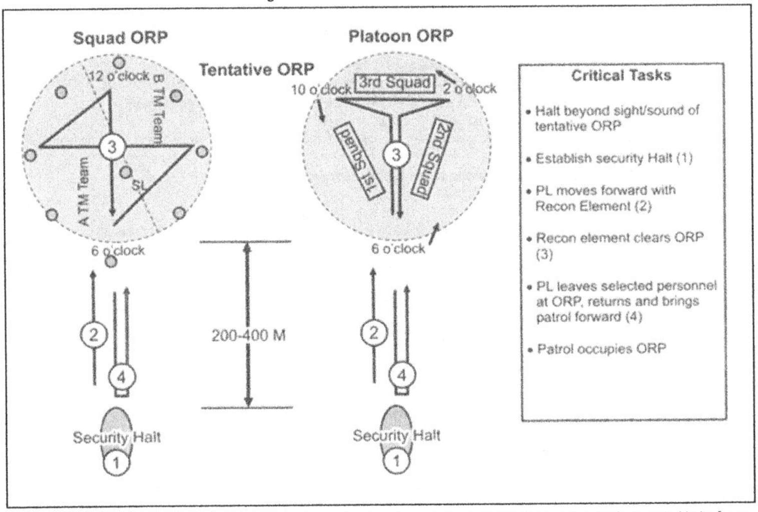


d. **Patrol Base**. A patrol base is a security perimeter that is set up when a squad or platoon conducting a patrol halts for an extended period. Patrol bases should not be occupied for more than a 24-hour period (except in emergency). A patrol never uses the same patrol base twice.

(1) ***Use***. Patrol bases are typically used--

- To avoid detection by eliminating movement.
- To hide a unit during a long detailed reconnaissance.
- To perform maintenance on weapons, equipment, eat and rest.
- To plan and issue orders.
- To reorganize after infiltrating on an enemy area.
- To establish a base from which to execute several consecutive or concurrent operations.

(2) ***Site Selection***. The leader selects the tentative site from a map or by aerial reconnaissance. The site's suitability must be confirmed and secured before the unit moves into it. Plans to establish a patrol base must include selecting an alternate patrol base site. The alternate site is used if the first site is unsuitable or if the patrol must unexpectedly evacuate the first patrol base.

(3) ***Planning Considerations***. Leaders planning for a patrol base must consider the mission and passive and active security measures. A patrol base (PB) must be located so it allows the unit to accomplish its mission.

- Observation posts and communication with observation posts.
- Patrol or platoon fire plan.
- Alert plan.
- Withdrawal plan from the patrol base to include withdrawal routes and a rally point, rendezvous point, or alternate patrol base.
- A security system to make sure that specific Soldiers are awake at all times.
- Enforcement of camouflage, noise, and light discipline.
- The conduct of required activities with minimum movement and noise.
- Priorities of Work.

(4) ***Security Measures***.

- Select terrain the enemy would probably consider of little tactical value.
- Select terrain that is off main lines of drift.
- Select difficult terrain that would impede foot movement, such as an area of dense vegetation, preferably bushes and trees that spread close to the ground.
- Select terrain near a source of water.
- Select terrain that can be defended for a short period and that offers good cover and concealment.
- Avoid known or suspected enemy positions.
- Avoid built-up areas.
- Avoid ridges and hilltops, except as needed for maintaining communications.
- Avoid small valleys.
- Avoid roads and trails.

(5) ***Occupation***. See Figure 5-8.

(a) A PB is reconned and occupied in the same manner as an ORP, with the exception that the platoon will typically plan to enter at a 90 degree turn. The PL leaves a two-man OP at the turn, and the patrol covers any tracks from the turn to the PB.

(b) The platoon moves into the PB. Squad-sized patrols will generally occupy a cigar-shaped perimeter; platoon-sized patrols will generally occupy a triangle-shaped perimeter.

(c) The PL and another designated leader inspect and adjust the entire perimeter as necessary.

(d) After the PL has checked each squad sector, each SL sends a two-man R&S team to the PL at the CP. The PL issues the three R&S teams a contingency plan, reconnaissance method, and detailed guidance on what to look for (enemy, water, built-up areas or human habitat, roads, trails, or possible rally points).

(e) Where each R&S team departs is based on the PLs guidance. The R&S team moves a prescribed distance and direction, and reenters where the PL dictates.

- Squad-sized patrols do not normally send out an R&S team at night.
- R&S teams will prepare a sketch of the area to the squad front if possible.
- The patrol remains at 100 % alert during this recon.
- If the PL feels the patrol was tracked or followed, he may elect to wait in silence at 100 % alert before sending out R&S teams.
- The R&S teams may use methods such as the "*I*," the "*Box*," or the "*T*." Regardless of the method chosen, the R&S team must be able to provide the PL with the same information.
- Upon completion of R&S, the PL confirms or denies the patrol base location, and either moves the patrol or begins priorities of work.

(6) ***Passive (Clandestine) Patrol Base (Squad)***.

- The purpose of a passive patrol base is for the rest of a squad or smaller size element.
- Unit moves as a whole and occupies in force.
- Squad leader ensures that the unit moves in at a 90-degree angle to the order of movement.
- A claymore mine is emplaced on route entering patrol base.
- Alpha and Bravo teams sit back-to-back facing outward, ensuring that at least one individual per team is alert and providing security.

(7) ***Priorities of Work (Platoon and Squad)***. Once the PL is briefed by the R&S teams and determines the area is suitable for a patrol base, the leader establishes or modifies defensive work priorities in order to establish the defense for the patrol base. Priorities of work are not a laundry list of tasks to be completed; to be effective, priorities of work must consist of a task, a given time, and a measurable performance standard. For each priority of work, a clear standard must be issued to guide the element in the successful accomplishment of each task. It must also be designated whether the work will be controlled in a centralized or decentralized manner. Priorities of work are determined IAW METT-TC. Priorities of Work may include, but are not limited to the following tasks:

(a) *Security (Continuous).*

- Prepare to use all passive and active measures to cover all of the perimeter all of the time, regardless of the percentage of weapons used to cover that all of the terrain.
- Readjust after R&S teams return, or based on current priority of work (such as weapons maintenance).
- Employ all elements, weapons, and personnel to meet conditions of the terrain, enemy, or situation.
- Assign sectors of fire to all personnel and weapons. Develop squad sector sketches and platoon fire plan.
- Confirm location of fighting positions for cover, concealment, and observation and fields of fire. SLs supervise placement of aiming stakes and claymores.
- Only use one point of entry and exit, and count personnel in and out. Everyone is challenged IAW the unit SOP.
- Hasty fighting positions are prepared at least 18 inches deep (at the front), and sloping gently from front to rear, with a grenade sump if possible.

(b) *Withdrawal Plan*. The PL designates the signal for withdrawal, order of withdrawal, and the platoon rendezvous point and/or alternate patrol base.

(c) *Communication (Continuous)*. Commo must be maintained with higher headquarters, OP's, and within the unit. May be rotated between the patrol's RTOs to allow accomplishment of continuous radio monitoring, radio maintenance, act as runners for PL, or conduct other priorities of work.

(d) *Mission Preparation and Planning*. The PL uses the patrol base to plan, issue orders, rehearse, inspect, and prepare for future missions.

(e) *Weapons and Equipment Maintenance*. The PL ensures that machine guns, weapon systems, commo equipment, and night vision devices (as well as other equipment) are maintained. These items are not disassembled at the same time for maintenance (no more than 33 percent at a time), and weapons are not disassembled at night. If one machine gun is down, then security for all remaining systems is raised.

(f) *Water Resupply*. The PSG organizes watering parties as necessary. The watering party carries canteens in an empty rucksack or duffel bag, and must have commo and a contingency plan prior to departure.

(g) *Mess Plan*. At a minimum, security and weapons maintenance are performed prior to mess. Normally no more than half the platoon eats at one time. Rangers typically eat 1 to 3 meters behind their fighting positions.

- Rest/sleep plan management. The patrol conducts rest as necessary to prepare for future operations.
- Alert Plan and Stand-to. The PL states the alert posture and the stand-to time. He develops the plan to ensure all positions are checked periodically, OP's are relieved periodically, and at least one leader is always alert. The patrol typically conducts stand-to at a time specified by unit SOP (such as 30 minutes before and after BMNT or EENT).
- Resupply. Distribute or cross-load ammunition, meals, equipment, etc.
- Sanitation and personal hygiene. The PSG and medic ensure a slit trench is prepared and marked. All Soldiers will brush teeth, wash face, shave, wash hands, armpits, groin, feet, and darken (brush shine) boots daily. The patrol will not leave trash behind.

Figure 5-8. PATROL BASE

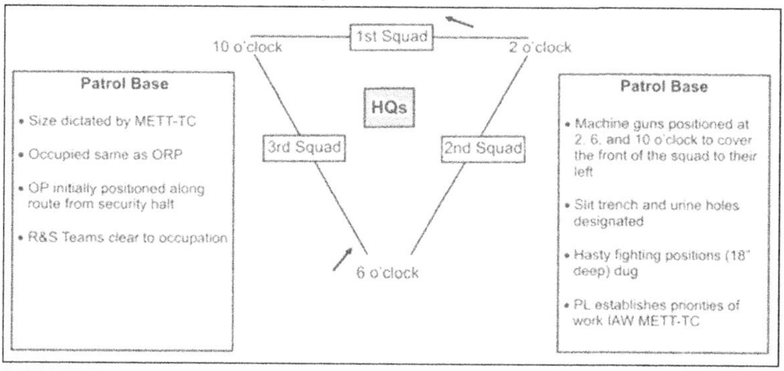


5-6. **MOVEMENT TO CONTACT**. The MTC is one of the five types of offensive operations. An MTC gains or regains contact with the enemy. Once contact is made, the unit develops the situation. Normally a platoon conducts an MTC as part of a larger force.

a. **Techniques**. The two techniques of conducting a movement to contact are search and attack and approach march.

(1) ***Search and Attack (S&A)***. This technique is used when the enemy is dispersed, is expected to avoid contact, disengage or withdraw, or you have to deny his movement in an area. The search and attack technique involves the use of multiple platoons, squads, and fire teams coordinating their actions to make contact with the enemy. Platoons typically try to find the enemy and then fix and finish him. They combine patrolling techniques with the requirement to conduct hasty or deliberate attacks once the enemy has been found.

(a) *Planning Considerations*.

- The factors of METT-TC.
- The requirement for decentralized execution.
- The requirement for mutual support.
- The length of operations.
- Minimize Soldier's load to improve stealth and speed.
- Resupply and MEDEVAC.
- Positioning key leaders and equipment.
- Employment of key weapons.
- Requirement for patrol bases.
- Concept for entering the zone of action.
- The concept for linkups while in contact.

(b) *Critical Performance Measures*.

- The platoon locates the enemy without being detected.
- Once engaged, fixes the enemy in position and maneuvers against the enemy.
- Maintains security throughout actions to avoid being flanked.

(2) ***Approach March***. The concept of the approach march is to make contact with the smallest element, allowing the commander the flexibility of destroying or bypassing the enemy. A platoon uses the approach march method as part of a larger unit. It can be tasked as the advance guard, move as part of the main body, or provide flank or rear security for the company or battalion. They may also receive on-order missions as part of the main body.

(a) *Fundamentals*. These basics are common to all movements to contact.

- Make enemy contact with smallest element possible.
- Rapidly develop combat power upon enemy contact.

- Provide all-round security for the unit.
- Support higher unit's concept.
- Reports all information rapidly and accurately and strives to gain and maintain contact with the enemy.
- Requires decentralized execution.

(b) *Planning Considerations*. The following issues should be considered heavily for MTC operations:

- Factors of METT-TC.
- Reduced Soldier's load.

(c) *Critical Performance Measures*.

- PL selects the appropriate movement formation based on likelihood of enemy contact.
- Maintains contact, once contact is made, until ordered to do otherwise.

b. **Task Standards**. The platoon moves NLT the time specified in the order, the platoon makes contact with the smallest element possible, and the main body is not surprised by the enemy. Once the platoon makes contact, it maintains contact. The platoon destroys squad and smaller-sized elements, and fixes elements larger than a squad. The platoon maintains sufficient fighting force capable of conducting further combat operations. Reports of enemy locations and contact are forwarded. If not detected by the enemy, the PL initiates a hasty attack. The platoon sustains no casualties from friendly fire. The platoon is prepared to initiate further movement within 25 minutes of contact, and all personnel and equipment are accounted for.

NOTES

# Chapter 6
# BATTLE DRILLS

Infantry battle drills describe how platoons and squads apply fire and maneuver to commonly encountered situations. They require leaders to make decisions rapidly and to issue brief oral orders quickly.

## Section I. INTRODUCTION

This section defines and describes the format for battle drills.

6-1. **DEFINITION.** FM 7-1 defines a battle drill as "*a collective action rapidly executed without applying a deliberate decision-making process.*"

a. Characteristics of a battle drill are—

- They require minimal leader orders to accomplish and are standard throughout the Army.
- Sequential actions are vital to success in combat or critical to preserving life.
- They apply to platoon or smaller units.
- They are trained responses to enemy actions or leader's orders.
- They represent mental steps followed for offensive and defensive actions in training and combat.

b. A unit's ability to accomplish its mission often depends on Soldiers and leaders to execute key actions quickly. All Soldiers and their leaders must know their immediate reaction to enemy contact, as well as follow-up actions. Drills are limited to situations requiring instantaneous response; therefore, Soldiers must execute drills instinctively. This results from continual practice. Drills provide small units with standard procedures essential for building strength and aggressiveness.

(1) They identify key actions that leaders and Soldiers must perform quickly.

(2) They provide for a smooth transition from one activity to another; for example, from movement to offensive action to defensive action.

(3) They provide standardized actions that link Soldier and collective tasks at platoon level and below. (Soldiers perform individual tasks to CTT or SDT standard.)

(4) They require the full understanding of each individual and leader, and continual practice.

6-2. **FORMAT.** Drills include a ***title***, a ***situation*** to cue the unit or leader to start the drill, the ***required actions*** in sequence, and supporting illustrations. Where applicable, drills are cross-referenced with material in other chapters, or other drills, or both. Training standards for battle drills are in the mission training plan (MTP).

## Section II. DRILLS

This section provides the battle drills for small units.

### BATTLE DRILL 1
### REACT TO CONTACT

**SITUATION:** A squad or platoon receives fires from enemy individual or crew-served weapons.

**REQUIRED ACTIONS:** See Figure 6-1.

1. Soldiers immediately assume the nearest covered positions.
2. Soldiers return fire immediately on reaching the covered positions.
3. Squad/team leaders locate and engage known or suspected enemy positions with well-aimed fire, and they pass information to the platoon/squad leader.
4. Fire team leaders control the fire of their Soldiers by using standard fire commands (initial and supplemental) containing the following elements:
    - Alert
    - Direction
    - Description of target
    - Range
    - Method of fire (manipulation and rate of fire)
    - Command to commence firing

5. Soldiers maintain contact (visual or oral) with the Soldiers on their left and right.

6. Soldiers maintain contact with their team leaders and indicate the location of enemy positions.

7. The leaders (visual or oral) check the status of their personnel.

8. The squad/team leaders maintain visual contact with the platoon/squad leader.

9. The platoon/squad leader moves up to the squad/fire team in contact and links up with its leader.
   a. The platoon leader brings his RTO, platoon FO, the nearest squad's squad leader, and one machine gun team.
   b. The squad leader of the trail squad moves to the front of his lead fire team.
   c. The PSG also moves forward with the second machine gun team and links up with the platoon leader, ready to assume control of the base-of-fire element.

10. The platoon/squad leader determines whether or not his unit must move out of an EA.

11. The platoon/squad leader determines whether his unit can gain and maintain suppressive fires with the element already in contact (based on the volume and accuracy of enemy fires against the element in contact). The platoon/squad leader assesses the situation. He identifies—
   - The location of the enemy position and obstacles.
   - The size of the enemy force engaging the unit in contact. The number of enemy automatic weapons, the presence of any vehicles, and the employment of indirect fires are indicators of the enemy strength.
   - Vulnerable flanks.
   - Covered and concealed flanking routes to the enemy position.

12. The platoon/squad leader determines the next COA, such as fire and movement, assault, breach, knock out bunker, enter and clear a building or trench.

13. The platoon/squad leader reports the situation to the company commander/platoon leader and begins to maneuver his unit.

14. Platoon leader directs platoon FO to call for and adjust indirect fires (mortars or artillery). Squad leaders relay requests through the platoon leader. The platoon/squad leader in conjunction with the platoon FO maintains accurate battle tracking of all friendly elements to facilitate quick clearance of fires.

15. Leaders relay all commands and signals from the platoon chain of command.

16. The PSG positions the base of fire element to observe and to provide supporting fires.

***Note:*** Once the platoon has executed the React to Contact drill, the platoon leader quickly assesses the situation, for example, enemy size and location. He picks a COA. The platoon leader reports the situation to the company commander.

Figure 6-1. REACT TO CONTACT

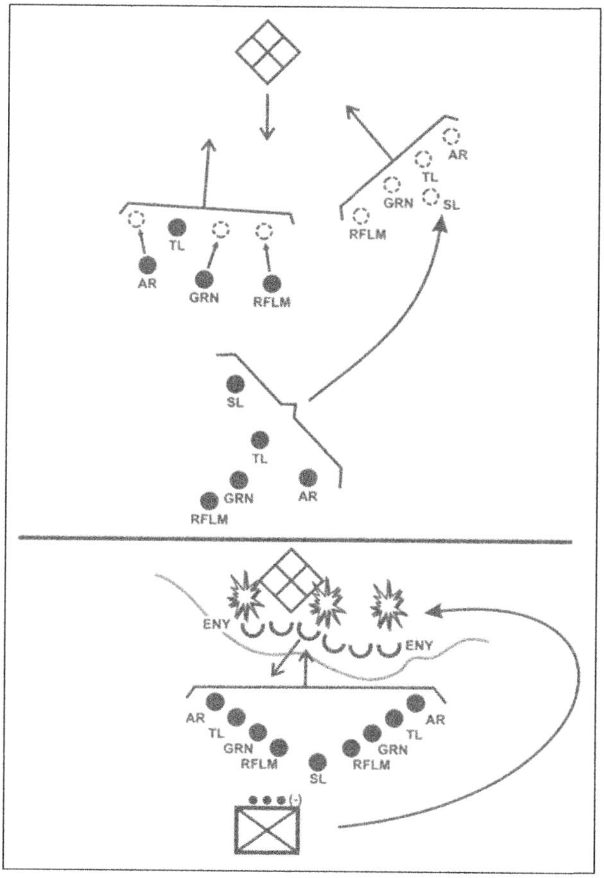

## BATTLE DRILL 2
## BREAK CONTACT

**SITUATION:** The squad or platoon is under enemy fire and must break contact.

**REQUIRED ACTIONS:** See Figure 6-2.

1. The platoon/squad leader directs fire support for the disengagement. He accomplishes this by—
   - Directing one squad/fire team in contact to support the disengagement of the remainder of the unit.
   - Considering the use of indirect fires for breaking contact.
   - Clearing the location, task, purpose, and method of conducting the fire mission with the platoon FO.

2. The platoon/squad leader orders a distance and direction, a terrain feature, or last ORP for the movement of the first squad/fire team. In conjunction with the platoon FO, the leader maintains accurate battle tracking of all friendly elements to facilitate quick clearance of fires.

3. The base of fire squad/team continues to suppress the enemy. The platoon/squad leader directs the platoon FO to execute the fire mission, if needed.

4. The moving squad/team assumes the overwatch position. The squad/team should use M203 grenade launchers, throw fragmentation and concussion grenades, and use smoke grenades to mask movement. The platoon/squad leader directs the platoon FO to execute smoke mission to screen friendly elements movement, if needed.

5. The moving squad/team takes up the designated position and engages the enemy position.

6. The platoon leader directs the base-of-fire element to move to its next location. Based on the terrain and the volume and accuracy of the enemy's fire, the moving fire team or squad may need to use fire and movement techniques.

7. The platoon/squad continues to bound away from the enemy until—
   - It breaks contact. It must continue to suppress the enemy as it does this.
   - It passes through a higher-level SBF position.
   - Its squad/fire teams are in the assigned position to conduct the next mission.

8. The leader should consider changing his unit's direction of movement once contact is broken. This will reduce the ability of the enemy to place effective indirect fires on the unit.

9. If the platoon/squad becomes disrupted, Soldiers stay together and move to the last designated rally point.

10. The platoon/squad leaders account for Soldiers, report, reorganize as necessary, and continue the mission. With the platoon FO, the platoon/squad develops a quick fire plan to support the hasty defense or new route of march.

Figure 6-2. BREAK CONTACT

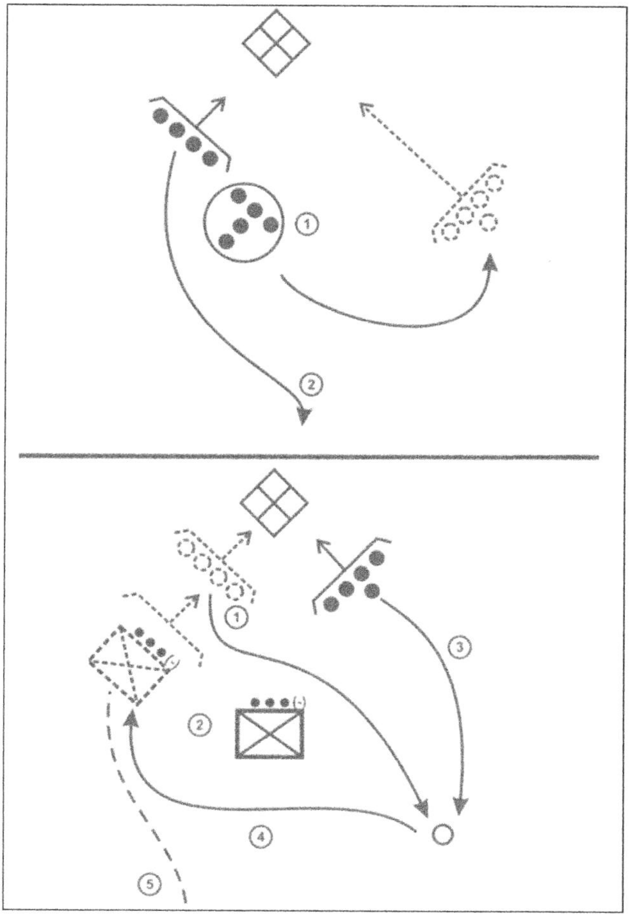

## BATTLE DRILL 3
## REACT TO AMBUSH

**SITUATION:** The squad or platoon enters a kill zone and the enemy initiates an ambush with a casualty-producing device and a high volume of fire.

**REQUIRED ACTIONS:** See Figure 6-3.

1. Near ambush (within hand-grenade range).
   a. Soldiers in the kill zone return fire immediately. How they go about this depends on the terrain. If—
      - Cover is not available, Soldiers immediately, without order or signal, assume prone position and throw concussion or fragmentation and smoke grenades.
      - Cover is available, without order or signal, Soldiers seek the nearest covered position, assume the prone position, and throw fragmentation or concussion and smoke grenades.

   b. Immediately after the explosion of the concussion or fragmentation grenades, Soldiers in the kill zone return fire and assault through the ambush position using fire and movement.
   c. Soldiers not in the kill zone identify the enemy's position. Fire is shifted as the personnel in the kill zone begin to assault.
   d. Soldiers in the kill zone continue the assault to eliminate the ambush or until contact is broken.
   e. The platoon/squad conducts consolidation and reorganization.

2. Far ambush (out of hand-grenade range). Soldiers receiving fire immediately return fire, take up covered positions, and suppress the enemy by—
   - Destroying or suppressing enemy crew-served weapons.
   - Sustaining suppressive fires.

3. Soldiers (squad/teams) not receiving fires move by a covered and concealed route to a vulnerable flank of the enemy position and assault using fire and movement techniques.

4. Soldiers in the kill zone continue suppressive fires and shift fires as the assaulting squad/team fights through the enemy position.

5. The platoon FO calls for and adjusts indirect fires as directed by the platoon leader. On order, he shifts or ceases fires to isolate the enemy position or to attack them with indirect fires as they retreat.

6. The platoon/squad leader reports, reorganizes as necessary, and continues the mission.

Figure 6-3. REACT TO AMBUSH

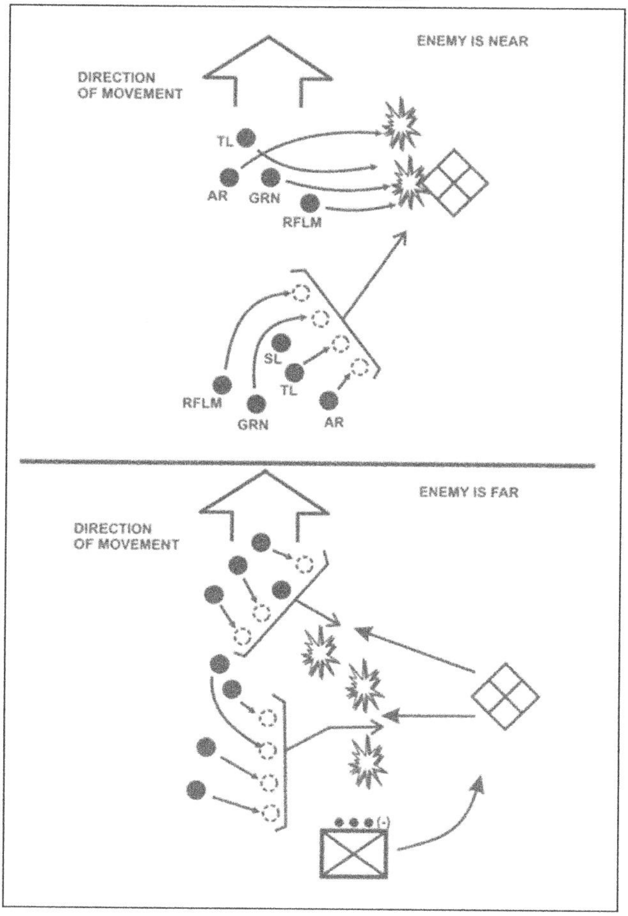

## BATTLE DRILL 4
## KNOCK OUT BUNKERS

**SITUATION:** While moving as part of a larger force, the platoon identifies the enemy in bunkers.

**REQUIRED ACTIONS:** See Figures 6-4 and 6-5.

1. The platoon initiates contact.
   a. The squad in contact establishes a base of fire.
   b. The platoon leader, his RTO, platoon FO, and one machine gun team move forward to link up with the squad leader of the squad in contact.
   c. The PSG moves forward with the second machine gun team and assumes control of the base-of-fire squad.
   d. The base-of-fire squad—
      - Destroys or suppresses enemy crew-served weapons first.
      - Obscures the enemy position with smoke (M203).
      - Sustains suppressive fires at the lowest possible level.
   e. The platoon leader directs platoon FO to call for and adjust indirect fires. The platoon leader in conjunction with the platoon FO maintains accurate battle tracking of all friendly elements to facilitate quick clearance of fires.
2. The platoon leader determines that he can maneuver by identifying—
   - Enemy bunkers, other supporting positions, and any obstacles.
   - Size of the enemy force engaging the platoon. The number of enemy automatic weapons, the presence of any vehicles, and the employment of indirect fires are indicators of enemy strength.
   - A vulnerable flank of at least one bunker.
   - A covered and concealed flanking route to the flank of the bunker.
3. The platoon leader determines which bunker is to be assaulted first and directs one squad (not in contact) to knock it out.
4. If necessary, the PSG repositions a squad, fire team, or machine gun team to isolate the bunker as well as to continue suppressive fires.
5. The assaulting squad, with the platoon leader and his RTO, move along the covered and concealed route and take action to knock out the bunker. The following occurs.
   a. On the platoon leader's signal, the support squad ceases or shifts fires to the opposite side of the bunker from which the squad is assaulting.
   b. At the same time, the platoon FO shifts indirect fires to isolate enemy positions.
6. The assaulting squad leader reports to the platoon leader and reorganizes his squad.
7. The platoon leader—
   - Directs the supporting squad to move up and knock out the next bunker.
   - Directs the assaulting squad to continue and knock out the next bunker.
   - Rotates squads as necessary.
8. The platoon leader reports, reorganizes as necessary, and continues the mission. The company follows up the success of the platoon attack and continues to assault enemy positions.

Figure 6-4. KNOCK OUT A BUNKER (SQUAD)

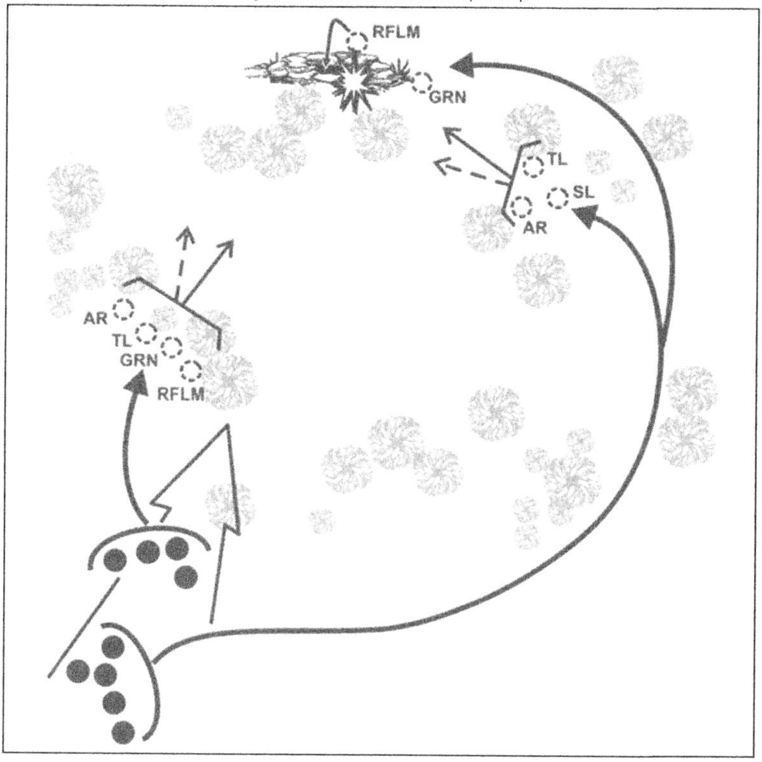

Figure 6-5. KNOCK OUT BUNKERS (PLATOON)

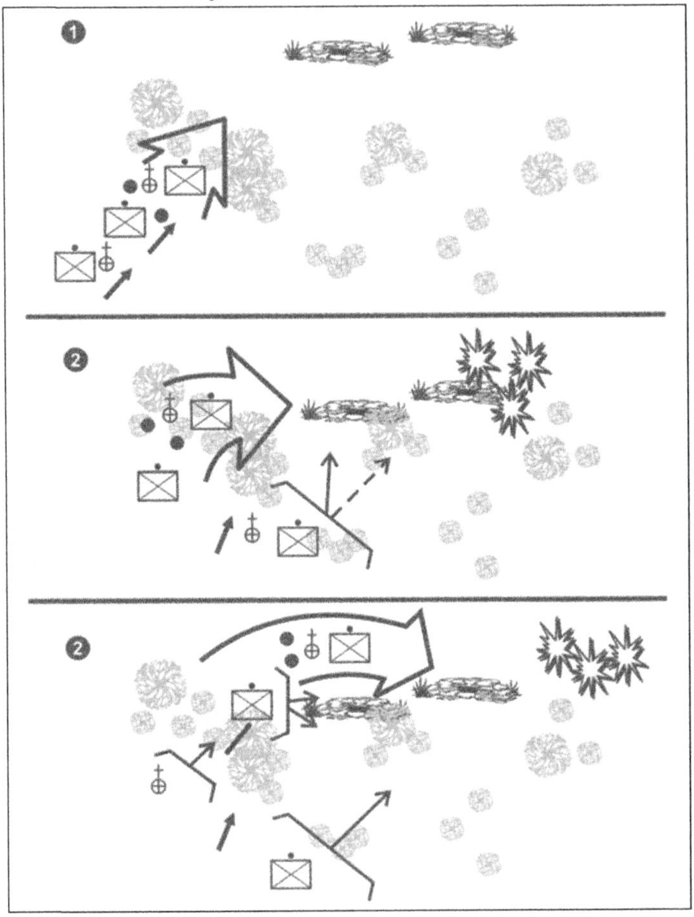

## BATTLE DRILL 5
## ENTER BUILDING/CLEAR ROOM

**SITUATION:** Operating as part of a larger force (in daylight or darkness), the squad is tasked to participate in clearing a building. The platoon leader directs the squad to enter the building or to clear a room. An entry point or breach has already been identified or will be created before initiating the entry. For a detailed discussion of urban entry breaching techniques, see FM 3.06-11. Enemy forces and noncombatants may or may not be present in the room and or building to be cleared.

**REQUIRED ACTIONS:** See Figures 6-6 through 6-14.

1. Platoon and squad leaders must consider the task and purpose they have been given and the method they are to use to achieve the desired results. They must operate IAW the ROE and must be aware of the affects that platoon weapons will have on the type and composition of the buildings.

2. To seize or gain control of a building may not always require committing troops into the structure or closing with the enemy. Before initiating this action and exposing members of the clearing squad to direct enemy contact and risking casualties, the platoon leader should consider/direct employment of all organic, crew-served, and supporting weapon systems onto the objective area to suppress and neutralize the threat, providing the mission, purpose, building composition, and ROE permit.

3. When conducting urban operations, Soldiers must be equipped at all times with a properly-mounted and immediately-useable NVD or light source to illuminate the immediate area.

***Note:*** The following discussion assumes that only the platoon's organic weapons are supporting the infantry squad. Urban situations may require precise application of firepower. This situation is especially true of an urban environment in which the enemy is mixed with noncombatants. Noncombatants may be found in the room, which can restrict the use of fires and reduce the combat power available to a squad leader. His squad may have to operate in no-fire areas. ROE can prohibit the use of certain weapons until a specific hostile action takes place. All Soldiers must be aware of the ROE. Leaders must include the specific use of weapons in their planning for precision operations in urban terrain. Leaders should always consider the use of snipers or designated marksmen to apply precise fires to the objective.

4. Clearing team members must approach the entry point quickly, quietly, and in standard order. The squad leader must ensure he is in a position to control the actions of both fire teams. This approach preserves the element of surprise and allows for quick entry and domination of the room. If a breach is required, the order may be slightly modified based on the breach technique (FM 3-06.11). The members of the fire team are assigned numbers one through four. The rifleman is # 1 and the grenadier is # 3. If one member of the clearing team is armed with the SAW rather than an M16 rifle or carbine, he should be designated # 4. The team leader is normally the # 2 man, because he will have the most immediate decision to make as he enters the room.

5. The entire team enters the room as quickly and smoothly as possible and clears the doorway immediately. If possible, the team moves from a covered or concealed position already in their entry order. Ideally, the team arrives and passes through the entry point without having to stop. If the team must stop to effectively "*stack*" outside the entry point, it must do so only momentarily, and it must provide cover.

6. The door is the focal point of anyone in the room. It is known as the "*fatal funnel*," because it focuses attention at the precise point where the individual team members are the most vulnerable. Moving into the room quickly reduces the chance of anyone being hit by enemy fire directed at the doorway.

7. For this battle to be effectively employed, each member of the team must know his sector of fire and how his sector overlaps and links with the sectors of the other team members. Team members do not move to the points of domination and then engage their targets. Rather, they engage targets as they calmly and quickly move to their designated points. Engagements must not slow movement to their points of domination. Team members may shoot from as short a range as 1 to 2 inches. They engage the most immediate threat first and then the less immediate threats in sector. Immediate threats are personnel who—

- Are armed and prepared to return fire immediately.
- Block movement to the position of domination.
- Are within arm's reach of a clearing team member.
- Are within 3 to 5 feet of the breach point.

8. The squad leader designates the assault team and identifies the location of the entry point for them.

9. The squad leader positions the follow-on assault team to provide overwatch and supporting fires for the initial assault team.

10. Assault team members use available cover and concealment, and move as close to the entry point as possible.

a. If an explosive breach or a ballistic breach is to be performed by a supporting element, the assault team remains in a covered position until the breach is made. They may provide overwatch and fire support for the breaching element if necessary.

b. All team members must signal one another that they are ready before the team moves to the entry point.

c If stealth is a consideration, team members avoid the use of verbal signals, which may alert the enemy and remove the element of surprise.

d. Assault team members must move quickly from the covered position to the entry point, minimizing the time they are exposed to enemy fire.

11. The assault team enters through the entry point or breach. Unless a grenade will be thrown prior to entry, the team should avoid stopping outside the point of entry.

a. If required, the # 2 man throws a grenade of some type (fragmentary, concussion, stun) into the room before entry.

b. The use of grenades should be consistent with the ROE and building structure. The grenade should be cooked off before being thrown, if applicable to the type of grenade used.

c. If stealth is not a factor, the thrower should sound off with a verbal indication ("*Frag out,*" "*Concussion out,*" "*Stun out*") that a grenade of some type is being thrown. If stealth is a factor, only visual signals are given as the grenade is thrown.

**CAUTION**

If walls and floors are thin, fragments from fragmentation grenades and debris created by concussion grenades can injure Soldiers outside the room. If the structure has been stressed by previous explosive engagements, the use of these grenades could cause it to collapse. Leaders must determine the effectiveness of these types of grenades compared to possibilities of harm to friendly troops.

12. On the signal to go, or after the grenade detonates, the assault team moves through the entry point (Figure 6-6 through Figure 6-9) and quickly takes up positions inside the room that allow it to completely dominate the room and eliminate the threat. Unless restricted or impeded, team members stop movement only after they have cleared the door and reached their designated point of domination. In addition to dominating the room, all team members are responsible for identifying possible loopholes and mouseholes in the ceiling, walls, and floor.

***Note:*** Where enemy forces may be concentrated and the presence of noncombatants is highly unlikely, the assault team can precede their entry by throwing a fragmentation or concussion grenade (structure dependent) into the room, followed by bursts of automatic small-arms fire by the # 1 man as he enters. Carefully consider the ROE and building composition before employing this method.

13. The # 1 and # 2 men are initially concerned with the area directly to their front, then along the wall on either side of the door or entry point. This area is in their path of movement, and it is their primary sector of fire. Their alternate sector of fire is from the wall they are moving toward, back to the opposite far corner.

14. The # 3 and # 4 men start at the center of the wall opposite their point of entry and clear to the left if moving toward the left, or to the right if moving toward the right. They stop short of their respective team member (either the # 1 man or the # 2 man).

15. The team members move toward their points of domination, engaging all threat or hostile targets in sequence in their sector. Team members must exercise fire control and discriminate between hostile and noncombatant room occupants. (The most practical way to do this is to identify whether or not the target has a weapon in his or her hands.) Shooting is done without stopping, using reflexive shooting techniques. Because the Soldiers are moving and shooting at the same time, they must move using the careful hurry. Figure 6-10, shows all four team members at their points of domination for a room with a center door and their overlapping sectors of fire.

16. The first man (rifleman) enters the room and eliminates the immediate threat. He has the option of going left or right, normally moving along the path of least resistance to one of two corners. When using a doorway as the point of entry, the team uses the path of least resistance, which they determine initially based on the way the door opens. If the door opens inward, the first man plans to move away from the hinges. If the door opens outward, he plans to move toward the hinged side. Upon entering, his direction is influenced by the size of the room, the enemy situation, and furniture or other obstacles that hinder or channel his movement.

17. The direction each man moves in should not be preplanned unless the exact room layout is known. Each man should go in a direction opposite the man in front of him (Figure 6-6). Every team member must know the sectors and duties of each position.

18. As the first man goes through the entry point, he can usually see into the far corner of the room. He eliminates any immediate threat and continues to move along the wall if possible and to the first corner, where he assumes a position of domination facing into the room.

**Figure 6-6. FIRST MAN ENTERS A ROOM, FOLLOWED BY TEAM LEADER**

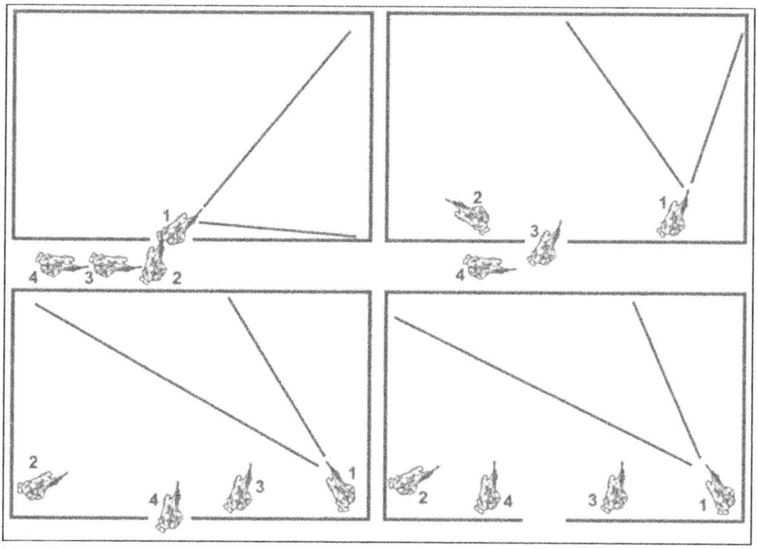


***Note:*** Team members must always stay within 1 meter of the wall. If a team member finds his progress blocked by some object that will force him more than 1 meter from the wall, he must either step over it (if able) or stop where he is and clear the rest of his sector from where he is. If this action creates dead space in the room, the team leader directs which clearing actions to take once other members of the team have reached their points of domination.

19. The second man (normally the team leader), entering almost simultaneously with the first, moves in the opposite direction, following the wall (Figure 6-7). The second man must clear the entry point, clear the immediate threat area, clear his corner, and move to a dominating position on his side of the room. The second man must also immediately determine if he is entering a center door or corner door and act accordingly (Figure 6-7 and Figure 6-10).

Figure 6-7. SECOND MAN (TEAM LEADER) ENTERS A ROOM

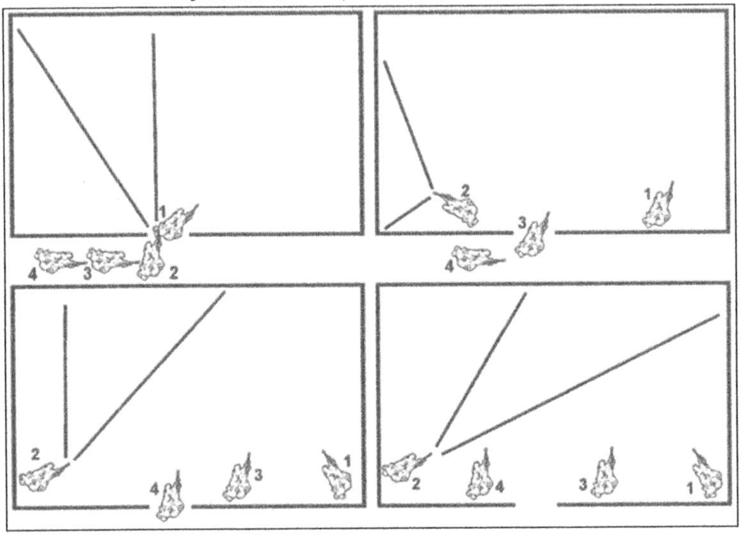


20. The third man (normally the grenadier) simply goes opposite of the second man inside the room, moves at least 1 meter from the entry point, and takes a position that dominates his sector (Figure 6-8).

Figure 6-8. THIRD MAN ENTERS A ROOM

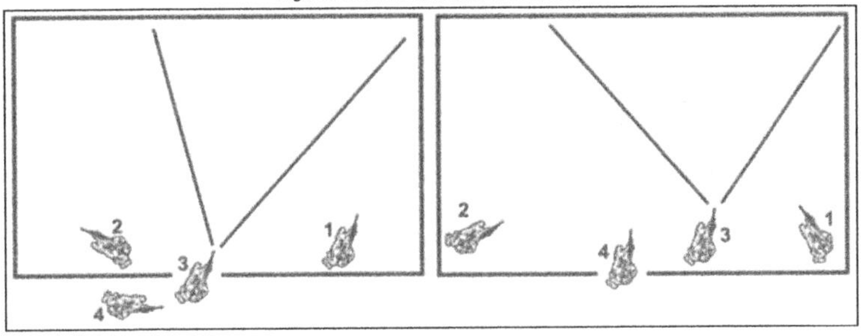

21. The fourth man (normally the SAW gunner) moves opposite of the third man, clears the doorway by at least 1 meter, and moves to a position that dominates his sector (Figure 6-9).

**Figure 6-9. FOURTH MAN IN A ROOM**

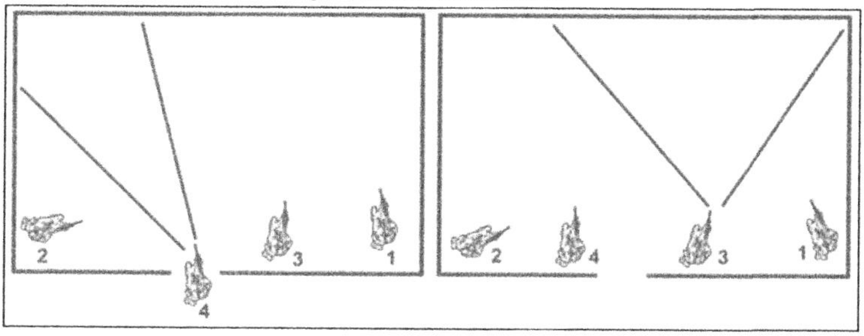


***Note:*** If the path of least resistance takes the first man to the left, then all points of domination are the mirror image of those shown in the diagrams.

22. Points of domination should not be in front of doors or windows so team members are not silhouetted to the outside of the room (Figure 6-10). No movement should mask the fire of any of the other team members.

**Figure 6-10. POINTS OF DOMINATION AND SECTORS OF FIRE, CENTER DOOR VERSUS CORNER DOOR**

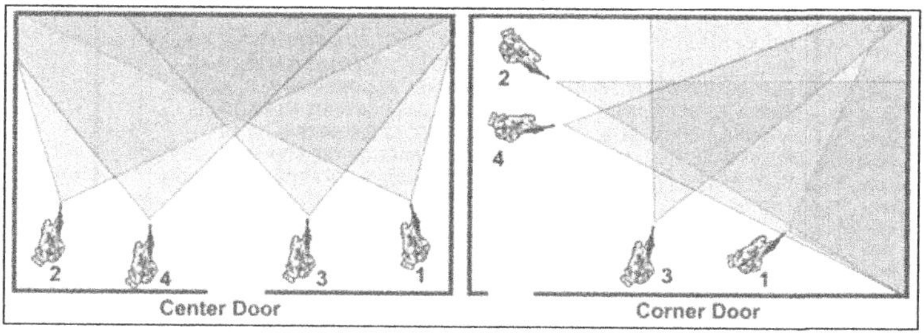


23. On order, any member of the assault team may move deeper into the room overwatched by the other team members. The team leader must control this action.

24. Once the room is cleared, the team leader signals to the squad leader that the room has been cleared. The squad leader marks the room IAW unit SOP (see FM 3.06-11). The squad leader determines whether or not his squad can continue to clear through the building.

25. The squad reorganizes as necessary. Leaders redistribute the ammunition. The squad leader reports to the platoon leader when the room is clear.

***Note:*** If at any point a team member experiences a weapon malfunction in the presence of enemy combatants, he has to make an immediate decision. If the enemy is outside immediate danger distance from the Soldier, then the Soldier should drop to one knee, which indicates that he has experienced a weapon malfunction and prevents fratricide by ensuring that the next Soldier's fields of fire are clear. Once on a knee, the Soldier must remain there until the team leader directs him to stand up. If the kneeling Soldier corrects his weapon malfunction, he can continue to engage targets from his kneeling position. If, however, the enemy Soldier is so close that he presents an immediate threat, and if dropping to a knee would expose the US Soldier to immediate harm, then the US Soldier must try to subdue or disable the enemy Soldier. He can strike the enemy with the muzzle of his weapon or a bayonet to the face, throat, or chest. Or, he could grapple with the enemy soldier and take him to the ground immediately to clear the other team members' fields of fire. Once they have cleared their sectors and eliminated any other threats in the room, the other members of the clearing team assist.

26. Although this battle drill is effective, leaders might have to modify it for the situation. Some reasons and methods for modifying the technique are shown in Table 6-1.

**Table 6-1. REASONS FOR AND METHODS OF MODIFYING ENTRY TECHNIQUES**

| REASON | METHOD |
|---|---|
| Objective rooms are consistently small | Clear in teams of 2 or 3 |
| Shortage of personnel | Clear in teams of 2 or 3 |
| Enemy poses no immediate threat | 2 or 3 men search rooms to ensure no enemy or noncombatants are present |
| No immediate threat, speed is critical | 1 to 3 men visually search each room |

27. When full four-man teams are not available for room clearing, two- and three-man teams can be used. If the # 1 or # 2 man discovers that the room is very small, he can yell, "*Short room*" or "*Short*," which tells the # 3 or # 4 man (whoever following the # 1 or # 2 man) to stay outside the room. Figures 6-11 and 6-12 show the points of domination and sectors of fire for a two-man clearing team. Figures 6-13 and 6-14 show the actions for a three-man team.

**DANGER**

**RICOCHETS POSE A HAZARD. ALL SOLDIERS MUST BE AWARE OF THE TYPE OF WALL CONSTRUCTION OF THE ROOM BEING CLEARED. THE WALLS OF AN ENCLOSED ROOM PRESENT MANY RIGHT ANGLES. COMBINED WITH HARD SURFACES SUCH AS CONCRETE, A BULLET MAY CONTINUE TO RICOCHET AROUND A ROOM UNTIL ITS ENERGY IS SPENT. AFTER HITTING THREAT PERSONNEL, BALL AMMUNITION MAY PASS THROUGH THE BODY AND RICOCHET. BODY ARMOR AND THE KEVLAR HELMET PROVIDE SOME PROTECTION FROM THIS HAZARD.**

Figure 6-11. POINTS OF DOMINATION AND SECTORS OF FIRE (TWO-MAN TEAM, CENTER DOOR)

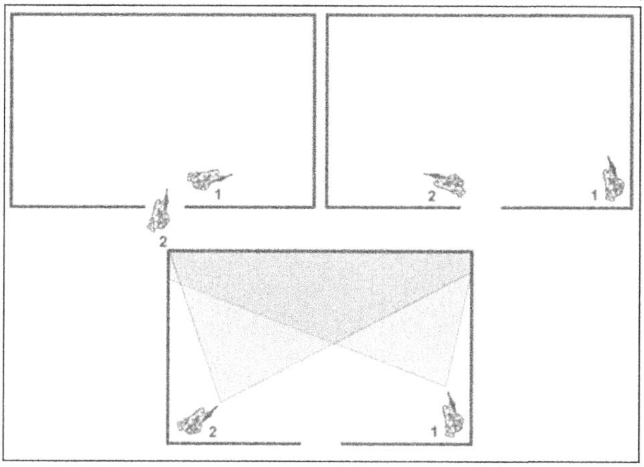

Figure 6-12. POINTS OF DOMINATION AND SECTORS OF FIRE (TWO-MAN TEAM, CORNER DOOR)

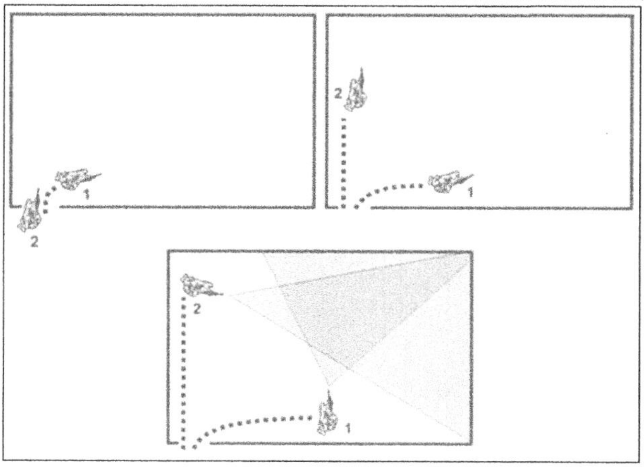

Figure 6-13. POINTS OF DOMINATION AND SECTORS OF FIRE (THREE-MAN TEAM, CENTER DOOR)

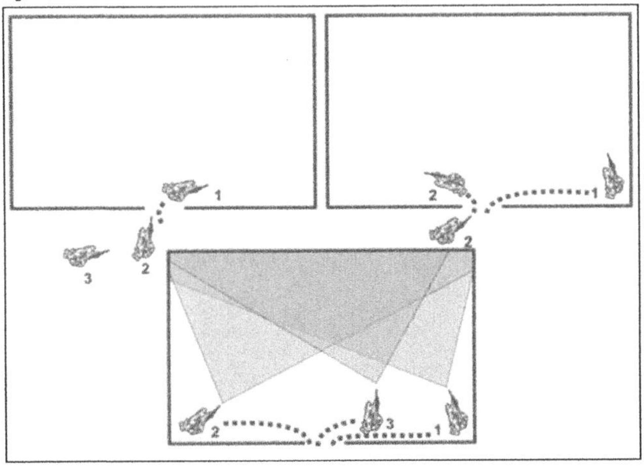


Figure 6-14. POINTS OF DOMINATION AND SECTORS OF FIRE (THREE-MAN TEAM, CORNER DOOR)

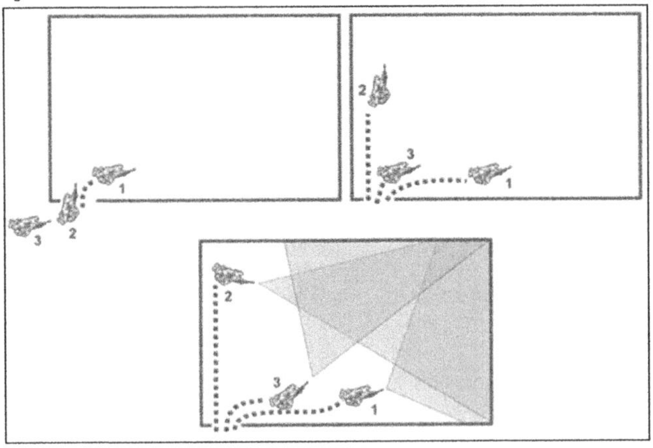

28. While moving through a building Soldiers may encounter the following architectural building features.

a. **Multiple Team/Multiple Rooms**. Figure 6-15 shows this.

**STEP 1:** First team enters and clears Room 1.

**STEP 2:** Squad leader determines the direction from which the second clearing team must enter Room 1, based on the location of the Room 2 entry point.

**STEP 3:** First team collapses inward to allow the second team to move into the room.

**STEP 4:** Second team 'stacks left' and prepares to enter Room 2.

**Figure 6-15. MULTIPLE TEAMS / MULTIPLE ROOMS**

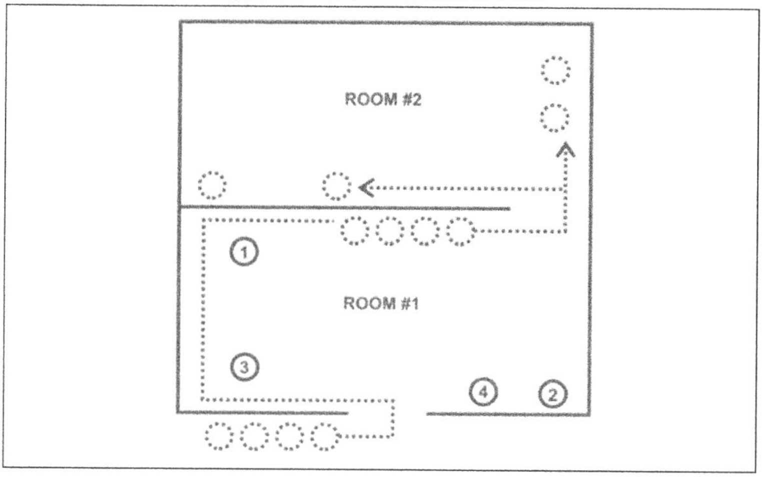

b. **Open Stairwell**. This is a gap between flights of stairs that allows a person to look up and down between the flights. Figure 6-16 shows an open stairwell.

**STEP 1:** The # 1 man pulls security on the highest point he can see / engage.

**STEP 2:** The # 2 man moves up the stairs on the inside with the # 3 man to a point that he can see / engage the next landing, where he turns around and continues to move up to the next landing.

**STEP 3:** The # 3 man moves up the stairs with the # 2 man on the outside and engages the threat on the immediate landing.

**STEP 4:** The # 4 man moves up the stairs with the # 1 man, on the squeeze, the # 2 man turns around to engage the next landing.

**STEP 5:** The flow continues with the # 2 man picking up the sector of the # 1 man had. The # 3 man picks up where the # 2 man was. The # 4 man picks up where the # 3 man was. The # 1 man picks up where the # 4 man was.

***Note:*** Most stairwells require a second team.

**Figure 6-16. OPEN STAIRWELL**

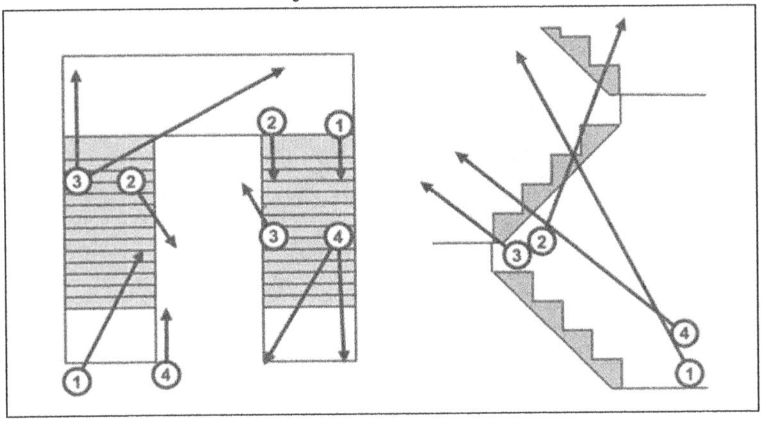

c. **Closed Stairwell**. This is any stairwell with walls separating the flights of stairs. Figure 6-17 shows a closed stairwell.

**STEP 1:** The # 1 man checks high to ensure there is no opening on the landing or between the stairs.

**STEP 2:** The # 2 man pulls long security to the next bend or landing.

**STEP 3:** The # 1 or # 3 man moves up the steps with the # 2 man. As they approach the corner, the # 2 man signals his presence by tapping the # 1 man on the shoulder.

**STEP 4:** Keying off the # 1 man's movement, the other men simultaneously break around the corner.

**STEP 5:** If no fire is received, the # 2 man moves to the opposite wall. Both men continue to move up until they reach their objective.

**STEP 6:** The # 3 and # 4 men will continue to move 3 to 4 steps behind.

***Note:*** Avoid getting locked into a security position such as inside a stairwell. Also, avoid spreading yourselves thin or getting separated by more than one floor of stairs.

**Figure 6-17. CLOSED STAIRWELL**

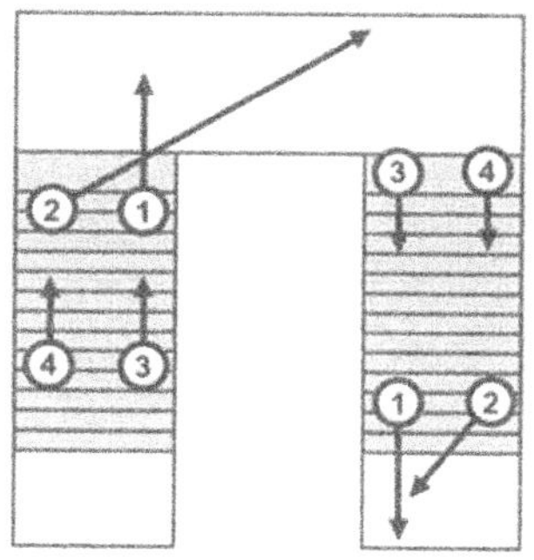

d. **Hallway Movement**. Clearing team(s) move down the hallway using the frontal security (cross cover) technique (Figure 6-18).

Figure 6-18. HALLWAY MOVEMENT

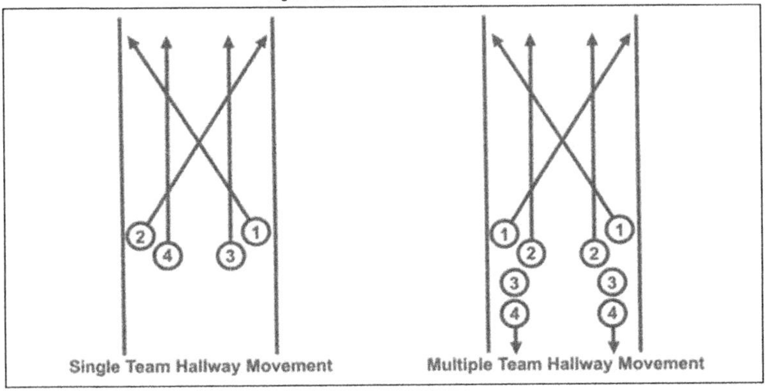

e. **T-Shaped Intersection**. See Figure 6-19. This technique can incorporate the dynamic corner clear (Figure 6-20).

**STEP 1:** Each # 1 man goes to a knee covering his sector.

**STEP 2:** On a predetermined signal, each two-man team will break the corner, picking up their sectors of fire.

**Figure 6-19. T-SHAPED INTERSECTION**

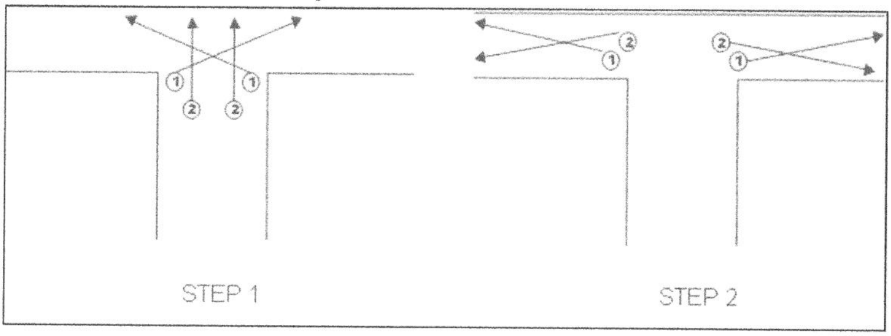

f. **Dynamic Corner**. (See Figure 6-20.)

**STEP 1:** The #1 and #2 man, as they approach the corner, have to clear. Do not slow down.

**STEP 2:** The #2 man will tap the #1 man on the shoulder about 2-3 meters away from the corner letting the #1 man know the #2 man is with him.

**STEP 3:** Keying off of the #1 man's movement, they both break the corner simultaneously.

**STEP 4:** The #1 man goes low to a knee, the #2 man stays high.

**STEP 5:** If the Rangers are not receiving fire the #2 man rabbits, or moves, to the far side.

**STEP 6:** The #1 and #2 man take up sectors of fire.

**STEP 7:** The #3 and #4 man take long security in the direction of movement.

**Figure 6-20. DYNAMIC CORNER**

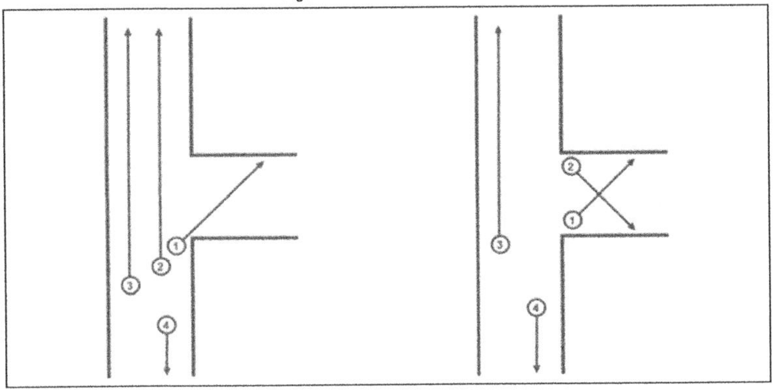

g. **Three-Way Intersection**. (See Figure 6-21.)

**STEP 1:** The #1 through #4 man will use one of the corner clearing techniques to clear the intersection.

**STEP 2:** When the intersection is secure, the #5 and #6 man move through the intersection and enter either more hallway or a room. If they are entering more hallway, the #1 through #4 man provide cross coverage as the #5 and #6 man move through the intersection into more hallway. If they are moving into a room, the #5 and #6 man move to one side of the door and signal the #1 and #3 or #2 and #4 man to enter the stack depending on which side of the door they stack on. The remaining members of the team, not in the stack, will continue to provide security down the hallway.

**STEP 3:** If the #5 and #6 man move into more hallway the #1 through #4 man will enter the stack and proceed down the hallway with the #3 and #4 man providing rear security.

**Figure 6-21. THREE WAY INTERSECTION**

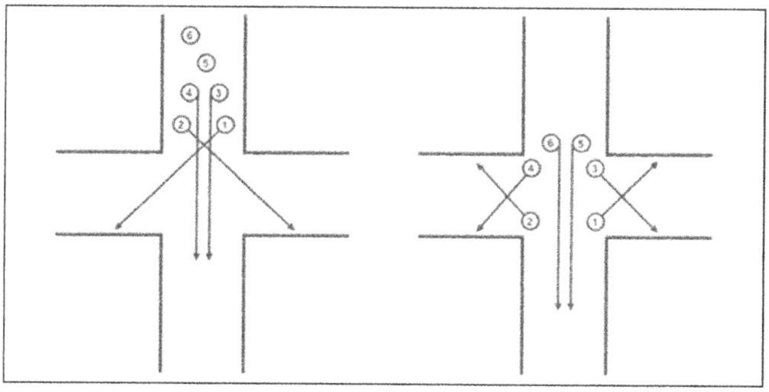

**BATTLE DRILL 6**
**ENTER/CLEAR A TRENCH**

**SITUATION:** The platoon is attacking as part of a larger force and identifies the enemy in a trench line. The platoon deploys and establishes a base of fire. The platoon leader determines that he has sufficient combat power to maneuver and assault the trench line.

**REQUIRED ACTIONS:** See Figure 6-22 and Figure 6-23.

1. The platoon leader directs one squad to enter the trench and secure a foothold.

2. The platoon leader designates the entry point of the trench line and the direction of movement once the platoon begins clearing.

3. The PSG positions Soldiers and machine guns to suppress the trench and isolate the entry point.

4. Platoon leader directs platoon FO to initiate fire mission, if necessary, in support of assault. They maintain accurate battle tracking of all friendly elements to facilitate quick clearance of fires. The platoon FO ceases or shifts fires to isolate the OBJ as assault team advances.

5. The assaulting squad executes actions to enter the trench and establish a foothold. The squad leader signals to the platoon leader that the foothold is secure, and the follow on elements can move into the trench. The squad leader remains at the entry point and marks it. The platoon follows the success of the seizure of the foothold with the remainder of the platoon as part of the platoon actions to clear a trench line.

6. The platoon leader moves into the trench with the assaulting squad.

7 The platoon leader directs one of the base-of-fire squads to move into the trench and begin clearing it in the direction of movement from the foothold.

8. The base-of-fire element repositions as necessary to continue suppressive fires.

9. The assaulting squad passes the squad that has secured the foothold and executes actions to take the lead and clear the trench.

a. The squad leader designates a lead fire team and a trail fire team.

b. The lead fire team and the squad leader move to the forward most secure corner or intersection. The squad leader tells the team securing that corner or intersection that his squad is ready to continue clearing the trench. The trail fire team follows, maintaining visual contact with the last Soldier of the lead team.

***Notes:*** 1. The fire support element must be able to identify the location of the lead fire team in the trench at all times.

2. Throughout this battle drill, the team leader positions himself at the rear of the fire team to have direct control (physically, if necessary) of his Soldiers. Other Soldiers in the fire team rotate the lead to change magazines and prepare grenades. Rotating the lead provides constant suppressive fires down the trench and maintains the momentum of the attack as the squad clears the trench.

c. The lead fire team passes the element securing the foothold. The following then occurs.

(1) The lead Soldier of the fire team moves abreast of the Soldier securing the corner or intersection, taps him, and announces, "*Taking the lead.*"

(2) The Soldier securing the corner or intersection acknowledges that he is handing over the lead by shouting, "*Okay!*" He allows the fire team to pass him.

d. The lead fire team starts clearing in the direction of movement. They arrive at a corner or intersection. The following then occurs.

(1) Allowing for cook-off (2 seconds maximum) and shouting, "*Frag out,*" the second Soldier prepares and throws a grenade around the corner.

(2) Upon detonation of the grenade, the lead Soldier moves around the corner firing three-round bursts and advancing as he fires. The entire fire team follows him to the next corner or intersection.

e. The squad leader—

- Follows behind the team.
- Ensures that the trailing fire team moves up and is ready to pass the lead at his direction.
- Rotates fire teams as necessary to keep his Soldiers fresh and to maintain the momentum of the attack.
- Requests indirect fires, if required, through the platoon leader. The squad leader also directs the employment of the M203 to provide immediate suppression against positions along the trench line.
- Ensures fire teams maintain sufficient interval to prevent them from being engaged by the same enemy fires.

f. At each corner or intersection, the lead fire team performs the same actions previously described.

g. If the lead Soldier finds that he is nearly out of ammunition before reaching a corner or intersection, he announces, "*Ammo,*" and then--

(1) The lead Soldier stops and moves against one side of the trench, ready to let the rest of the team pass. He continues to aim his weapon down the trench in the direction of movement.

(2) The next Soldier ensures that he has a full magazine, moves abreast of the lead Soldier, taps him, and announces, "*Taking the lead.*"

(3) The lead Soldier acknowledges that he is handing over the lead by shouting, "*Okay.*" Positions rotate and the squad continues forward.

h. The trailing fire team secures intersections and marks the route within the trench as the squad moves forward. The trailing fire team leader ensures that follow-on squads relieve his buddy teams to maintain security.

i. The squad leader reports the progress of the clearing operation. The base-of-fire element must be able to identify the location of the lead fire team in the trench at all times.

10. The platoon leader rotates squads to keep the Soldiers fresh and to maintain the momentum of the assault.

11. The PSG calls forward ammunition resupply and organizes teams to move it forward into the trench.

12. The base-of-fire element ensures that all friendly forces move into the trench only through the designated entry point to avoid fratricide.

13. The platoon leader reports to the company commander that the trench line is secured, or he is no longer able to continue clearing. If trench line is secured, then platoon leader directs platoon FO to develop a fire plan to support the defense of the platoon position.

**DANGER**

**FIRE TEAMS: MAINTAIN SUFFICIENT INTERVALS TO AVOID ENGAGEMENT BY THE SAME ENEMY FIRES.**

**Figure 6-22. ENTER A TRENCH (SQUAD)**

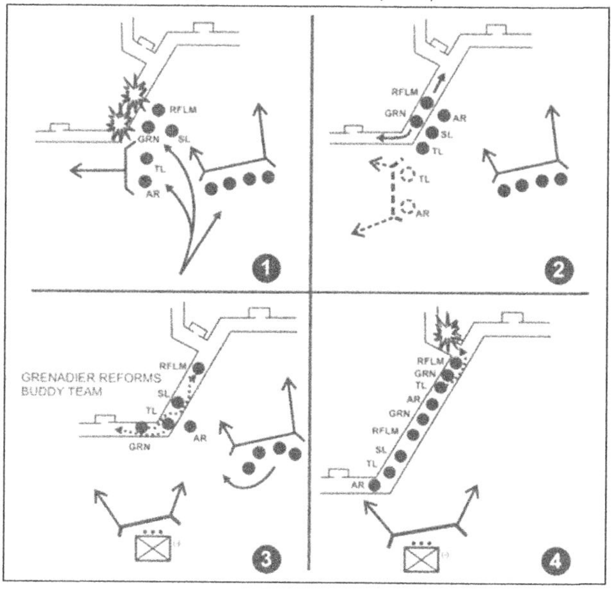

Figure 6-23. CLEAR A TRENCH LINE (PLATOON)

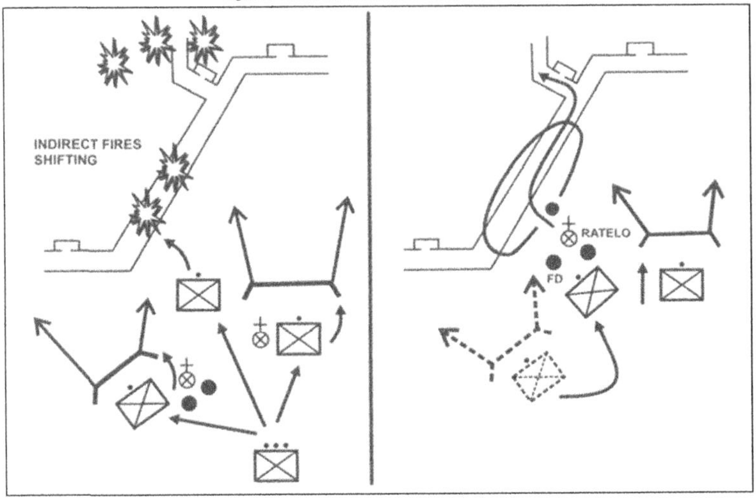


**BATTLE DRILL 7**
**CONDUCT INITIAL BREACH OF A MINED WIRE OBSTACLE (PLATOON)**

**SITUATION:** The platoon is operating as part of a larger force. The lead squad identifies a wire obstacle reinforced with mines that cannot be bypassed and enemy positions on the far side of the obstacle.

**REQUIRED ACTIONS:** See Figure 6-24 and Figure 6-25.

1. The squad in contact reacts to contact.
2. The platoon gains suppressive fires. The following then occurs.
   a. The squad in contact establishes a base-of-fire position.
   b. The platoon leader, his RTO, platoon FO, and the squad leader of the next squad with one machine gun team move forward to link up with the squad leader of the squad in contact.
3. The platoon leader determines that he can maneuver by identifying—
   - The obstacle and enemy positions.
   - The size of the enemy force engaging the squad, for example, the number of enemy automatic weapons, the presence of any vehicles, and the employment of indirect fires are indicators of enemy strength.
   - A breach point.
   - A covered and concealed route to the breach point.
4. The platoon leader directs the squad in contact to support the movement of another squad to the breach point. He indicates the base-of-fire position and the route to it, the enemy position to be suppressed, and the breach point and route the rest of the platoon will take to it. The platoon leader also clears the location, task, purpose, and method of conducting the fire mission with the platoon FO.

5. On the platoon leader's signal, the base-of-fire squad—
- Destroys or suppresses enemy weapons that are firing effectively against the platoon.
- Obscures the enemy position with smoke.
- Continues suppressive fires at the lowest possible level.

6. The platoon leader designates one squad as the breach squad and the remaining squad as the assault squad once the breach has been made. The assault squad may add its fires to the base-of-fire squad. Normally, it follows the covered and concealed route of the breach squad and assaults through immediately after the breach is made.

7. The base-of-fire squad moves to the breach point and establishes a base of fire.

8. The PSG moves forward to the base-of-fire squad with the second machine gun and assumes control of the squad.

9. The platoon leader leads the breach and assault squads along the covered and concealed route.

10. The platoon FO calls for and adjusts indirect fires as directed by the platoon leader. The platoon leader in conjunction with the platoon FO maintains accurate battle tracking of all friendly elements to facilitate quick clearance of fires.

11. The breach squad executes actions to breach the obstacle.

a. The squad leader directs one fire team to support the movement of the other fire team to the breach point.

b. The squad leader identifies the breach point.

c. The base-of-fire element continues to provide suppressive fires and isolates the breach point.

d. The breaching fire team, with the squad leader, moves to the breach point using the covered and concealed route. From there, the following takes place:

(1) The squad leader and breaching fire team leader employ smoke grenades to obscure the breach point. The platoon base-of-fire element shifts direct fires away from the breach point and continues to suppress key enemy positions. The platoon FO ceases indirect fires or shifts them beyond the obstacle.

(2) The breaching fire team leader positions himself and the automatic rifleman on one flank of the breach point to provide close-in security.

(3) The grenadier and rifleman of the breaching fire team probe for mines and cut the wire obstacle, marking their path as they proceed. If available, a bangalore is preferred.

(4) Once the obstacle has been breached, the breaching fire team leader and the automatic rifleman move to the far side of the obstacle and take up covered and concealed positions with the rifleman and grenadier. The team leader signals to the squad leader when they are in position and ready to support.

(5) The squad leader signals the base-of-fire team leader to move his fire team up and through the breach. He then moves through the obstacle and joins the breaching fire team, leaving the grenadier (or anti-armor specialist) and rifleman of the supporting fire team on the near side of the breach to guide the rest of the platoon through.

(6) Using the same covered and concealed route as the breaching fire team, the base-of-fire team moves through the breach and takes up covered and concealed positions on the far side.

12. The squad leader reports the situation to the platoon leader and posts guides at the breach point.

13. The platoon leader leads the assault squad through the breach in the obstacle and positions them beyond the breach to support the movement of the remainder of the platoon or assaults the enemy position covering the obstacle. Platoon leader directs the platoon FO to shift indirect fires off the enemy position.

14. The breaching squad continues to widen the breach to allow vehicles to pass through.

15. The platoon leader reports the situation to the company commander, and then he directs his base-of-fire squad to move up and through the obstacle. The platoon leader appoints Soldiers to guide the company through the breach point.

16. The PSG brings the remaining elements forward and through the breach on the platoon leader's command.

17. The company follows up the success of the platoon as it conducts the breach and continues the assault against the enemy positions.

Figure 6-24. CONDUCT INITIAL BREACH OF A MINED WIRE OBSTACLE (SQUAD)

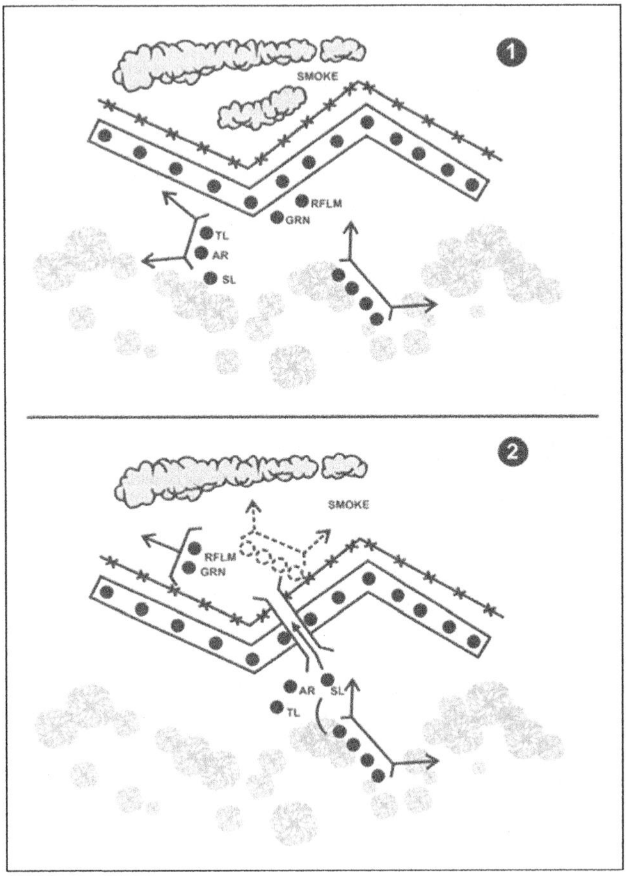

Figure 6-25. CONDUCT INITIAL BREACH OF A MINED WIRE OBSTACLE (PLATOON)

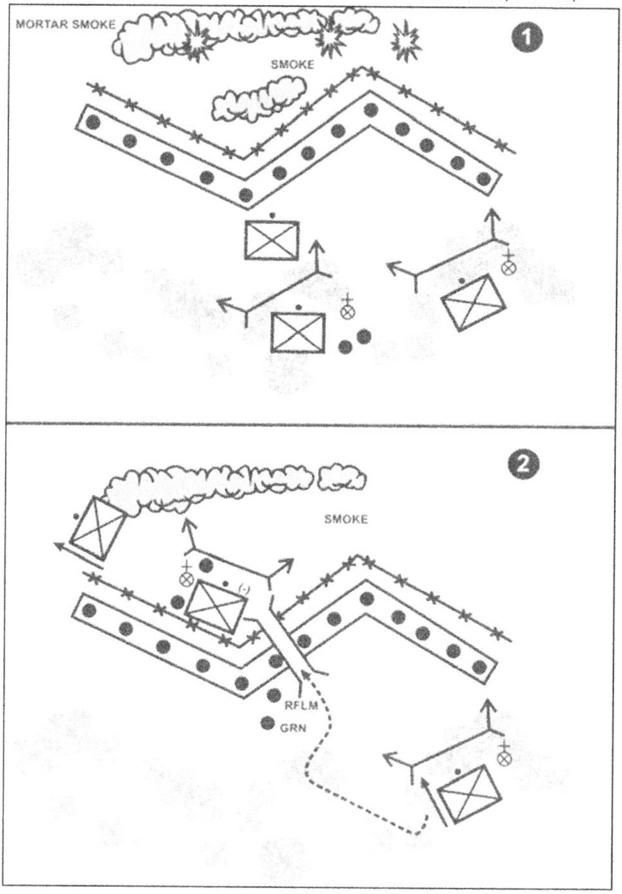

## BATTLE DRILL 8
## REACT TO INDIRECT FIRE

**SITUATION:** A platoon or squad while moving or at the halt (not dug in) receives indirect fire.

**REQUIRED ACTIONS:** NA

Any Soldier announces, "*Incoming.*" Then, if the platoon/squad is moving, the following takes place:

1. The platoon/squad leader gives direction and distance for the platoon/squad to move to a rally point by ordering direction and distance, for example, "*Four o'clock, three hundred meters.*"

2. Platoon/squad members move rapidly along the direction and distance to the rally point.

3. At the rally point, the leader immediately accounts for personnel and equipment, and forms the platoon/squad for a move to an alternate position.

4. The senior leader submits a SITREP to higher headquarters.

5. If the platoon/squad is halted (not dug in) or is preparing to move because they hear incoming artillery, the following takes place:

   a. The platoon/squad leader gives direction and distance for the platoon/squad to move to a rally point by ordering direction and distance, for example, "*Four o'clock, three hundred meters.*"

   b. Platoon/squad members secure all mission-essential equipment and ammunition and move rapidly along the direction and distance to the rally point.

   c. At the rally point, the leader immediately accounts for personnel and equipment, and forms the platoon/squad for a move to an alternate position.

   d. The senior leader submits a SITREP to higher headquarters.

***Note:*** If platoon/squad members are in defensive (dug in) positions, then members will remain in those positions if appropriate. Senior leader submits SITREP.

NOTES

# Chapter 7
# COMMUNICATIONS

The basic requirement of combat communications is to provide rapid, reliable, and secure interchange of information. Communications are a vital aspect in successful mission accomplishment. The information in this chapter helps the Ranger squad/platoon maintain effective communications and correct any radio antenna problems.

## Section I. INTRODUCTION

This section discusses military frequency modulated radios and automated net control devices (ANCDs).

7-1. **MILITARY FREQUENCY MODULATED RADIOS**. Military frequency modulated (FM) radios are composed of a receiver and a transmitter. Together, the receiver-transmitter has many capabilities and features that enable you to perform your mission more effectively. The radio can operate in Single Channel mode or frequency hopping mode. The radio has about 2,320 SC Channels and includes voice and digital communication. The operating voltage for the manpack radio is 13.5 volts from the primary battery. The range of the manpack radio is 5-10 KM on Hi power. This range is based upon line of sight and is derived from averages achieved under normal atmospheric and weather conditions. Ranges depend upon location, sighting, weather, and surrounding noise level, among other factors. the automated net control device (ANCD) is an essential tool. It replaces the signal operating instructions booklet. The ANCD has many capabilities. The main purpose of the ANCD for your mission is to obtain signal operating instructions electronically. The following discusses assembly, operation, and troubleshooting for the AN/PRC-119; provides step-by-step instructions to obtain SOI from an ANCD; and explains how to repair broken antennas:

a. **Manpack Radio Assembly**. To assemble a manpack radio you must first check and install a battery.
   (1) Inspect the battery box for dirt or damage.
   (2) Stand RT on front panel guards.
   (3) Check battery life condition (you will be using the rechargeable BB-390 batteries).
   (4) Place battery in box.
   (5) Close battery cover, and secure using latches.
   (6) Return radio to upright position.
   (7) If used battery was installed, enter the battery life condition into radio by performing the following:
      (a) Set **FCTN** to **LD**.
      (b) Press **BAT**, then **CLR**.
      (c) Enter number recorded on side of battery.
      (d) Press **STO**.
      (e) Set **FCTN** to **SQ ON**.

b. **Antenna**.
   (1) Inspect whip antenna connector on antenna and on radio for damage.
   (2) Screw whip antenna into base.
   (3) Hand tighten.
   (4) Carefully mate antenna base with RT ANT connector.
   (5) Hand tighten.
   (6) Position antenna as needed by bending goose neck.

***Note:*** Keep antenna straight, if possible. If the antenna is bent to a horizontal position, it may be necessary to turn the radio in order to receive and transmit messages.

c. **Handset**.
   (1) Inspect the handset for damage.
   (2) Push handset on **AUD/DATA** and twist clockwise to lock in place.

d. **Field Pack**.
   (1) Place radio transmitter (RT) in field pack with antenna on the left shoulder.
   (2) Fold top flap of field over RT and secure flap to field pack using straps and buckles.

e. **Presets**.
   (1) Set **CHAN** to **1**.
   (2) Set **MODE** to **SC**.
   (3) Set **RF PWR** to **HI**.

(4) Set **VOL** to mid range.
(5) Set **DIM** full clockwise.
(6) Set **FCTN** to **LD**.
(7) Set **DATA RATE** to **OFF**.

f. **Single Channel Loading Frequencies**.
(1) Obtain Ranger SOI.
(2) Set **FCTN** to **LD**.
(3) Set **MODE** to **SC**.
(4) Set **CHAN** to **MAN**, **Cue**, or desired channel (**1** to **6**) where frequency is to be stored.
(5) Press **FREQ** (display will show "**00000**," or the frequency the RT is currently set to).
(6) Press **CLR** (display will show five lines).
(7) Enter the number of the new frequency.
(8) If you make a mistake with a number, press **CLR**.
(9) Press **STO** (display will blink).
(10) Set **FCTN** to **SQ ON**.

g. **Clearing of Frequencies**.
(1) Set **MODE** to **SC**.
(2) Set **CHAN** to **MAN**, **Cue** or the channel whose frequency is to be cleared.
(3) Press **FREQ**.
(4) Press **CLR**.
(5) Press **Load**; then press **STO**.
(6) Set **FCTN** to **SQ ON**.

h. **Scanning of Multiple Frequencies**.
(1) Load all desired frequencies using "Single Channel Loading Frequencies" instructions (see 7-2h).
(2) Set **CHAN** to **CUE**.
(3) Set **SC** to **FH**.
(4) Set **FCTN** to **SQ ON**.
(5) Press **STO** (display will say **SCAN**).
(6) Press **8**. You will now be able to scan more than one frequency.

7-2. **AUTOMATED NET CONTROL DEVICE**. The RTO retrieves all necessary COMSEC and information from the ANCD. To retrieve the SOI from an ANCD:

a. Press the **ON** button on the ANCD keypad.
b. Press the letter lock button to unlock the keys on the ANCD.
c. Press **Main Menu** key.
d. Once you press the main menu key, this is what should show on the screen.

**Appl Date Time Setup**
**Util Bit.**

e. Using the arrow keys, scroll over to **Appl** (as shown above) and press the enter button on the keypad.

f. Once you have entered into the **Appl** (applications) menu using the arrow keys scroll over to **SOI** and press the enter button on the key pad. For example--

**RDS SOI Radio.**
**(Appl)** <--*(Indicates that you are in the applications menu).*

**SOI menu will look like this:**
**Qref Group Net Sufx Pyro**
**Tmpd Set C/S Find Memo**

g. Using the arrow keys scroll over to **Set** and press the enter button on the keypad.
h. Press enter on choose once in the **Set** menu.

> *EXAMPLE:*
> SELECT:
> **Choose Send Receive**.

i. Once in the **Choose** menu press the enter key on set **1-5** or **6-10**.

j. After you choose your selection out of the **Set** menu, the ANCD will automatically return back to the **SOI** menu.

k. Using the keypad scroll over to **Tmpd** and enter the number out of the set you need for that day. After this operation, the ANCD will automatically return back to the **SOI** menu.

l. Using the arrow keys on the keypad scroll over to **Net** and press the enter button on the keypad. The Net menu should look like the following example:

> *EXAMPLE:*
> T05 <--(time period) 1 <--(platoon) / A <--(company) /A3Q <--(prefix).
> C 60.000.

***Note:*** When in **Net** menu, this is where you will obtain your frequency and the first part of your call sign.

m. After you obtain the information you need out of **Net**, press the **Abort** button on the keypad. This will return you to the **SOI** menu.

n. Using the keypad, scroll over to **Sufx** and press the **Enter** button.

o. Once you are in the **Sufx** menu, use the up and down arrow keys on the ANCD to locate your two-digit designator. These two digits will go at the end of your prefix; your prefix plus the two digits equal your call sign.

p. Once you are done, press the **OFF** button on the ANCD to end operation.

## Section II. ANTENNAS

This section discusses repair techniques, construction and adjustment, field-expedient antennas, antenna length and orientation, and improvement of marginal communications.

7-3. **REPAIR**. Antennas are sometimes broken or damaged, causing either a communications failure or poor communications. If a spare antenna is available, the damaged antenna is replaced. When there is no spare, the Ranger squad/platoon may have to construct an emergency antenna. The following paragraphs contain suggestions for repairing antennas and antenna supports and the construction and adjustment of emergency antennas.

> **DANGER**
>
> **SERIOUS INJURY OR DEATH CAN RESULT FROM CONTACT WITH THE RADIATING ANTENNA OF A MEDIUM-POWER OR HIGH-POWER TRANSMITTER. TURN THE TRANSMITTER OFF WHILE ADJUSTING THE ANTENNA.**

a. **Whip Antennas**. When a whip antenna is broken into two sections, the part of the antenna that is broken off can be connected to the part attached to the base by joining the sections.

(1) Use the method shown in A, Figure 7-1 when both parts of the broken whip are available and usable.

(2) Use the method in B, Figure 7-1, when the part of the whip that was broken off is lost, or when the whip is so badly damaged that it cannot be used.

(3) To restore the antenna to its original length, a piece of wire is added that is nearly the same length as the missing part of the whip. The pole support is then lashed securely to both sections of the antenna. The two antenna sections are cleaned thoroughly to ensure good contact before connecting them to the pole support. If possible, the connections are soldered.

b. **Wire Antennas**. Emergency repair of a wire antenna may involve the repair or replacement of the wire used as the antenna or transmission line; or the repair or replacement of the assembly used to support the antenna.

(1) When one or more wires of an antenna are broken, the antenna can be repaired by reconnecting the broken wires. To do this, lower the antenna to the ground, clean the ends of the wires, and twist the wires together. Whenever possible, solder the connection.

(2) If the antenna is damaged beyond repair, construct a new one. Make sure that the lengths of the wires on the substitute antenna match the length of the original.

(3) Antenna supports may also require repair or replacement. A substitute item may be used in place of a damaged support and, if properly insulated, can be of any material of adequate strength. If the radiating element is not properly insulated, field antennas may be shorted to ground and be ineffective. Many commonly found items can be used as field-expedient insulators.

(4) The best of these items are plastic or glass to include plastic spoons, buttons, bottle necks, and plastic bags. Though less effective than plastic or glass but still better than no insulator at all are wood and rope. The radiating element--the actual antenna wire-should touch only the antenna terminal and should be physically separated from all other objects, other than the supporting insulator. (Figure 7-2 shows how to make emergency insulators.)

**Figure 7-1. USE OF BROKEN ANTENNA**

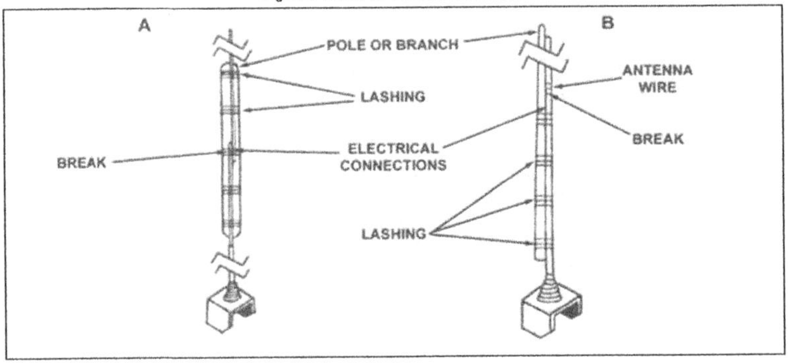


**Figure 7-2. METHODS OF CONSTRUCTING EMERGENCY INSULATORS**

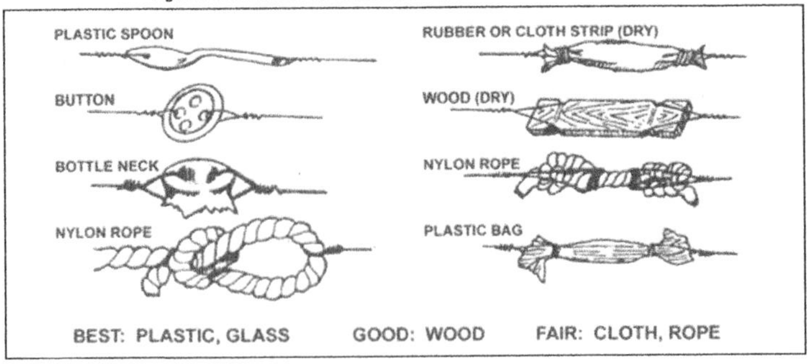

7-4. **CONSTRUCTION AND ADJUSTMENT**. Ranger squad/platoons may use the following methods to construct and adjust antennas:

a. **Construction**. The best kinds of wire for antennas are copper and aluminum. In an emergency, however, Rangers use any type of wire that is available.

(1) The exact length of most antennas is critical. The emergency antenna should be the same length as the antenna it replaces.

(2) Antennas supported by trees can usually survive heavy wind storms if the trunk of a tree or a strong branch is used as a support. To keep the antenna taut and to prevent it from breaking or stretching as the trees sway, the Ranger attaches a spring or old inner tube to one end of the antenna. Another technique is to pass a rope through a pulley or eyehook. The rope is attached to the end of the antenna and loaded with a heavyweight to keep the antenna tightly drawn.

(3) Guidelines used to hold antenna supports are made of rope or wire. To ensure the guidelines will not affect the operation of the antenna, the Ranger cuts the wire into several short lengths and connects the pieces with insulators.

b. **Adjustment**. An improvised antenna may change the performance of a radio set. The following methods can be used to determine if the antenna is operating properly:

(1) A distant station may be used to test the antenna. If the signal received from this station is strong, the antenna is operating satisfactorily. If the signal is weak, the Ranger adjusts the height and length of the antenna and the transmission line to receive the strongest signal at a given setting on the volume control of the receiver. This is the best method of tuning an antenna when transmission is dangerous or forbidden.

(2) In some radio sets, the Ranger uses the transmitter to adjust the antenna. First, he sets the controls of the transmitter to normal; then, he tunes the system by adjusting the antenna height, the antenna length, and the transmission line length to obtain the best transmission output.

7-5. **FIELD-EXPEDIENT OMNIDIRECTIONAL ANTENNAS**. Vertical antennas are omnidirectional. The omnidirectional antenna transmits and receives equally well in all directions.

a. **Vertical Antennas**. Most tactical antennas are vertical. For example, the man-pack portable radio uses a vertical whip and so do the vehicular radios in tactical vehicles. A vertical antenna can be made by using a metal pipe or rod of the correct length, held erect by means of guidelines. The lower end of the antenna should be insulated from the ground by placing it on a large block of wood or other insulating material. A vertical antenna may also be a wire supported by a tree or a wooden pole (Figure 7-3).

**Figure 7-3. VERTICAL ANTENNA MADE FROM A WIRE SUPPORTED BY A TREE OR WOODEN POLE**

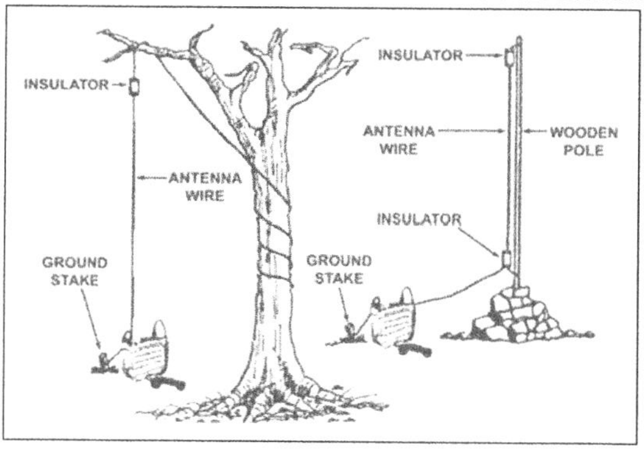

b. **Modification of Connection on Shorter Vertical Antennas**. For short vertical antennas, the pole may be used without guidelines (if properly supported at the base). If the length of the vertical mast is not long enough to support the wire upright, it may be necessary to modify the connection at the top of the antenna (Figure 7-4; see FM 24-18).

**Figure 7-4. MODIFIED CONNECTION AT TOP OF ANTENNA**

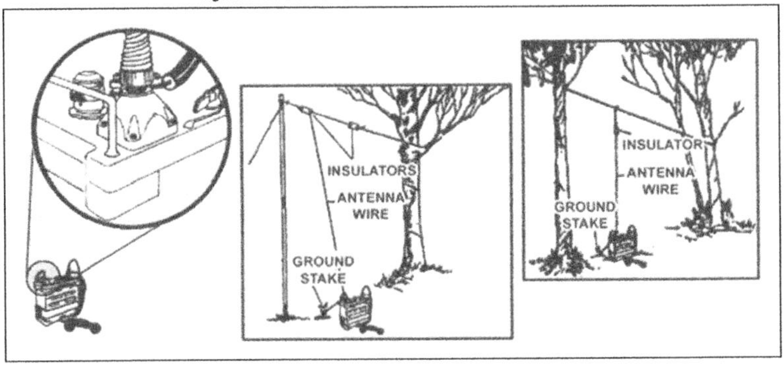

c. **End-Fed, Half-Wave Antenna**. An emergency, end-fed half-wave antenna (Figure 7-5) can be constructed from available materials such as field wire, rope, and wooden insulators. The electrical length of this antenna is measured from the antenna terminal on the radio set to the far end of the antenna. The best performance can be obtained by constructing the antenna longer than necessary and then shortening it, as required, until the best results are obtained. The ground terminal of the radio set should be connected to a good earth ground for this antenna to function efficiently.

**Figure 7-5. END-FED, HALF-WAVE ANTENNA**

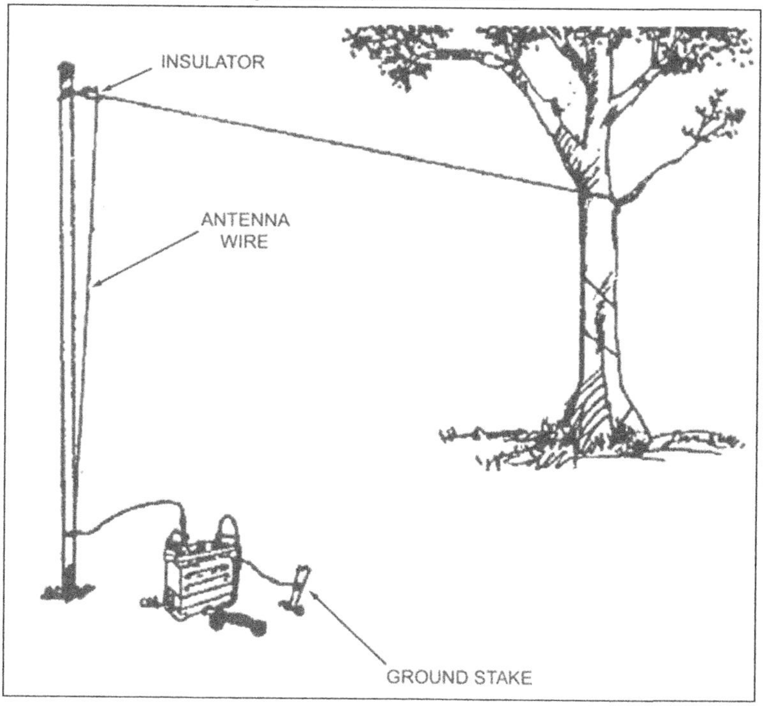

d. **Center-Fed, Doublet Antenna**. The center-fed doublet is a half-wave antenna consisting of two quarter wavelength sections on each side of the center (Figure 7-6). Doublet antennas are directional broadside to their length, which makes the vertical doublet antenna omnidirectional. This is because the radiation pattern is doughnut-shaped and bidirectional.

**Figure 7-6. CENTER-FED, DOUBLET ANTENNA**

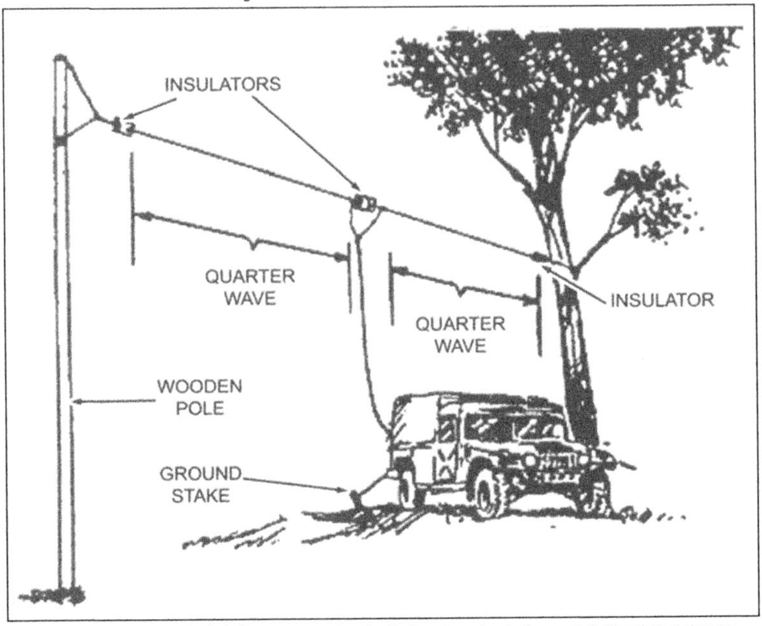


(1) Compute the length of a half-wave antenna by using the formula previously discussed. Cut the wires as close as possible to the correct length; this is very important.

(2) Uses transmission line for conducting electrical energy from one point to another and for transferring the output of a transmitter to an antenna. Although it is possible to connect an antenna directly to a transmitter, the antenna is usually located some distance away.

(3) Support center-fed half-wave FM antennas entirely with pieces of wood. (A horizontal antenna of this type is shown in A, Figure 7-7 and a vertical antenna in B, Figure 7-7.) Rotate these antennas to any position to obtain the best performance.

(a) If the antenna is erected vertically, bring out the transmission line horizontally from the antenna for a distance equal to at least one-half of the antenna's length before it is dropped down to the radio set.

**Figure 7-7. CENTER-FED ANTENNAS SUPPORTED WITH WOOD**

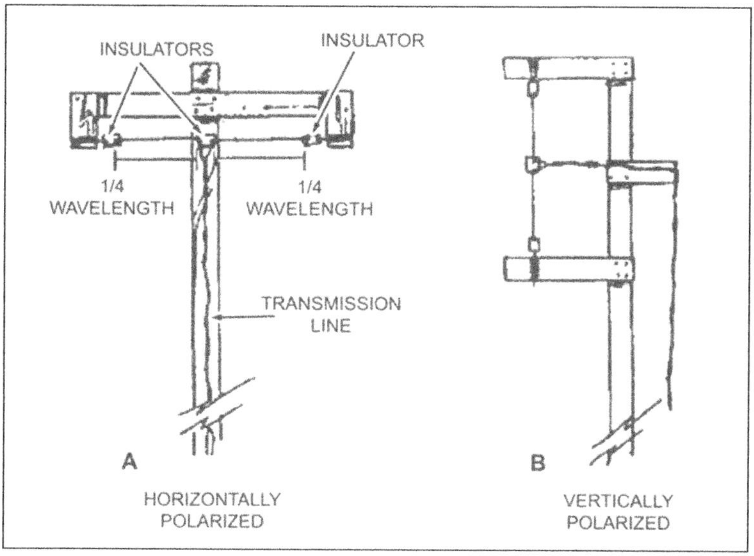

(b) The half-wave antenna is used with FM radios (Figure 7-8). It is effective in heavily wooded areas to increase the range of portable radios. Connect the top guidelines to a limb or pass it over the limb and connect it to the tree trunk or a stake.

**Figure 7-8. HALF-WAVE ANTENNA USED WITH FM RADIO**

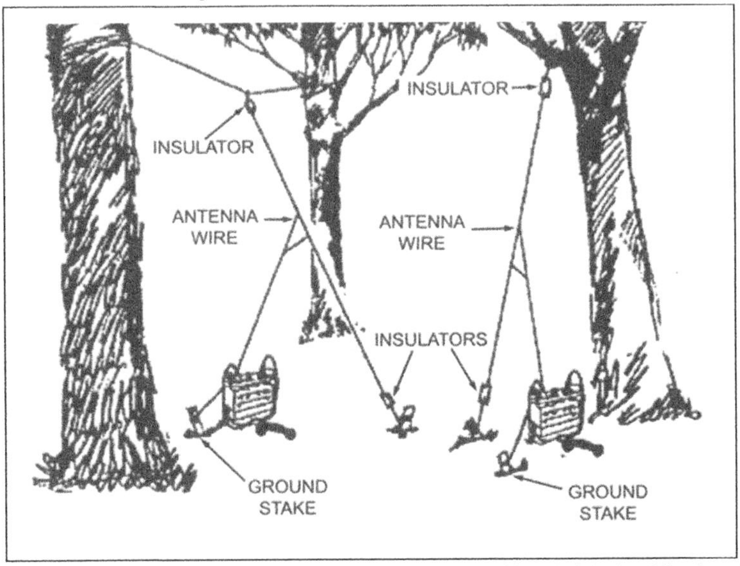


c. **Expedient 292-Type Antenna**. The expedient 292-type antenna was developed for use in the jungle and, if used properly, can increase the team's ability to communicate. In its entirety, the antenna is bulky, heavy, and not acceptable for sniper team operations. The team can, however, carry only the mast head and antenna sections, mounting these on wood poles or hanging them from trees; or, the team can make a complete expedient 292-type antenna (Figure 7-9), using WD-1, wire, and other readily available material. The team can also use almost any plastic, glass, or rubber objects for insulators. Dry wood is acceptable when nothing else is available. (See Figure 7-2 for types of insulators that may be used.) The following describes how to make this antenna:

(1) Use the quick-reference table (Table 7-2) to determine the length of the elements (one radiating and three ground planes) for the frequency that will be used. Cut these elements (A, Figure 7-9) from WD-1 field wire (or similar wire). Cut spacing sticks (B, Figure 7-9) the same length. Place the ends of the sticks together to form a triangle and tie the ends with wire, tape, or rope. Attach an insulator to each corner. Attach a ground-plane wire to each insulator. Bring the other ends of the ground-plane wires together, attach them to an insulator (C, Figure 7-9), and tie securely. Strip about 3 inches of insulation from each wire and twist them together (D, Figure 7-9).

(2) Tie one end of the radiating element wire to the other side of insulator C and the other end to another insulator (D, Figure 7-9). Strip about 3 inches of insulation from the radiating element at insulator C.

(3) Cut enough WD-1 field wire to reach from the proposed location of the antenna to the radio set. Keep this line as short as possible, because excess length reduces the efficiency of the system. Tie a knot at each end to identify it as the *hot* lead. Remove insulation from the *hot* wire and tie it to the radiating element wire at insulator C. Remove insulation from the other wire and attach it to the bare ground-plane element wires at insulator C. Tape all connections and do not allow the radiating element wire to touch the ground-plane wires.

(4) Attach a rope to the insulator on the free end of the radiating element and toss the rope over the branches of a tree. Pull the antenna as high as possible, keeping the lead-in routed down through the triangle. Secure the rope to hold the antenna in place.

(5) At the radio set, remove about 1 inch of insulation from the *hot* lead and about 3 inches of insulation from the other wire. Attach the *hot* line to the antenna terminal (doublet connector, if so labeled). Attach the other wire to the metal case-the handle, for example. Be sure both connections are tight or secure.

(6) Set up correct frequency, turn on the set, and proceed with communications.

**Figure 7-9. COMPLETE EXPEDIENT 292-TYPE ANTENNA**

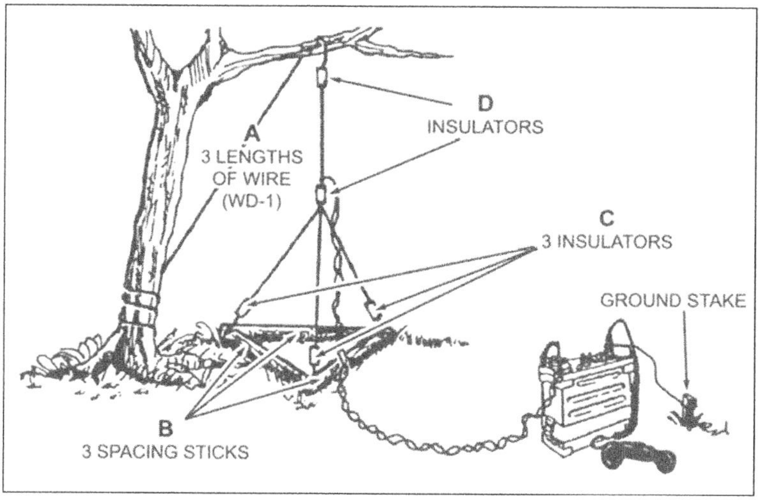

Table 7-2. QUICK REFERENCE TABLE

| OPERATING FREQUENCY IN MHz | ELEMENT LENGTH (RADIATING ELEMENTS AND GROUND-PLANE ELEMENTS) | |
|---|---|---|
| 30 | 2.38 m | (7 ft 10 in) |
| 32 | 2.23 m | (7 ft 4 in) |
| 34 | 2.1 m | (6 ft 11 in) |
| 36 | 1.98 m | (6 ft 6 in) |
| 38 | 1.87 m | (6 ft 2 in) |
| 40 | 1.78 m | (5 ft 10 in) |
| 43 | 1.66 m | (5 ft 5 in) |
| 46 | 1.55 m | (5 ft 1 in) |
| 49 | 1.46 m | (4 ft 9 in) |
| 52 | 1.37 m | (4 ft 6 in) |
| 55 | 1.3 m | (4 ft 3 in) |
| 58 | 1.23 m | (4 ft 0 in) |
| 61 | 1.17 m | (3 ft 10 in) |
| 64 | 1.12 m | (3 ft 8 in) |
| 68 | 1.05 m | (3 ft 5 in) |
| 72 | 0.99 m | (3 ft 3 in) |
| 76 | 0.94 m | (3 ft 1 in) |

7-6. **FIELD-EXPEDIENT DIRECTIONAL ANTENNAS**. The vertical half-rhombic antenna (Figure 7-10) and the long-wire antenna (Figure 7-11) are two field-expedient directional antennas. These antennas consist of a single wire, preferably two or more wavelengths long, supported on poles at a height of 3 to 7 meters (10 to 20 feet) above the ground. The antennas will, however, operate satisfactorily as low as 1 meter (about 3 feet) above the ground--the radiation pattern is directional. The antennas are used mainly for either transmitting or receiving high-frequency signals.

**Figure 7-10. VERTICAL HALF-RHOMBIC ANTENNA**

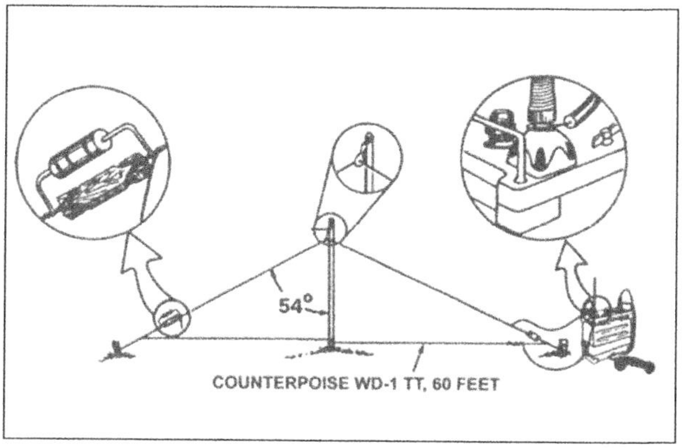


**Figure 7-11. LONG-WIRE ANTENNA**

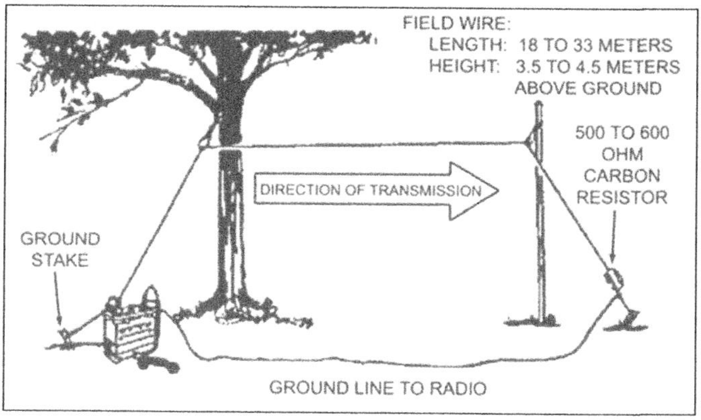

a. The vee antenna (Figure 7-12) is another field-expedient directional antenna. It consists of two wires forming a V with the open area of the "V" pointing toward the desired direction of transmission or reception. To make construction easier, the legs should slope downward from the apex of the "V." This is called a sloping vee antenna (Figure 7-13). The angle between the legs varies with the length of the legs to achieve maximum performance. (Use Table 7-1 to determine the angle and the length of the legs.)

**Figure 7-12. VEE ANTENNA**

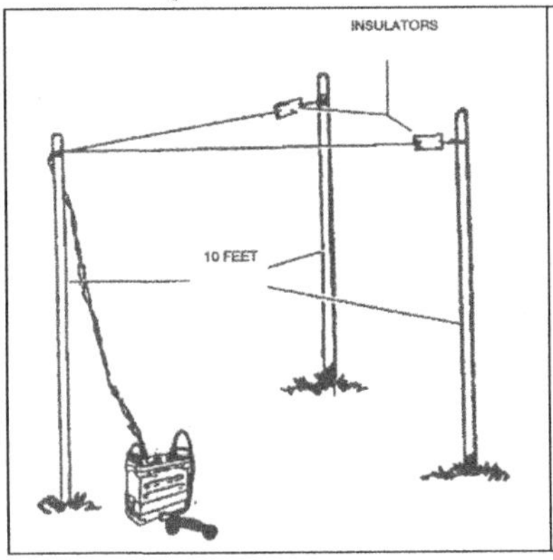


**Figure 7-13. SLOPING VEE ANTENNA**

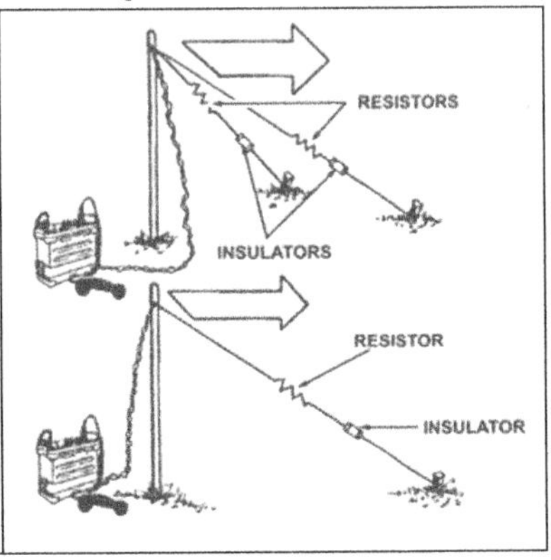


b. When the antenna is used with more than one frequency or wavelength, use an apex angle that is midway between the extreme angles determined by the chart. To make the antenna radiate in only one direction, add noninductive terminating resistors from the end of each leg (not at the apex) to ground. (See TM 11-666.)

7-7. **ANTENNA LENGTH**. The length of an antenna must be considered in two ways: both a physical and an electrical length. These two lengths are never the same. The reduced velocity of the wave on the antenna and a capacitive effect (known as end *effect*) make the antenna seem longer electrically than it is physically. The contributing factors are the ratio of the diameter of the antenna to its length and the capacitive effect of terminal equipment, such as insulators and clamps, used to support the antenna.

a. To calculate the physical length of an antenna, use a correction of 0.95 for frequencies between 3.0 and 50.0 MHz The figures given below are for a half-wave antenna.

b. Use the following formula to calculate the length of a long-wire antenna (one wavelength or longer) for harmonic operation:

***EXAMPLE:***

If N = the number of half-wavelengths in the total length of the antenna = 3,
and if the frequency in MHz = 7, then--

$$\text{Length (meters)} = \frac{150(N-0.05)}{\text{Frequency in MHz}} = \frac{150(3-.05)}{7} =$$

$$\frac{150 \times 2.95}{7} = \frac{442.50}{7} = 63.2 \text{ meters}$$

7-8. **ANTENNA ORIENTATION**. If the azimuth of the radio path is not provided, the azimuth should be determined by the best available means. The accuracy required in determining the azimuth of the path depends on the radiation pattern of the directional antenna. In transportable operation, the rhombic and V antennas may have such a narrow beam as to require great accuracy in azimuth determination. The antenna should be erected for the correct azimuth. Great accuracy is not required in erecting broad-beam antennas. Unless a line of known azimuth is available at the site, the direction of the path is best determined by a magnetic compass.

7-9. **IMPROVEMENT OF MARGINAL COMMUNICATIONS**. Under certain situations, it may not be feasible to orient directional antennas to the correct azimuth of the desired radio path. As a result, marginal communications may suffer. To improve marginal communications, the following procedure can be used:

a. Check, tighten, and tape cable couplings and connections.
b. Return all transmitters and receivers in the circuit.
c. Ensure antennas are adjusted for the proper operating frequency.
d. Change the heights of antennas.
e. Move the antenna a short distance away and in different locations from its original location.

NOTES

## Chapter 8

# ARMY AVIATION

Army aviation and infantry units can be fully integrated with other members of the combined arms team to form powerful and flexible air assault task forces that can project combat power throughout the entire depth, width, and breadth of the modern battlefield with little regard for terrain barriers. These combat operations are deliberate, precisely planned, and vigorously executed. They are assigned to strike the enemy when and where he is most vulnerable (See page 2-28, *Air Movement Annex* and page 2-37, *Army Aviation Coordination Checklist*).

8-1. **REVERSE PLANNING SEQUENCE**. Successful air assault execution is based on a careful analysis of METT-TC and detailed, precise reverse planning. Five basic plans that comprise the reverse planning sequence are developed for each air assault operation. The battalion is the lowest level that has sufficient personnel to plan, coordinate, and control air assault operations. When company size or lower operations are conducted, the bulk of the planning takes place at battalion or higher headquarters. They are:

a. **Ground Tactical Plan**. The foundation of a successful air assault operation is the commander's ground tactical plan. All additional plans must support this plan. The plan specifies actions in the objective area to ultimately accomplish the mission and address subsequent operations.

b. **Landing Plan**. The landing plan must support the ground tactical plan. This plan outlines a sequence of events that allows elements to move into the area of operations, and ensures that units arrive at designated locations at prescribed times prepared to execute the ground tactical plan.

c. **Air Movement Plan**. The air movement plan is based on the ground tactical and landing plans. It specifies the schedule and provides instructions for air movement of troops, equipment, and supplies from PZs to LZs.

d. **Loading Plan**. The loading plan is based on the air movement plan. It ensures that troops, equipment, and supplies are loaded on the correct aircraft. Unit integrity is maintained when aircraft loads are planned. Cross-loading may be necessary to ensure survivability of command and control assets, and that the mix of weapons arriving at the LZ is ready to fight.

e. **Staging Plan**. The staging plan is based on the loading plan and prescribes the arrival time of ground units (troops, equipment and supplies) at the PZ in the order of movement

8-2. **SELECTION AND MARKING OF PICKUP AND LANDING ZONES**

a. Small unit leaders should consider the following when selecting a PZ/LZ:

(1) ***Size***. Minimal circular landing point separation from other aircraft and obstacles is needed:

- OH-58D – 25 meters.
- UH-1, AH-1 – 35 meters.
- UH-60, AH-64 – 50 meters.
- Cargo helicopters – 80 meters.

(2) ***Surface Conditions***. Avoid potential hazards such as sand, blowing dust, snow, tree stumps, or large rocks.

(3) ***Ground Slope***.

- 0% - 6 % -- land upslope.
- 7% - 15% -- land sideslope.
- over 15% -- no touchdown (aircraft may hover).

(4) ***Obstacles***. An obstacle clearance ratio of 10 to 1 is used in planning approach and departure of the PZ and LZ, for example, a 10-foot tall tree requires 100 feet of horizontal distance for approach or departure. Obstacles will be marked with a red chemlight at night, or red panels in daytime. Avoid using markings if the enemy would see them.

(5) ***Approach/Departure***. Approach and departure are made into the wind and along the long axis of the PZ/LZ.

(6) ***Loads***. The greater the load, the larger the PZ/LZ must be to accommodate the insertion or extraction.

b. **Marking of PZs and LZs**.

(1) ***Day***. A ground guide will mark the PZ or LZ for the lead aircraft by holding an M16/M4 rifle over his head, by displaying a folded VS-17 panel chest high, or by other coordinated and identifiable means.

(2) ***Night***. The code letter "Y" (inverted "Y") is used to mark the landing point of the lead aircraft at night (Figure 8-1). Chemical lights or "beanbag" lights are used to maintain light discipline. A swinging chemlight may also be used to mark the landing point.

Figure 8-1. INVERTED "Y"

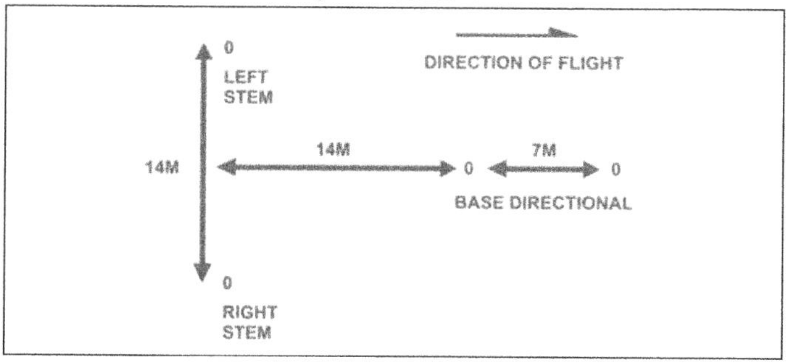


8-3. **AIR ASSAULT FORMATIONS**. Aircraft supporting an operation may use any of the following PZ/LZ configurations which are prescribed by the air assault task force (AATF) commander working with the air mission commander (AMC):

a. ***Heavy Left or Right***. Requires a relatively long, wide landing area; presents difficulty in pre-positioning loads; restricts suppressive fire by inboard gunners; provides firepower to front and flank (Figure 8-2).

Figure 8-2. HEAVY LEFT/HEAVY RIGHT

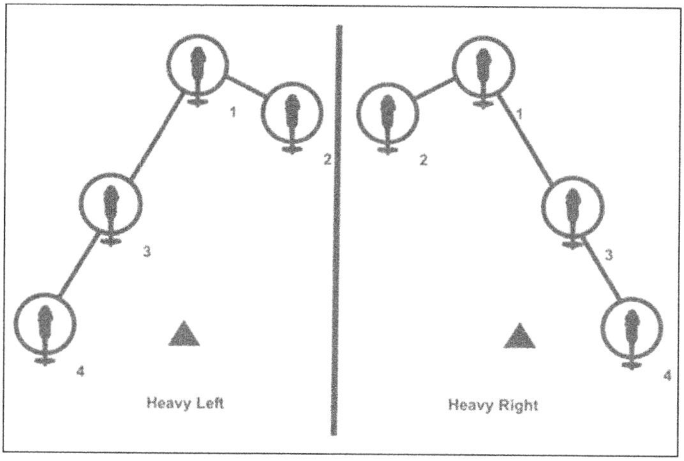

b. **Diamond**. Allows rapid deployment for all-round security; requires small landing area; presents some difficulty in pre-positioning loads; restricts suppressive fire of inboard gunners (Figure 8-3).

Figure 8-3. DIAMOND

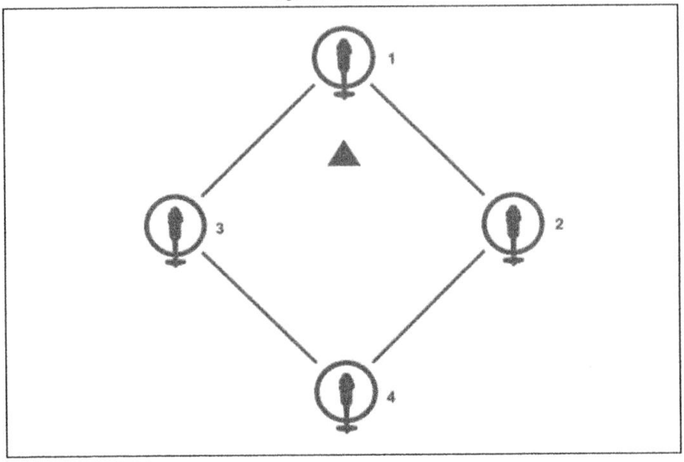


c. **Vee**. Requires a relatively small landing area; allows rapid deployment of forces to the front; restricts suppressive fire of inboard gunners; presents some difficulty in pre-positioning loads (Figure 8-4).

Figure 8-4. VEE

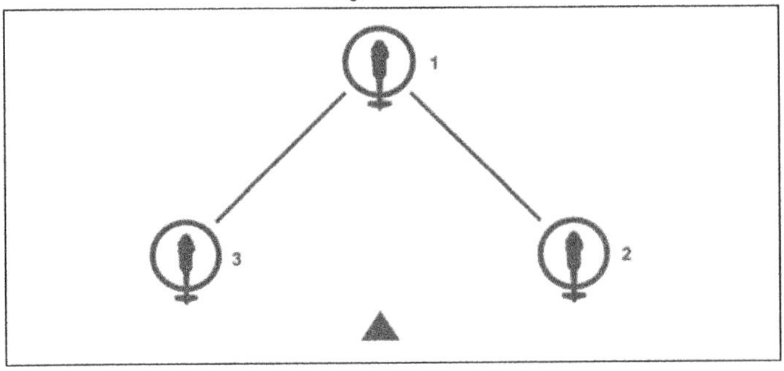

d. **Echelon Left or Right**. Requires a relatively long, wide landing area; presents some difficulty in pre-positioning loads; allows rapid deployment of forces to the flank; allows unrestricted suppressive fire by gunners (Figure 8-5).

**Figure 8-5. ECHELON LEFT/ECHELON RIGHT**

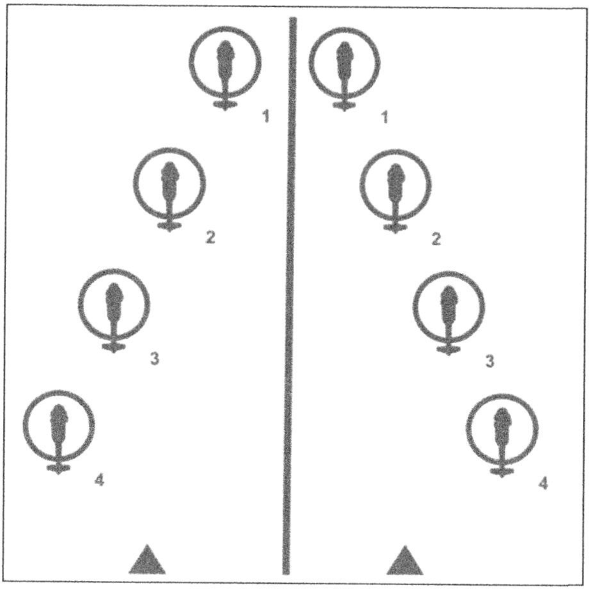

e. **Trail**. Requires a relatively small landing area; allows rapid deployment of forces to the flank; simplifies pre-positioning loads; allows unrestricted suppressive fire by gunners (Figure 8-6).

**Figure 8-6. TRAIL**

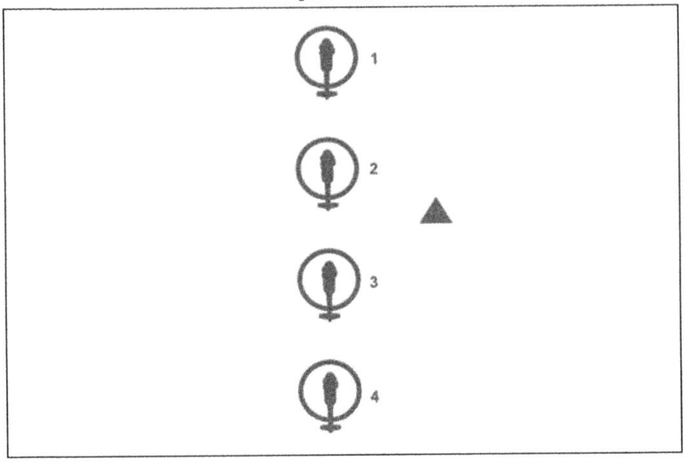

f. **Staggered Trail Left or Right**. Requires a relatively long, wide landing area; simplifies pre-positioning loads; allows rapid deployment for all round security; gunners' suppressive fire restricted somewhat (Figure 8-7).

**Figure 8-7. STAGGERED TRAIL LEFT/STAGGERED TRAIL RIGHT**

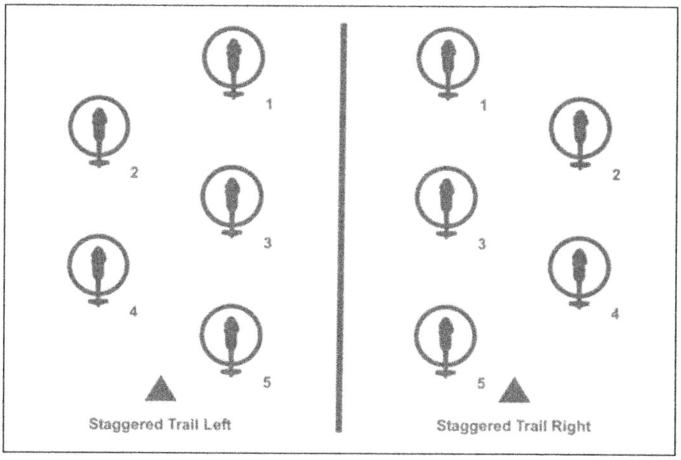

8-4. **PICKUP ZONE OPERATIONS**. Prior to arrival of aircraft, the PZ is secured, PZ control party is positioned, and the troops and equipment are positioned in platoon/squad assembly areas. In occupying a patrol/squad assembly area, the patrol/squad leader does the following. Figure 8-8 shows an example of a large, one-sided PZ. Figures 8-8 through 8-12 show loading/unloading procedures and techniques:

- Maintain all-around security of the assembly area.
- Maintain communications.
- Organize personnel and equipment into chalks and loads.
- Conduct safety briefing and equipment check of troops.

**Figure 8-8. LARGE, ONE SIDED PZ**

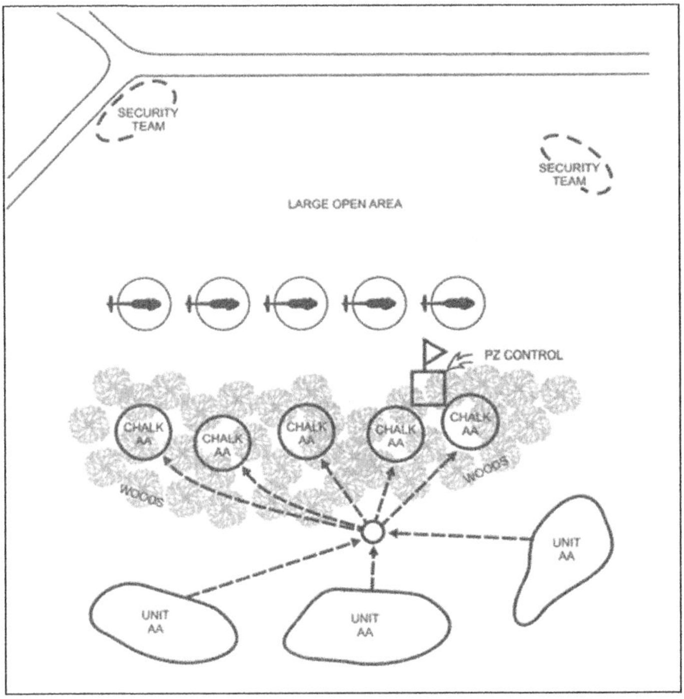

Figure 8-9. UH-60 LOADING SEQUENCE

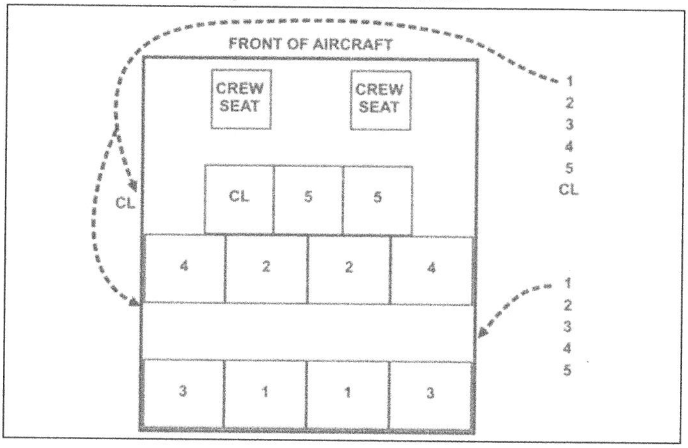


Figure 8-10. UNLOADING SEQUENCE

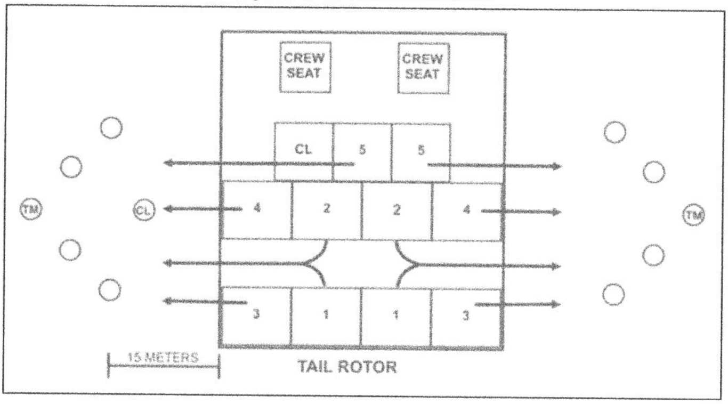

Figure 8-11. TACTICAL LOADING SEQUENCE

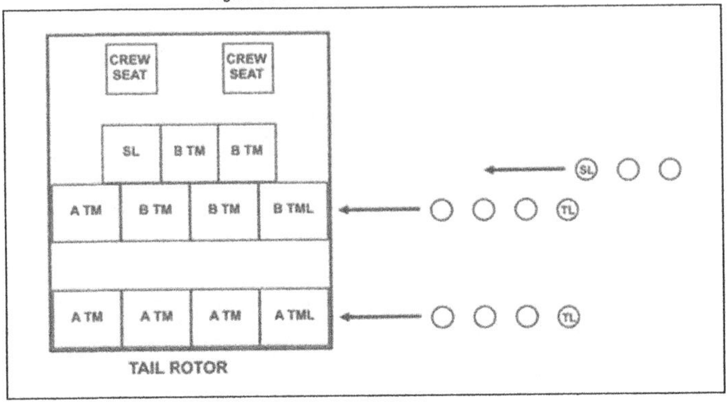


Figure 8-12. TACTICAL UNLOADING USING DOOR NEAREST COVER, CONCEALMENT

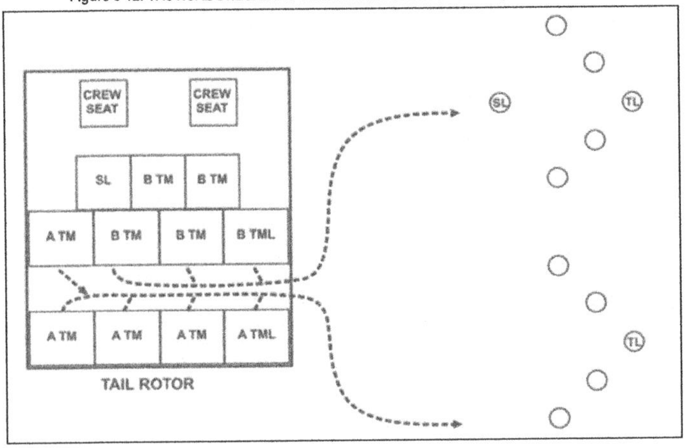

8-5. **SAFETY**. Safety is the primary concern of all leaders when operating in/around aircraft. The inclusion of aircraft into Ranger operations carries an inherent high-risk factor. The following guidelines are to be considered.

a. Approach the aircraft from 90 degrees to 45 degrees off the nose.

b. Weapons with blank firing adapters are pointed muzzles up.

c. Weapons loaded with live ammunition are muzzles down.

d. Wear the ballistic helmet.

e. When possible, conduct an air crew safety brief with all personnel. At a minimum, cover loading/off-loading, emergencies, and egress procedures.

f. Leaders need to carry a manifest and turn a copy into higher.

8-6. **CLOSE COMBAT ATTACK AVIATION**. Close combat attack is defined as a hasty or deliberate attack in support of units engaged in close combat. During CCA, armed helicopters engage enemy units with direct fire that impacts near friendly forces. Targets may range from a few hundred meters to a few thousand meters. Close combat attack is coordinated and directed by a team, platoon or company level ground unit soldiers using standardized CCA procedures in unit SOP's.

a. **Aircraft Capabilities and Limitations**.

(1) AH-64D (Apache):

- Air speed (knots): 164 max, 120 Cruise combat radius: 200 KM
- 30 mm chain gun: 1200 rds, 3,500 m range
- 2.75" rockets: 19 per pod (76), 3-5 km range
- Hellfire Missile: 8 per side (16), 5-8 km range

(2) Capabilities.

- MOBILITY - AO can cover corps or division area
- SPEED - 100-120 knots day/80-100 Knots night
- VERSATILITY - scout-weapons teams vs. pure attack
- LETHALITY - Attack bn can engage 288 targets
- VIDEO RECONNAISSANCE - Provide near real time intelligence

(3) ***Limitations***.

- Threat ID with FLIR
- Low ceilings (Clouds) <500' AGL degrades hellfire capability
- Combat service support consume large amounts of class III, V, and IX

(4) **OH-58D (Kiowa)**.

- Air speed (knots): 125 max, 100 cruise combat radius: 120 KM
- .50 cal MG: 500 rds, 2,000 m range
- 2.75" rockets: 7 per pod (14), 3-5 km range
- Hellfire: 2 per side (4), 5-8 km range
- Stinger: 2 per side (4), 4 km range

(5) ***Capabilities***.

- Mast-mounted sight
- Ability to designate targets while remaining masked
- Thermal imaging system (day/night)
- Laser designator / Aim1 laser
- Video image crosslink (VIXL)
- Moving map display
- Video

(6) ***Limitations***.

- Power limited
- Infrared crossover
- Battlefield obscurants
- Low ceilings - Hellfire
- Remote designation constraints
- Instrument meteorological conditions

c. **CCA Call-for-Fire Format**.

(1) IP/BP/ABF or friendly location________
- Grid/Lat Long

(2) Known point
- Terrain feature

(3) HDG to TGT ________ (magnetic)
- Specify from IP/BP/ABF or friendly location

(4) DST to TGT ________ (m) (Specify from IP/BP/ABF or friendly location.)

(5) TGT elevation________ (feet, mean sea level)

(6) TGT Description ________________

(7) TGT location ____________
- Grid
- Known point
- Terrain Feature

(8) Type TGT Mark ________ (day/night)

(9) location of friendly ________
- Omit if previously given
- Grid/Lat Long
- Known point
- Terrain feature

(10) Egress direction__________
- Cardinal direction not to over fly threats

d. **Marking Method Chart**. Table 8-1 shows this chart.

**Table 8-1. MARKING METHOD CHART**

| Method | D | N | NVG | NVS | Friendly | Target |
|---|---|---|---|---|---|---|
| Smoke | X | | | | X | X |
| Smoke IR | X | | X | | X | X |
| Signal Mirror | X | | | | X | |
| IR Laser | | | X | | X | X |
| Tracer | X | X | X | | | X |
| Glint Tape | | | | X | X | |
| CID | X | | | | X | |
| Strobe | | X | X | | X | |
| IR Strobe | | | X | | X | |
| IR Panel | | | | X | X | |
| Chem Light | | X | X | | X | |
| AN/PAQ-4 | | | X | | | X |
| VS-17 Panel | X | | | | X | |
| Spot Light | | X | X | | X | |
| MRE Heater | | | | X | X | |
| AN/PEQ-2 | | | X | | | X |
| Hydra 70 Illum | X | X | X | X | | X |
| Ref FM 3-04.111 Appendix Q for more detail | | | | | | |

e. **Risk-Estimate Distances**. REDs for CCA munitions are shown in Table 8-2.

**Table 8-2. RISK ESTIMATE DISTANCES FOR CCA MUNITIONS**

| *WEAPON* | *DANGER CLOSE* | *MAX EFF RANGE)* |
|---|---|---|
| 50 cal | 150 meters | 2,000 meters |
| 30 mm | 150 meters | 3,500 meters |
| 2.75 rocket | 175 meters | 8,000 meters |
| Hellfire Missile | 500 meters | 8,000 meters |
| Danger Close = Risk-Estimate Distances for 0.1% (1/10th of a percent) probability of incapacitation (PI) per FM 3-09.32. | | |

# Chapter 9
# WATERBORNE OPERATIONS

This chapter discusses rope bridges, poncho rafts, and watercraft.

9-1. **ROPE BRIDGE TEAM.** The Ranger patrol seldom has ready-made bridges, so they must know how to employ field-expedient stream crossing techniques.

a. **Personnel.**

(1) # 1 man................Lead safety swimmer and far side lifeguard.
(2) # 2 man................Rope puller, swims water obstacle pulling 150-foot rope, ties off rope on far side anchor point.
(3) # 3 man................Near side lifeguard is the last man to cross water obstacle.
(4) # 4 man................Bridge team commander (BTC; most knowledgeable person on team).
(5) # 5/6 men...........Rope tighteners.

b. **One-Rope Bridge (Wet Crossing).** Special equipment follows:

- Two snaplinks per piece of heavy equipment.
- Two snaplinks for every 120 feet of rope.
- One 14-foot utility rope per person (Swimmer safety line).
- Two snaplinks per person.
- One waterproof bag per RTO.
- Two B-7 life preservers.
- Three flotation work vests.
- Two 150-foot nylon ropes.

c. **Planning**. A stream crossing annex is prepared in conjunction with the unit's operation order. Special organization is accomplished at this time. For a platoon size patrol, a squad is normally given the task of providing the bridge team. The squad leader designates the most technically proficient soldier in the squad as the bridge team commander (BTC).

d. **Rehearsals and Inspections**.

(1) Rehearse the entire stream crossing, emphasizing--

- Security and actions on enemy contact.
- Actual construction of the rope bridge within the 8-minute time standard on dry land.
- Individual preparation.
- Order of crossing.
- All signals and control measures.
- Reorganization.

(2) Conduct rehearsals as realistically as possible.

(3) Ensure personnel are proficient in the mechanics of a stream crossing operation.

(4) Inspect for equipment completeness, correct rigging and preparation, personnel knowledge and understanding of the operation.

(5) Actions of the # 4 man (BTC) during the preparation phase.

(a) Rehearse the bridge team.
(b) Account for all equipment in the bridge kit.
(c) Ensure 150-foot rope is coiled.

e. **Execution Phase.**

(1) Steps for the establishment and conduct of bridge stream crossing:

(a) Leader halts short of the river, local security is established, and a recon is conducted of the area for the presence of the enemy and for crossing site suitability/ necessity. He directs the BTC to construct the bridge.

(b) Security is established up and downstream while unit leader briefs BTC on anchor points. The unit leader counts individuals across.

(c) The bridge team begins to establish the rope bridge while unit members begin individual preparation.

(d) The BTC is responsible for construction of one-rope bridge and selection of the near side anchor point as well as the far side anchor point if visibility permits. He will tie a swimmer's safety line to anchor himself into the bridge. Tying an end of the line bowline around the waist secured with an overhand knot, and on the free running end, an end of line bowline with an overhand knot. A snaplink will be placed on the loop portion of the end of line bowline, which will extend no further than arms length away from the soldier (standard waterborne uniform). This will ensure that the soldier is never more than one arms length away from the rope bridge should he lose his grip.

(e) Noise and light discipline are enforced and security is maintained.

(f) Individual soldiers put a snaplink in their end of the line bowline and the front sight assembly of every M-4/M16 or M203. M240 gunners put a snaplink through the front sight assembly and rear swivel of their M240 MG. RTOs (and others with heavy rucksacks) place an additional snaplink on their rucksack frame, top center.

(2) The responsibilities of the bridge team while constructing the rope bridge are as follows:

(a) # 1 man (Lead safety swimmer and far side lifeguard). He grounds his rucksack (with snaplink through top of frame) to the rear of the near side anchor point. He wears equipment in the following order (body out). He carries a safety line to assume duties of far side lifeguard:

-- Waterborne uniform (top button buttoned, pants unbloused)
-- B7 life preserver or engineer work vest
-- LBV
-- Weapon (across the back)

(b) # 1 man enters the water upstream from the # 2 man and stays an arms length away from the # 2 man on the upstream side. He identifies the far side anchor point upon exiting the water and once the # 2 man has exited the water moves to his far side lifeguard position downstream of the rope bridge with knotted safety line on wrist, LBV LBV/weapon grounded, and work vest held in throwing hand. He continues to wear the B7.

(c) # 2 man (rope puller) in waterborne uniform (same as # 1 man) wears his equipment in the following order. He grounds his rucksack (with snaplink through top of frame) to the rear of the near side anchor point. His duties are to swim across the water obstacle pulling the rope. He ties off the rope on the anchor point identified by the # 1 man with a round turn and two half hitches with a quick release. The direction of the round turn is the same direction as the flow of water (current) to facilitate exit off the rope bridge:

- Work vest.
- LBV.
- Weapon (across the back).
- Swimmer safety line.

(d) # 3 man (near side lifeguard) in the same waterborne uniform as the far side lifeguard. The # 3 man positions himself on the downstream side of the bridge before the # 1/2 men enter the water. He (# 1 man) grounds his rucksack (with snaplink through top of frame) on rear of near-side anchor point. After the PSG crosses and verifies the headcount, the # 1 man unties the quick release at the near side anchor point. The # 3 man reties his safety line into an Australian rappel seat, hooks the end of line bowline into his snaplink, and connects his snaplink to the one on the end-of-line bowline . # 3 man is the last pulled across the water obstacle. Before crossing the water obstacle, he dons his equipment in the following order:

- B7.
- Work vest.
- LBV.
- Weapon.

(e) # 4 man (BTC). He is in the standard waterborne uniform with LBV and sling rope tied in safety line (round the waist bowline with end of line bowline no more than one arms length). He is responsible for construction of rope bridge and organization of bridge team. He is also responsible for back feeding the rope and tying end of line bowlines. He designates the near side anchor point, ties the Figure Eight slip of the transport tightening system, and hooking all personnel to the rope bridge. He ensures that the transport tightening knot is on the upstream side of the rope bridge. He ensures that all individuals are in the waterborne uniform, hooked into the rope facing the current with the safety line routed through the trailing shoulder of the individual's LBV and rucksack. He ensures that the weapon is hooked onto the rope. He controls the flow of traffic on the bridge. He is responsible for crossing with the # 1 man's rucksack. He is generally the next to the last man to cross (follows PSG who is keeping head count).

(f) # 5 and # 6 men (rope tighteners) wear the waterborne uniform with LBV and safety line. They tighten the transport tightening knot. They also take the rucksacks of the # 2 and # 3 men across. Once they reach the far side, # 5 and # 6 men pull the last man (# 3 man) across.

(g) The rucksacks of # 1/2/3 men are transported across by # 4/5/6 men. The rucksacks of the # 1/2/3 men are hooked into the rope by the snaplink through the top of the frames and the # 4/5/6 men pull them across. The weapons of # 4/5/6 men are attached between the # 4/5/6 men and the rucksack that they are pulling across the bridge.

(h) BTC rehearses the bridge team during the planning sessions and directs construction and emplacement. The unit leader selects the crossing site which complements the tactical plan.

- # 3 man positions himself downstream of crossing site.
- # 1 man enters water upstream of # 2. He stays one arm's length from the # 2 man and is prepared to render any assistance to the # 2 man. Both swim in conjunction upstream to compensate for the current. BTC feeds rope out of rucksack positioned on the downstream side of near side anchor point.

(i) The # 1 man exits and identifies the far side anchor point (if BTC cannot identify it for the # 2 man). # 2 man exits on the upstream side of the far side anchor point. The rope is now routed to facilitate movement onto and off the bridge.

(j) Radios and heavy equipment are waterproofed and rigged. All individuals don waterborne uniform and tie safety lines. PSG moves to anchor point and maintains accountability through headcount.

(k) The # 2 man signals the BTC that the rope is temporarily attached to the far side anchor point, and the BTC pulls out excess slack and ties the transport tightening system using a Figure Eight slip. The BTC signals the # 2 man to pull the knot twelve to fifteen feet from the near side anchor point. After this is done the # 2 two man ties round turns 18 to 24 inches off the water with the remaining rope and secures the rope, to itself, with a snaplink. The # 2 man signals the BTC and the pulling team (# 4/5/6) tightens the bridge, pulling the transport tightening system as close as possible to the near side anchor point.

(l) The # 1 man moves downstream and assumes his duties as the far side lifeguard. The bridge team commander will tie off the rope with a round turn and two half hitches around the near side anchor point. The BTC will place himself on the upstream side of the bridge (facing downstream). He begins to hook individuals into the rope inspecting them for safety.

***Note:*** Any Ranger identified as a weak swimmer will cross with no other personnel on the rope bridge. The weak swimmer crosses individually to allow the near and far side lifeguards to focus their attention exclusively on the weak swimmer and not be distracted by other personnel crossing the bridge.

(m) The # 2 man moves to the upstream side of the rope bridge, assists personnel off the rope on the far side, and maintains head count. The # 5 and # 6 men cross with the # 1 and # 2 men's rucksacks..

(n) The BTC maintains the flow of traffic ensuring that no more than three men are on the bridge at any one time (one hooking up, one near the center, and one being unhooked). Once the PSG has accounted for everyone on the near side, he withdraws left and right (L/R) security and sends them across. PSG follows security across. The # 3 man hooks the BTC (with # 3's ruck) onto the rope. Once the BTC has crossed, the # 3 man unhooks near side anchor point and the BTC unties far side anchor point. The # 3 man ties an Australian rappel seat with snaplink to the front, hooks onto the snaplink that is in the end of the line bowline on the 120-foot rope, and then signals the # 4/5/6 men to take in slack. The # 3 man extends arms in front of his head, slightly upstream to fend off debris and is pulled across by the # 4/5/6. All individuals (except # 1/2/3 and RTOs) wear rucksacks across. The # 4/5/6 men hook the rucksacks of the # 1/2/3 men onto the bridge by the snaplink. All the men cross facing upstream.

(o) Once the far side headcount, weapons and equipment are verified (between PSG and # 5 man), personnel reorganize, and then they continue the mission.

(p) Personnel with heavy equipment:

**M240** All major groups are tied together with 1/4-inch cord. An anchor line bowline runs through the rear swivel, down the left side of gun. Tie a round turn through the trigger guard. Route the cord down the right side and tie off two half hitches around the forearm assembly with a round turn and two half hitches through the front sight posts. The remainder of the working end is tied off with an end of the rope bowline approximately one foot from the front sight post large enough to place leading hand through. The M240 is secured to the bridge by snaplinks on the front sight post and rear swivel. The M240 is pulled across by the trailing arm of the M240 gunner.

**AN-PRC/II9s** These are waterproofed before crossing a one-rope bridge. Once far side FM commo is established, the near side RTO breaks down and waterproofs his radio, and prepares to cross the bridge. He puts a snaplink in the top center of the rucksack frame (same as for # 1/2/3 men). The BTC will hook the rucksack to the rope.

***Note:*** Using two snaplinks binds the load on the rope. Adjust arm straps all the way out. RTO pulls radio across the rope bridge.

9-2. **PONCHO RAFT.** Normally a poncho raft is constructed to cross rivers and streams when the current is not swift. A poncho raft is especially useful when the unit is still dry and the platoon leader desires to keep the individuals equipment dry.

a. **Equipment Requirements**.

(1) Two serviceable ponchos.

(2) Two weapons (poles can be used in lieu of weapons).

(3) Two rucksacks per team.

(4) 10 feet of utility cord per team.

(5) One sling rope per team.

b. **Conditions**. Poncho rafts are used to cross water obstacles when any or all of the following conditions are found:

(1) The water obstacle is too wide for 150-foot rope.

(2) No sufficient near or far shore anchor points are available to allow rope bridge construction.

(3) Under no circumstances will poncho rafts be used as a means to cross a water obstacle if an unusually swift current is present.

c. **Choosing a Crossing Site**. Before a crossing site is used, a thorough reconnaissance of the immediate area is made. Analyzing the situation using METT-TC, the patrol leader chooses a crossing site that offers as much cover and concealment as possible and has entrance and exit points that are as shallow as possible. For speed of movement, it is best to choose a crossing site that has near and far shore banks that are easily traversed by an individual Ranger.

d. **Execution Phase**. Steps for the construction of a poncho raft:

(1) Pair off the unit/patrol in order to have the necessary equipment.

(2) Tie off the hood of one poncho and lay out on the ground with the hood up.

(3) Weapons are then placed in the center of the poncho, approximately 18 inches apart, muzzle to butt.

(4) Next, rucksacks and LBV are placed between the weapons with the two individuals placing their rucksacks as far apart as possible.

(5) The two will then start to undress (bottom to top), first with their boots, taking the laces completely out for subsequent use as tie downs if necessary).

(6) The boots are then placed over muzzle/butt of weapon toe in.

(7) Members continue to undress, folding each item neatly and placing it on top of their boots.

(8) Once all of the equipment is placed between the two weapons or poles, the poncho is snapped together. The snapped portion of the poncho is then lifted into the air and tightly rolled down to the equipment. Start at the center and work out to the end of the raft creating pigtails at the end. This is accomplished much easier if done by both soldiers together. The pigtailed ends are then folded in toward the center top of the raft and tied off with a single boot lace.

(9) The other poncho is then laid out on the ground with the hood up and the first poncho with equipment is placed in its center. The second poncho is then snapped, rolled and tied in the same manner as the first poncho. The third and fourth boot laces (or utility cord) are then tied around the raft approximately one foot from each end for added security. The poncho raft is now complete.

***Note:*** The patrol leader must analyze the situation using METT-TC and make a decision on the uniform for crossing the water obstacle such as weapons inside the poncho raft or slung across the back, remaining dressed or stripped down with clothes inside raft.

9-3. **WATERCRAFT.** Use of inland and coastal waterways may add flexibility, surprise, and speed to tactical operations. Use of these waterways will also increase the load carrying capacity of normal dismounted units. Watercraft are employed in reconnaissance and assault operations.

a. **Inflation Method**. Inflatable with foot pumps using four separate valves located on the inside of the buoyancy tubes. Each of the four valves are used to section off the assault boat into eight separate airtight compartments. To pump air into the boat, turn all valves into the "orange" or "inflate" section of the valve. Once the assault boat is filled with air, turn all valves onto the "green" or "navigation" section. This will section the assault boat into eight separate compartments.

b. **Characteristics**.

- Overall length: 15 feet, 5 inches.
- Overall width: 6 feet, 3 inches.
- Weight: 265 pounds.
- Maximum payload: 2,756 pounds.
- Crew : 1 coxswain + either 10 paddlers or a 65 HP short-shaft outboard motor.

c. **Preparation**.

(1) ***Rubber Boat***.

- All rubber boats will have a twelve-foot bowline secured to the front starboard D-Ring. This rope will be tied with an anchor-line bowline, and the knot will be covered with 100-MPH tape.
- All rubber boats will have a fifteen foot center line tied to the rear floor D-Ring. The same procedure for securing the bowline will be used for the centerline.
- All rubber boats will be filled to 240 millibars of air, and checked to ensure that all valve caps are tight, and set in the NAVIGATE position.
- All rubber boats will be equipped with one foot pumps. These will be placed in the front pouches of the rubber boat. If no pouches are present, the foot pumps will be placed on the floor.
- All rubber boats will be inspected with the maintenance chart.

(2) ***Personnel and Equipment***.

- All personnel will wear work vest or kapoks (or another suitable positive flotation device).
- LBV is worn over the work vest, unbuckled at the waist.
- Individual weapon is slung across the back, muzzle pointed down and facing toward the inside of the boat.
- Crew served weapons, radios, ammunition and other bulky equipment is lashed securely to the boat to prevent loss if the boat should overturn. Machine guns with hot barrels are cooled prior to being lashed inside the boats.
- Radios and batteries are waterproofed.
- Pointed objects are padded to prevent puncturing the boat.

NOTES

d. **Positions.** Assign each individual a specific boat position (Figure 9-1).

e. **Duties.**

(1) Designate a commander for each boat, (normally coxswain)

(2) Designate a navigator (normally a leader within the platoon) - observer team as necessary.

(3) Crew is positioned as shown in Figure 9-2.

(4) Duties of the Coxswain.

(a) Responsible for control of the boat and actions of the crew.

(b) Supervises the loading, lashing and distribution of equipment.

(c) Maintains the course and speed of the boat.

(d) Gives all commands.

(5) # 2 paddler (long count) is responsible for setting the pace.

(6) # 1 paddler is the observer, stowing and using the bowline unless another observer is assigned.

f. **Embarkation and Debarkation Procedures.**

(1) When launching, the crew will maintain a firm grip on the boat until they are inside it: similarly, when beaching or debarking, they hold on to the boat until it is completely out of the water. Loading and unloading is done using the bow as the entrance and exit point.

(2) Keep a low center of mass when entering and existing the boat to avoid capsizing. Maintain 3 points of contact at all times.

(3) The long count is a method of loading and unloading by which the boat crew embarks or debarks individually over the bow of the boat. It is used at river banks, on loading ramps, and when deep water prohibits the use of the short count method.

(4) The short count is a method of loading or unloading by which the boat crew embarks or debarks in pairs over the sides of boat while the boat is in the water. It is used in shallow water allowing the boat to be quickly carried out of the water. The short count method of organization is primarily used during surf operations.

(5) Beaching the boat is a method of debarking the entire crew at once into shallow water and quickly carrying the boat out of the water.

g. **Commands.** Commands are issued by the coxswain to ensure the boat is transported over land and controlled in the water. All crew members learn and react immediately to all commands issued by the coxswain. T he various commands are as follows:

(1) "Short Count...count off," Crew counts off their position by pairs, such as 1,2,3,4,5 (passenger # 1, # 2, if applicable) coxswain.

(2) "Long Count...count off," Crew counts off the position by individual, such as 1,2,3,4,5,6,7,8,9,10, (Passenger # 1, # 2, if applicable), coxswain.

(3) "Boat Stations," Crew takes positions along side the boat.

(4) "High Carry...Move," (used for long distance move overland).

(a) On the preparatory command of "High Carry," the crew faces the rear of the boat and squats down grasping carrying handles with the inboard hand.

(b) On the command "Move," the crew swivels around, lifting the boat to the shoulders so that the crew is standing and facing to the front with the boat on their inboard shoulders.

(c) Coxswain guides the crew during movement.

(5) "Low Carry...Move," (used for short distance moves overland).

(a) On preparatory command of "Low Carry," the crew faces the front of the boat, bent at the waist, and grasps the carrying handles with their inboard hands.

(b) On the command of "Move," the crew stands up straight raising the boat approximately six to eight inches off the ground.

(c) Coxswain guides the crew during movement.

(6) "Lower the Boat---Move," Crew lowers the boat gently to the ground using the carrying handles.

(7) "Give Way Together," Crew paddles to front with # 2 setting the pace.

(8) "Hold," Entire crew keeps paddles straight downward motionless in the water stopping the boat.

(9) "Left side hold (Right)," Left crew holds, right continues with previous command.

(10) "Back paddle," Entire crew paddles backward propelling the boat to the rear.

Figure 9-1. BOAT POSITIONS

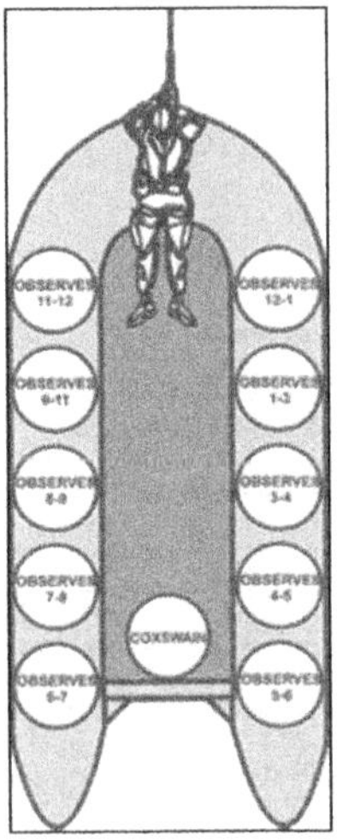


Figure 9-2. CREW POSITIONS, LONG COUNT AND SHORT COUNT

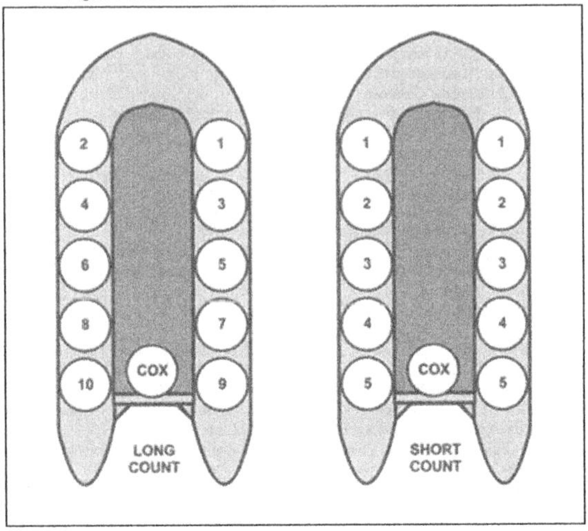


(11) "Back Paddle Left" (Right), Left crew back paddles causing the boat to turn left, right crew continues with previous command.

(12) "Rest Paddles," Crew members place paddles on their laps with blades outboard. This command may be given in pairs such as "# 1s, rest paddles").

h. **Securing of Landing Site.**

(1) If the landing site cannot be secured prior to the waterborne force landing, some form of early warning, such as scout swimmers, is considered. These personnel swim to shore from the assault boats and signal the boats to land. All signals and actions are rehearsed prior to the actual operation.

(2) If the patrol is going into an unsecured landing site, it can provide security by having a security boat land, reconnoiter the landing site and then signal to the remaining boats to land. This is the preferred technique.

(3) The landing site can be secured by force with all the assault boats landing simultaneously in a line formation. While this is the least desirable method of securing a landing site, it is rehearsed in the event the tactical situation requires its use.

(4) Arrival at the debarkation point.

(a) Unit members disembark according to leaders order (Figure 9-3).

(b) Local security is established.

(c) Leaders account for personnel and equipment.

(d) Unit continues movement.

- Soldiers pull security initially with work vest on.
- Coxswains and two men unlash and de-rig rucksacks.
- Soldiers return in buddy teams to secure rucksack and drop off work vest.
- Boats are camouflaged/cached if necessary prior to movement.

Figure 9-3. DEBARKATION

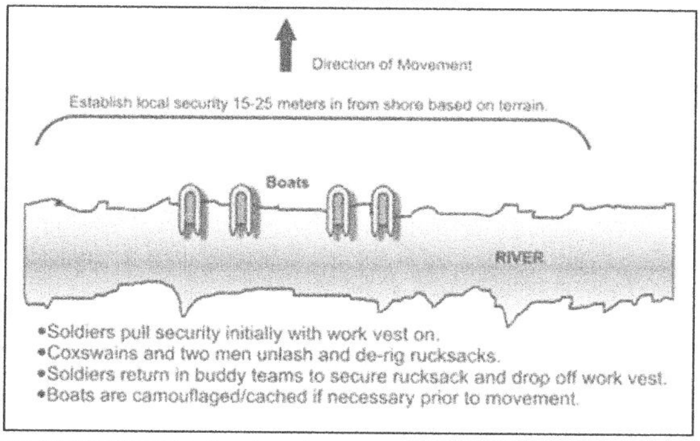


i. **Capsize Drill.** The following commands and procedures are used for capsize drills or to right an over-turned boat,

(1) ***PREPARE TO CAPSIZE***. This command alerts the crew and they raise paddles above their heads, with the blade pointed outward. Before capsizing, the coxswain will conduct a long count.

(2) ***PASS PADDLES***. All paddles are passed back and collected by the # 9/10 men.

(3) ***CAPSIZE THE BOAT***. All personnel slide into the water except the # 3/5/7 men. The # 1 man secures the bowline. The three men in the water grasp the capsize lines (ensuring the lines are routed under the safety lines) and stand on the buoyancy tubes opposite the capsize lines anchor points. The boat is then turned over by the # 3/5/7 men, who lean back and straighten their legs while pulling back on the capsize lines. As the boat lifts off the water, the # 4 man grasps the center carrying handle and rides the boat over. Once the boat is over, the # 4 man helps the # 3/7 men back onto the boat. The # 5 man holds onto the center carrying handle and turns the boat over the same way. The # 5 man rides the boat back over and helps the rest of the crew into the boat. As soon as the boat is capsized, the coxswain commands a long count to ensure that no one sank or was trapped under the boat. Every time the boat is turned over, he conducts another long count.

j. **River Movement.**

(1) ***Characteristics of River***.

(a) Know local conditions prior to embarking on river movement.

(b) A bend is a turn in the river course.

(c) A reach is a straight portion of river between two curves.

(d) A slough is a dead end branch from a river. They are normally quite deep and can be distinguished from the true river by their lack of current

(e) Dead water is a part of the river, due to erosion and changes in the river course that has no current. Dead water is characterized by excessive snags and debris.

(f) An island is usually a pear-shaped mass of land in the main current of the river. Upstream portions of islands usually catch debris and are avoided.

(g) The current in a narrow part of a reach is normally greater than in the wide portion.

(h) The current is greatest on the outside of a curve; sandbars and shallow water are found on the inside of the curve.

(i) Sandbars are located at those points where a tributary feeds into the main body of a river or stream.

(j) Because the # 1 and # 2 men are sitting on the front left and right sides of the boat, they observe for obstacles as the boat moves downriver. If either notices an obstacle on either side of the boat, he notifies the coxswain. The coxswain then adjusts steering to avoid the obstacle.

(2) ***Navigation***. The patrol leader is responsible for navigation. The three acceptable methods of river navigation are--

(a) Checkpoint and general route. These methods are used when the drop site is marked by a well-defined checkpoint and the waterway is not confused by a lot of branches and tributaries. They are best used during daylight hours and for short distances.

(b) Navigator-observer method. This method is the most accurate means of river navigation and is used effectively in all light conditions.

(c) *Equipment Needed*.

- Compass
- GPS
- Photo map (1st choice)
- Topographical map (2nd choice)
- Poncho (for night use)
- Pencil/Grease pencil
- Flashlight (for night use)

(d) *Procedure*. Navigator is positioned in center of boat and does not paddle. During hours of darkness, he uses his flashlight under the poncho to check his map. The observer (or # 1 man) is at the front of the boat.

(a) The navigator keeps his map and compass oriented at all time.

(b) The navigator keeps the observer informed of the configuration of the river by announcing bends, sloughs, reaches, and stream junctions as shown on his map.

(c) The observer compares this information with the bends, sloughs, reaches, and stream junctions he actually sees. When these are confirmed, the navigator confirms the boat's location on his map.

(d) The navigator also keeps the observer informed of the general azimuths of reaches as shown on his map and the observer confirms these with actual compass readings of the river.

(e) The navigator announces only one configuration at a time to the observer and does not announce another until it is confirmed and completed.

(f) A strip-map drawn on clear acetate backed by luminous tape may be used. The drawing is to scale or a schematic. It should show all curves and the azimuth and distance of all reaches. It may also show terrain features, stream junctions, and sloughs.

k. **Formations.** Various boat formations are used (day and night) for control, speed, and security (Figure 9-4). The choice of which is used depends on the tactical situation and the discretion of the patrol leader. He should use hand and arm signals to control his assault boats. The formations are:

- Wedge
- Line
- File
- Echelon
- Vee

**Figure 9-4. FORMATIONS**

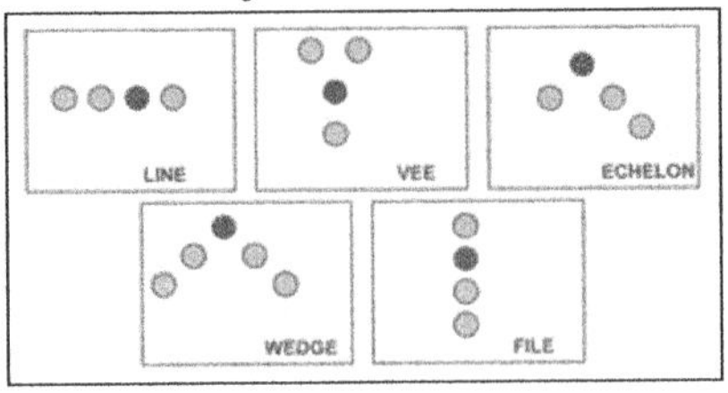

# Chapter 10
# MILITARY MOUNTAINEERING

In the mountains, commanders face the challenge of maintaining their units' combat effectiveness and efficiency. To meet this challenge, commanders conduct training that provides Rangers with the mountaineering skills necessary to apply combat power in a rugged mountain environment, and they develop leaders capable of applying doctrine to the distinct characteristics of mountain warfare.

10-1. **TRAINING**. Military mountaineering training provides tactical and fundamental mobility skills needed to move units safely and efficiently in mountainous terrain. Rangers should receive extensive training prior to executing combat operations in mountainous environments. Some of the areas to train are as follows:

- Characteristics of the mountain environment.
- Care and use of basic mountaineering equipment.
- Mountain bivouac techniques.
- Mountain communications.
- Mountain travel and walking techniques.
- Mountain navigation, hazard recognition and route selection.
- Rope management and knots.
- Natural and artificial anchors.
- Belay and rappel techniques.
- Installation construction and use such as rope bridges.
- Rock climbing fundamentals.
- Rope bridges and lowering systems.
- Individual movement on snow and ice.
- Mountain stream crossings (to include water survival techniques).

10-2. **DISMOUNTED MOBILITY**. Movement in more technical terrain demands specialized skills and equipment. Before Rangers can move in such terrain, a technical mountaineering team might have to secure the high ground.

10-3. **MOUNTAINEERING EQUIPMENT**. Mountaineering equipment is everything that allows the trained Ranger to accomplish many tasks in the mountains. The importance of this gear to the mountaineer is no less than that of the rifle to the Infantry Soldier.

a. **Ropes and Cords**. Ropes and cords are the most important pieces of mountaineering equipment. They secure climbers and equipment during steep ascents and descents. They are also used to install other ropes and to haul equipment. From WWII until the 1980's, US armed forces mainly used 7/16 *nylon laid rope*, often referred to as *green line* for all mountaineering operations. Since kernmantle ropes were introduced, other, more specialized ropes have in many cases replaced the all-purpose green line. Kernmantle ropes are made much like parachute cord, with a smooth sheath around a braided or woven core. Laid ropes are still used, but should be avoided whenever rope failure could cause injury or loss of equipment.

(1) ***Types***. Kernmantle rope is classified as *dynamic* or *static*:

(a) *Dynamic Ropes*. Ropes used for climbing are classified as *dynamic ropes*. These rope stretch or elongate 8 to 12 percent once subjected to weight or impact. This stretching is critical in reducing the impact force on the climber, anchors, and belayer during a fall by softening the catch. An 11-mm by 150-meter rope is standard for military use. Other sizes are available.

(b) *Static Ropes*. Static ropes are used in situations where rope stretch is undesired, and when the rope is subjected to heavy static weight. Static ropes should never be used while climbing, since even a fall of a few feet could generate enough impact force to injure climber and belayer, and/or cause anchor failure. Static Ropes are usually used when constructing rope bridges, fixed rope installations, vertical haul lines, and so on.

(2) ***Sling Ropes and Cordelettes***. A short section of static rope or static cord is referred to as a sling rope or cordelette. These are a critical piece of personal equipment during mountaineering operations. They are usually 7 to 8mm diameter and up to 21 feet long. The minimum Ranger standard is 8mm by 15 feet.

(3) ***Rope Care***. Ropes used daily should be used no longer that one year. An occasionally used rope can be used generally up to 4 to 5 years if properly cared for.

- Inspect ropes thoroughly before, during, and after use for cuts, frays, abrasion, mildew, and soft or worn spots.
- Never step on a rope or drag it on the ground unnecessarily.
- Avoid running rope over sharp or rough edges (pad if necessary).

- Keep ropes away from oil, acids, and other corrosive substances.
- Avoid running ropes across one another under tension (nylon to nylon contact will damage ropes).
- Do not leave ropes knotted or under tension longer than necessary.
- Clean in cool water, loosely coil, and hang to dry out of direct sunlight. Ultraviolet light rays harm synthetic fibers. When wet, hang rope to drip-dry on a rounded wooden peg, at room temperature (do not apply heat).

(4) ***Webbing and Slings***. Loops of tubular webbing or cord, called slings or runners, are the simplest pieces of equipment and some of the most useful. The uses for these simple pieces are endless, and they are a critical link between the climber, the rope, carabiners, and anchors. Runners are usually made from either 9/16- or 1-inch tubular webbing and are either tied or sewn by a manufacturer.

b. **Carabiners**. The carabiner is one of the most versatile pieces of equipment available in the mountains. This simple piece of gear is the critical connection between the climber, his rope, and the protection attaching him to the mountain. New metal alloys are stronger to hold hard falls; and, they are lighter than steel, which allows the Ranger to carry many at once. Consequently, steel carabiners are used less and less often. Basic carabiners come in several different shapes.

c. **Protection**. This is the generic term used to describe a piece of equipment, natural or artificial, that is used to construct an anchor. Protection, along with a climber, belayer, and climbing rope, forms the lifeline of the climbing team. The rope connects two climbers, and the protection connects them to the rock or ice. The two types of artificial protection are traditional (removable) and fixed (usually permanent) (Figures 10-1 and 10-2).

**Figure 10-1. EXAMPLES OF TRADITIONAL (REMOVABLE) PROTECTION USED ON ROCK**

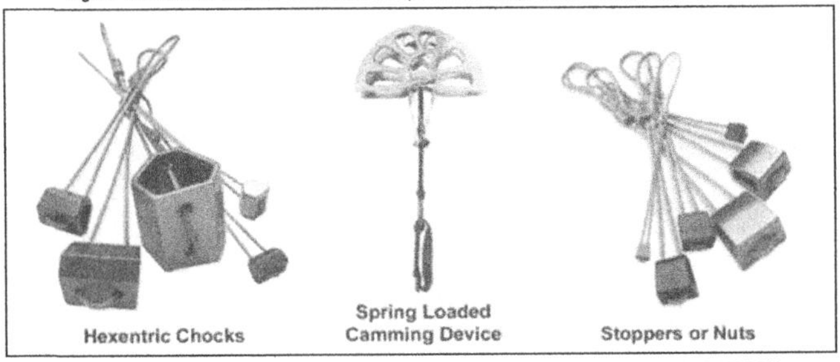


**Figure 10-2. EXAMPLES OF FIXED (PERMANENT OR SEMI) PROTECTION USED ON ROCK**

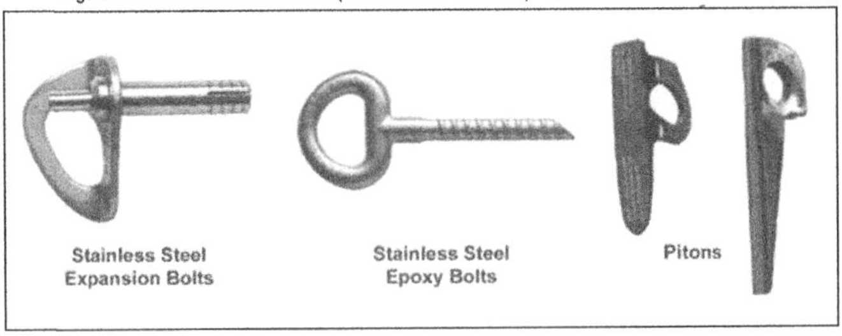

10-4. **ANCHORS**. Anchors form the base for all installations and roped mountaineering techniques. Anchors must be strong enough to support the entire weight of the load or of the impact placed upon them. Several pieces of artificial or natural protection may be incorporated together to make one multipoint anchor. Anchors are classified as artificial or natural:

a. **Artificial Anchors**. Artificial anchors are made from man-made materials. The most common ones use traditional or fixed protection (Figure 10-3).

**Figure 10-3. CONSTRUCTION OF THREE-POINT, PRE-EQUALIZED ANCHOR WITH FIXED ARTIFICIAL PROTECTION**

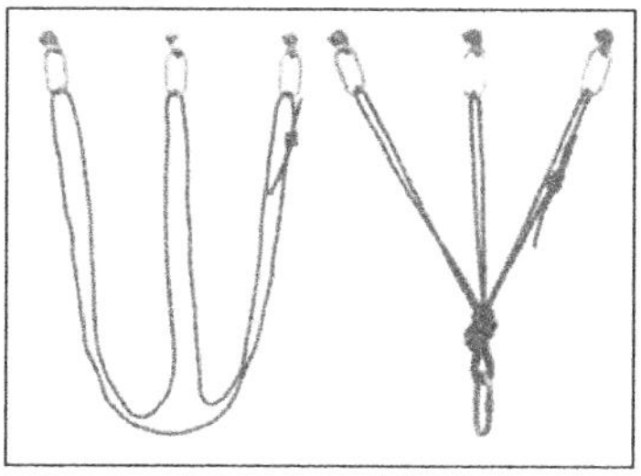

b. **Natural Anchors**. Natural anchors are usually very strong and often simple to construct with little equipment. Trees, shrubs, and boulders are the most common anchors. All natural anchors require is a method of attaching a rope. Regardless of the type of natural anchor used, it must support the weight of the load.

(1) ***Trees***. These are probably the most widely used of all anchors. In rocky terrain, trees usually have a very shallow root system. Check this by pushing or tugging on the tree to see how well it is rooted. Anchor as low as possible to keep the pull on the base of the tree, where it is nearer its own anchor and therefore stronger. Use padding on soft, sap-producing trees to keep sap off ropes and slings.

(2) ***Rock Projections and Boulders***. You can use these, but they must be heavy enough, and they must have a stable enough base to support the load.

(3) ***Bushes and Shrubs***. If no other suitable anchor is available, route a rope around the bases of several bushes. As with trees, place the anchoring rope as low as possible. Make sure all vegetation is healthy and well rooted to the ground.

(4) ***Tensionless Anchor***. Use this to anchor rope on high-load installations such as bridging. The wraps of the rope around the anchor (Figure 10-4) absorb the tension, keeping it off the knot and carabiner. Tie this anchor with at least four wraps around the anchor. A smooth anchor, such as a pipe or rail, might require several more wraps. Wrap the rope from top to bottom. Place a fixed loop into the end of the rope, and attach it loosely back onto the rope with a carabiner.

**Figure 10-4. TENSIONLESS NATURAL ANCHOR**

10-5. **KNOTS**.

a. **Square Knot**. This joins two ropes of equal diameter (Figure 10-5). It has two interlocking bights; the running ends exit on the same side of the standing portion of the rope. Each tail is secured with an overhand knot on the standing end. When you dress the knot, leave at least a 4-inch tail on the working end.

**Figure 10-5. SQUARE KNOT**

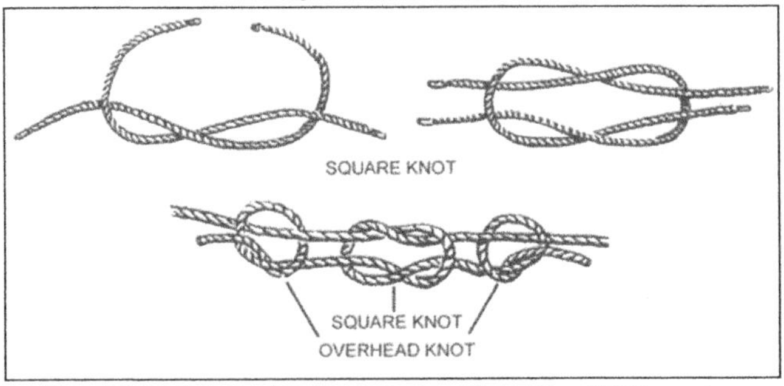


b. **Round-Turn with Two Half Hitches**. This is a constant tension anchor knot (Figure 10-6). The rope forms a complete turn around the anchor point (thus the name round turn), with both ropes parallel and touching, but not crossing. Both half hitches are tightly dressed against the round turn, with the locking bar on top. When you dress the knot, leave at least a 4-inch tail on the working end.

**Figure 10-6. ROUND-TURN WITH TWO HALF-HITCHES**

c. **End-of-the-Rope Clove Hitch**. The end-of-the-rope clove hitch (Figure 10-7) is an intermediate anchor knot that requires constant tension. Make two turns around the anchor. A locking bar runs diagonally from one side to the other. Leave no more than one rope width between turns of rope. Locking bar is opposite direction of pull. When you dress the knot, leave at least a 4-inch tail on the working end.

**Figure 10-7. END-OF-THE-ROPE CLOVE HITCH**

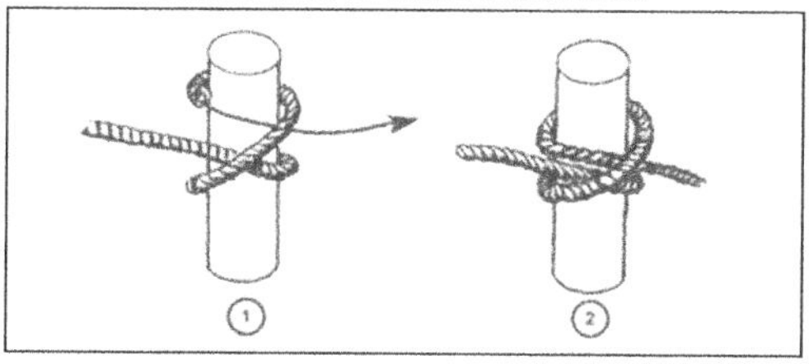


d. **Middle of the Rope Clove Hitch**. This knot (Figure 10-8) secures the middle of a rope to an anchor. The knot forms two turns around the anchor. A locking bar runs diagonally from one side to the other. Leave no more than one rope width between turns. Ensure the locking bar is opposite the direction of pull.

**Figure 10-8. MIDDLE-OF-THE-ROPE CLOVE HITCH**

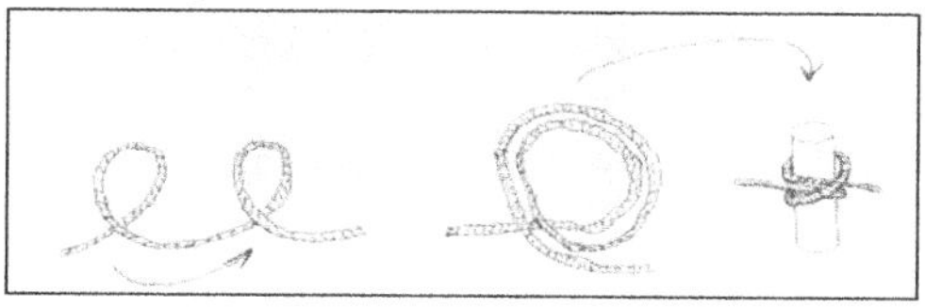

e. **Rappel Seat**. The rappel seat (Figure 10-9) is a rope harness used in rappelling and climbing. It can be tied for use with the left or right hand. Leg straps do not cross, and are centered on buttocks and tight. Leg straps form locking half hitches on rope around waist. A square knot is tied on the right hip and finished with two overhand knots. Tails must be even, no longer than 6 inches. A carabiner is inserted around all ropes, with the gate opening up and away. The carabiner must not contact the square or overhand knot. The rappel seat must be tight enough to prevent a fist from fitting between the rappeller's body and the harness.

**Figure 10-9. RAPPEL SEAT**

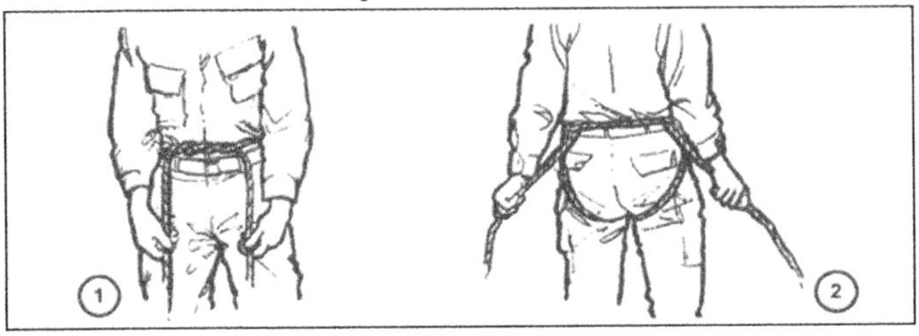


**Figure 10-9. RAPPEL SEAT (CONTINUED)**

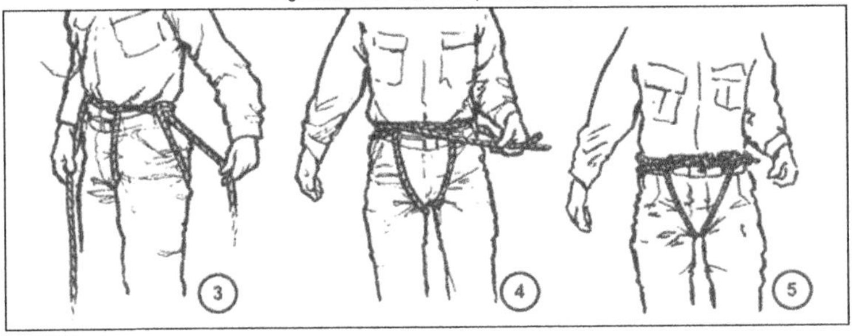

f. **Double Figure Eight**. Use a Figure Eight loop knot (Figure 10-10) to form a fixed loop in the end of the rope. It can be tied at the end of the rope or anywhere along the length of the rope. Figure Eight loop knots are formed by two ropes parallel to each other in the shape of a Figure Eight, no twists are in the Figure Eight. Fixed loops are large enough to insert a carabiner. When you dress the knot, leave at least a 4-inch tail on the working end.

**Figure 10-10. DOUBLE FIGURE EIGHT-LOOP KNOT**

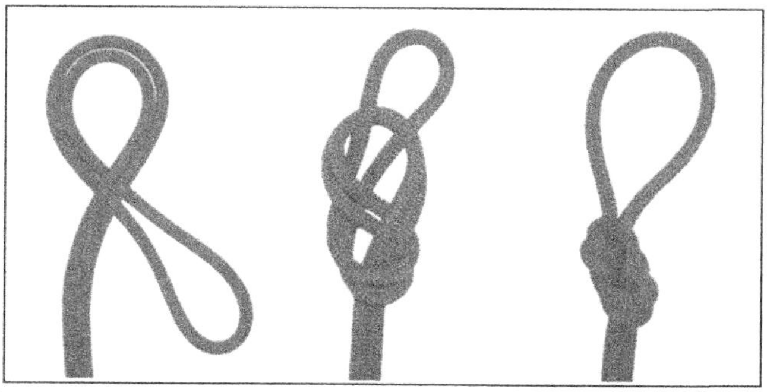

g. **Rerouted Figure Eight Knot**. This anchor knot also attaches a climber to a climbing rope. Form a Figure Eight in the rope, and pass the working end around an anchor. Reroute the end back through to form a double Figure Eight (Figure 10-11). Tie the knot with no twists. When you dress the knot, leave at least a 4-inch tail on the working end.

**Figure 10-11. REROUTED FIGURE EIGHT KNOT**

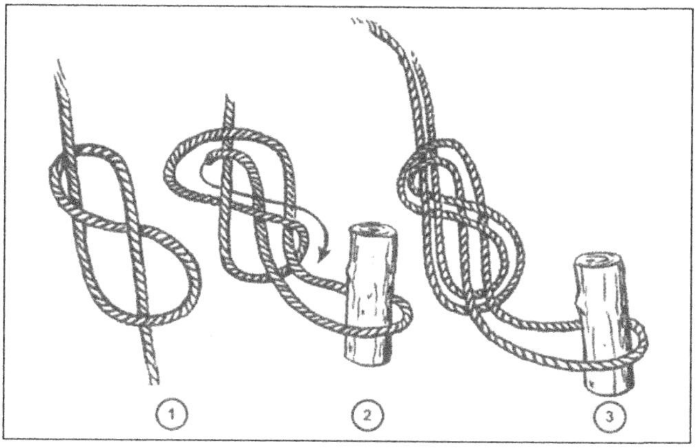

h. **Figure Eight Slip Knot**. The Figure Eight slip knot is used to form an adjustable bight in the middle of a rope. Knot is in the shape of a Figure Eight. Both ropes of the bight pass through the same loop of the Figure Eight. The bight is adjustable by means of a sliding section (Figure 10-12).

**Figure 10-12. FIGURE EIGHT SLIP KNOT**

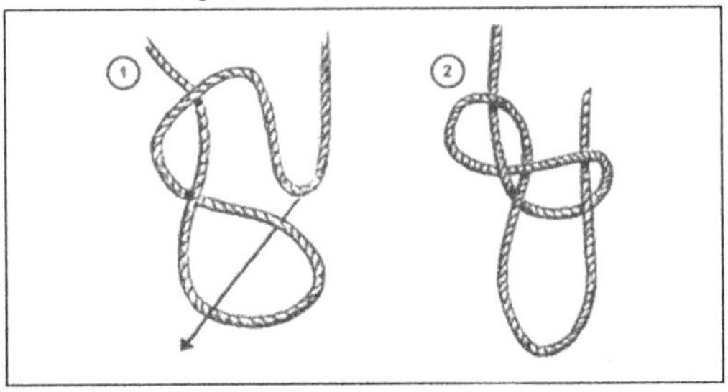


i. **End-of-the-Rope Prusik**. The End-of-the-Rope Prusik (Figure 10-13) attaches a movable rope to a fixed rope. The knot has two round turns, with a locking bar perpendicular to the standing end of the rope. Tie a bowline within 6 inches of the locking bar. When you dress the knot, leave at least a 4-inch tail on the working end.

**Figure 10-13. END-OF-THE-ROPE PRUSIK**

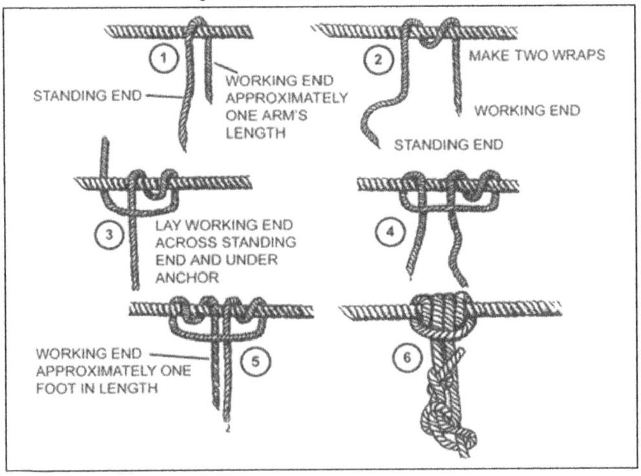

j. **Middle-of-the-Rope Prusik**. The middle-of-the-rope Prusik (Figure 10-14) attaches a movable rope to a fixed rope, anywhere along the length of the fixed rope. Make two round turns with a locking bar perpendicular to the standing end. Ensure the wraps do not cross and that the overhand knot is within 6 inches from the horizontal locking bar. Ensure the knot does not move freely on fixed rope.

**Figure 10-14. MIDDLE-OF-THE-ROPE PRUSIK**

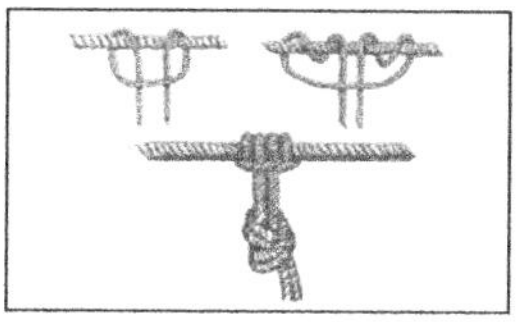

k. **Bowline on a Coil**. This knot (Figure 10-15) secures a climber to the end of a climbing rope. Make at least four parallel wraps around the body, between the hipbone and lower set of ribs. Keep all coils free of clothing, touch and are tight enough to ensure that a fist cannot be inserted between the wraps and the body. Three distinct coils show through the bight of the bowline. The rope coming off the bottom of the coils is on the right side, forward of the hip and forms the bight and the overhand knot. The rope coming off the top of the coils is on the left side, forward of the hips and forms the third and final coil showing through the bight of the bowline. Bowline is centered on the gig line and secured with an overhand knot with a minimum of 4 inch tail remaining after the knot is dressed.

**Figure 10-15. BOWLINE ON A COIL**

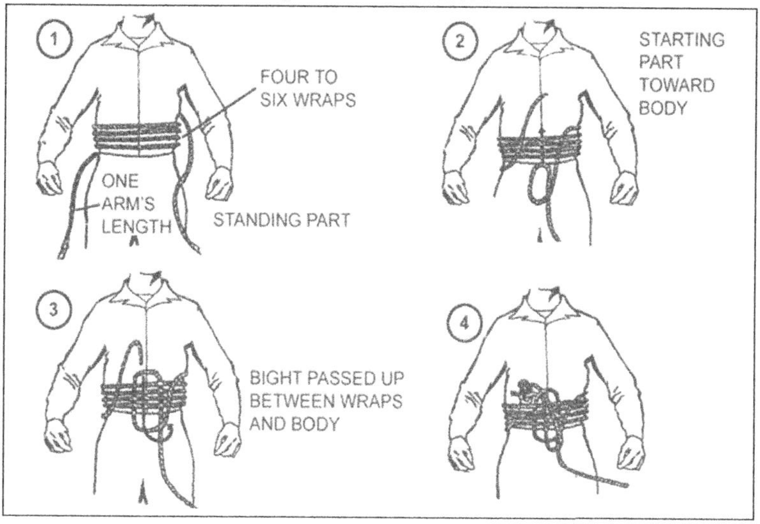

10-6. **BELAYS**. Belaying is any action taken to stop a climber's fall or to control the rate a load descends. The belayer also helps manage the rope for a climber by controlling the amount of rope that is taken out or in, or by controlling a rappeller's rate of descent. The belayer must be stably anchored to prevent him from being pulled out of his position, and losing control of the rope. The two types of belays are body and mechanical belays.

a. **Body Belay**. This belay (Figure 10-16) uses the belayer's body to apply friction by routing the rope around the belayer's body. Caution must be used when belaying from the body since the entire weight of the load may be placed on the belayer's body.

**Figure 10-16. BODY BELAY**

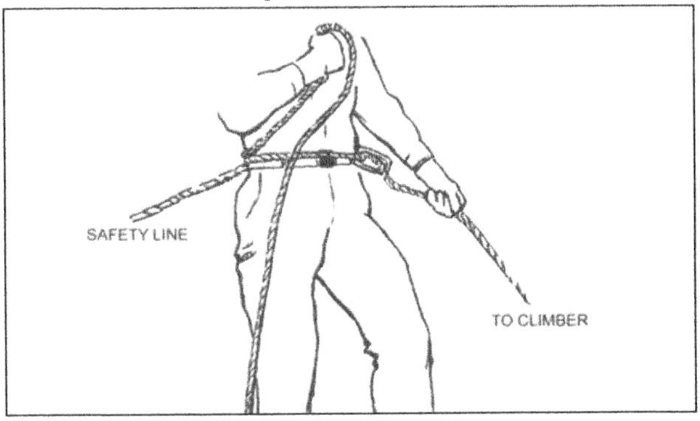


b. **Mechanical Belay**. This belay (Figure 10-17) uses mechanical devices to help the belayer control the rope, as in rappelling. A variety of devices are used to construct a mechanical belay for mountaineering.

**Figure 10-17. MECHANICAL BELAY**

(1) ***Munter Hitch***. One of the most often used belays, the munter hitch (Figure 10-18), requires very little equipment. The rope is routed through a locking, pear-shaped carabiner, then back on itself. The belayer controls the rate of descent by manipulating the working end back on itself with the brake hand.

**Figure 10-18. MUNTER HITCH**

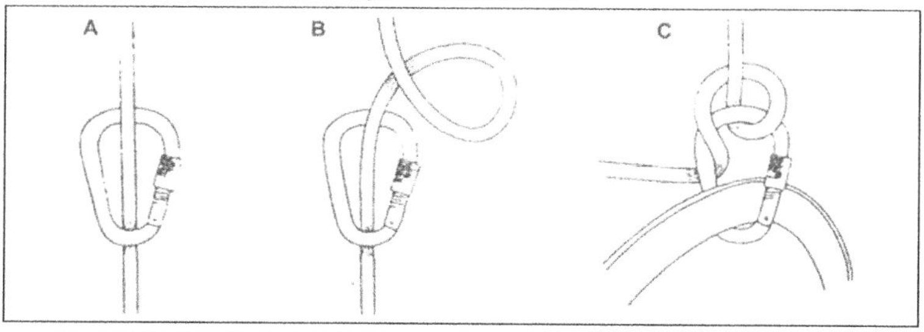


(2) ***Air Traffic Controller***. The ATC (air traffic controller) is a mechanical belay device (Figure 10-19) that locks down on itself when tension is applied in opposite directions. The belayer need apply very little force with his brake hand to control descent or arrest a fall.

**Figure 10-19. AIR TRAFFIC CONTROLLER**

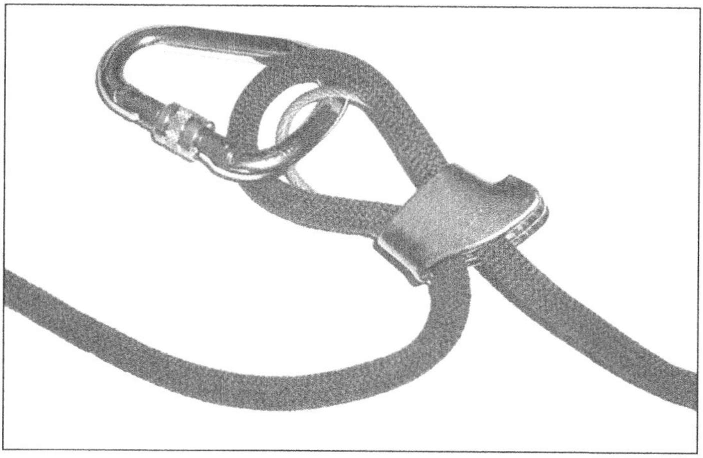

10-7. **CLIMBING COMMANDS**. Table 10-1 shows the sequence of commands used by climber and belayer.

**Table 10-1. SEQUENCE OF CLIMBING COMMANDS**

| *SEQUENCE* | *COMMAND* | *GIVEN BY* | *MEANING* |
|---|---|---|---|
| 1. | BELAY ON, CLIMB | Belayer | Alerts climber that belay is on and climber may climb. |
| 2. | CLIMBING | Climber | Alerts belayer that climber is climbing. |
| 3. | UP-ROPE | Climber | Belayer, remove excess slack in the rope. |
| 4. | BRAKE | Climber | Alerts belayer to immediately apply brake. |
| 5. | FALLING | Climber | Alerts belayer that climber is falling and that the belayer should immediately apply the brake and prepare to arrest the fall. |
| 6. | TENSION | Climber | Alerts belayer to remove all slack from climbing rope until rope is tight, then apply brake and hold position. |
| 7. | SLACK | Climber | Alerts belayer to pull slack into the climbing rope (Belayer may have to assist) |
| 8. | ROCK | Anyone | Alerts everyone that an object is about to fall near them. Belayer immediately applies brake. |
| 9. | POINT | Climber | Alerts belayer that the direction of pull on the climbing rope has changed in the event of a fall. |
| 10. | STAND-BY | Climber or Belayer | Alerts the other to hold position, stand by, I am not ready. |
| 11. | DO YOU HAVE ME? | Climber | Alerts belayer to prepare for a fall or to prepare to lower the climber. |
| 12. | I HAVE YOU | Belay | Alerts climber that the brake is on and the belayer is prepared for the climber to fall, or to lower him. |
| 13. | OFF-BELAY | Climber | Alerts belayer that climber is safetied in or it is safe to come off belay. |
| 14. | 3-METERS | Belayer | Alerts climber to the amount of rope between climber and belayer (May be given in feet or meters) |
| 15. | BELAY-OFF | Belayer | Alerts climber that belayer is off belay |

10-8. **ROPE INSTALLATIONS**. Rope installations may be constructed by teams to assist units in negotiating natural and manmade obstacles. installation teams consist of a squad sized element, with 2 to 4 trained mountaineers. Installation teams deploy early and prepare the AO for safe, rapid movement by constructing various types of mountaineering installations. Following construction of an installation, the squad, or part of it, remains on site to secure and monitor the system, assist with the control of forces across it, and make adjustments or repairs during its use. After passage of the unit, the installation team may then disassemble the system and deploy to another area as needed.

a. **Fixed Rope Installations**. A fixed rope is anchored in place to help Rangers move over difficult terrain. Its simplest form is a rope tied off on the top of steep terrain. As terrain becomes steeper or more difficult, fixed rope systems may require intermediate anchors along the route. Planning considerations are as follows:

- Does the installation allow you to bypass the obstacle?
- (Tactical considerations) Can obstacle be secured from construction through negotiation, to disassembly?
- Is it in a safe and suitable location? Is it easy to negotiate? Does it avoid obstacles?
- Are natural and artificial anchors available?
- Is the area safe from falling rock and ice?

b. **Vertical Hauling Line**. This installation (Figure 10-20) is used to haul men and equipment up vertical or near vertical slopes. It is often used in conjunction with the fixed rope.

**Figure 10-20. VERTICAL HAULING LINE**

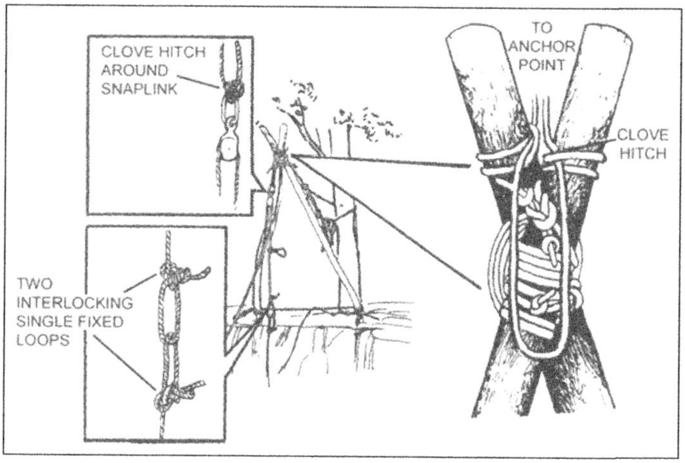


(1) ***Planning Considerations***.

- Does the installation allow you to bypass the obstacle?
- (Tactical considerations) Can you secure the installation from construction through negotiation, to disassembly?
- Does it have good loading and off-loading platforms? Are the platforms natural and easily accessible? Do they provide a safe working area?
- Does it allow sufficient clearance for load? Is there enough space between the slope and the apex of the A-frame to allow easy on-loading and off-loading of troops and equipment?
- Does it have an A-Frame for artificial height?
- Does it allow you to haul line in order to move personnel and equipment up and down slope?
- Does the A frame have a pulley or locking carabiner to ease friction on hauling line?
- Does it have a knotted hand line to help Rangers up the installation?
- Does it allow for Rangers top and bottom to monitor safe operation?

(2) ***Equipment***.

- Three 120-foot (50m) static ropes.
- Three 15-foot sling ropes for constructing A-frame.
- Two A-frame poles, 7 to 9 feet long, 4 to 6 inches in diameter (load dependent).
- Nine carabiners.
- One pulley with steel locking carabiner.

c. **Bridges**. Rope bridges are employed in mountainous terrain to bridge linear obstacles such as streams or rivers where the force of flowing water may be too great or temperatures are too cold to conduct a wet crossing.

(1) ***Construction***. The rope bridge is constructed using a static ropes. The max span that can be bridged is 50 percent of the ropes entire length for a dry crossing, 75 percent for a wet crossing. The ropes are anchored with an anchor knot on the far side

of the obstacle, and are tied off at the near end with a transport tightening system. Rope Bridge Planning Considerations are as follows:

- Does the installation allow you to bypass the obstacle?
- (Tactical considerations) Can you secure the installation from construction through negotiation, to disassembly?
- Is it in the most suitable location such as a bend in the river? Is it easily secured?
- Does it have near and far-side anchors?
- Does it have good loading and off-loading platforms?

(2) ***Equipment (1-Rope Bridge)***.

- One sling rope per-man.
- One steel locking carabineer.
- Two steel ovals.
- Two 120-foot static ropes.

(3) ***Construction Steps***. The first Ranger swims the rope to the far side and ties a tensionless anchor (Fig 10-4) between knee and chest level, with at least 6 to 8 wraps. The BTC ties a transport tightening system (Fig 10-21) to the near side anchor point. Then, he ties a Figure Eight slipknot and incorporates a locking half hitch around the adjustable bight. Insert 2 steel oval carabineers into the bight so the gates are opposite and opposed. The rope is then routed around the near side anchor point at waist level and dropped into the steel oval carabineers.

- A three-Ranger pulling team moves forward from the PLT. No more than three men should tighten the rope. Using more can overtighten the rope, bringing it near failure.
- Once the rope bridge is tight enough, the bridge team secures the transport-tightening system using two half hitches, without losing more than 4 inches of tension.
- Personnel cross using the commando crawl, rappel seat, or monkey crawl method (Figures 10-22 thru 24).

**Figure 10-21. TRANSPORT TIGHTENING SYSTEM**

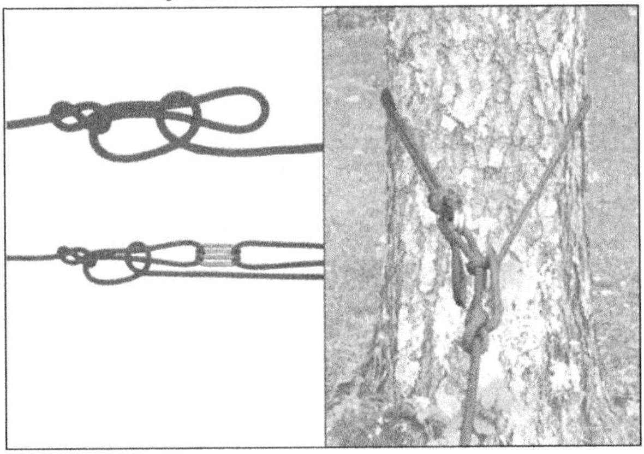

Figure 10-22. COMMANDO CRAWL METHOD

Figure 10-23. RAPPEL SEAT (TYROLEAN TRAVERSE) METHOD

Figure 10-24. MONKEY CRAWL METHOD

(4) ***Bridge Recovery***. Once all but two Rangers have crossed the rope bridge, the bridge team commander (BTC) chooses either the wet or dry method to dismantle it. If he chooses the dry method, he should he should first anchor his tightening system with the transport knot.

- The BTC back-stacks all of the slack coming out of the transport knot, then ties a fixed loop and places a carabiner into the fixed loop.
- The next to last Ranger to cross attaches the carabiner to his rappel seat or harness, and then moves across the bridge using the Tyrolean traverse method.
- The BTC then removes all knots from the system. The far side remains anchored. The rope should now only pass around the near side anchor.
- A three-man pull team, assembled on the far side, takes the end brought across by the next to last Ranger and pulls the rope tight again and holds it.
- The BTC then attaches himself to the rope bridge and moves across.
- Once across, the BTC breaks down the far side anchor, removes the knots, and then pulls the rope across. If the BTC chooses a wet crossing, any method can be used to anchor the tightening system.
- All personnel cross except the BTC or the strongest swimmer.
- The BTC then removes all knots from the system.
- The BTC ties a fixed loop, inserts a carabiner, and attaches it to his rappel seat or harness. He then manages the rope as the slack is pulled to the far side.
- The BTC then moves across the obstacle while being belayed from the far side.

d. **Suspension Traverse**. The suspension traverse is used to move personnel and equipment over rivers, ravines, chasms, and up or down a vertical obstacle. By combining the transport tightening system used during the Rope bridge, an A-Frame used for the Vertical haul Line (Figure 10-25), and belay techniques device, Units can construct a suspension traverse (Figures 10-26 and 10-27). Installation of a suspension traverse can be time-consuming and equipment-intensive. All personnel must be well trained and rehearsed in the procedures.

(1) ***Construction***. The suspension traverse must be constructed using a static ropes. The max span that can be bridged is generally 75 percent length of the shortest rope. Planning considerations are the same as rope bridge and vertical haul line combined.

(2) ***Equipment***.

- Three static installation ropes.
- Seven sling Ropes.
- Nine carabiners.
- One heavy duty double pulley.
- One locking carabiner.
- One canvas pad.

**Figure 10-25. ANCHORING THE TRAVERSE ROPE TO THE A-FRAME**

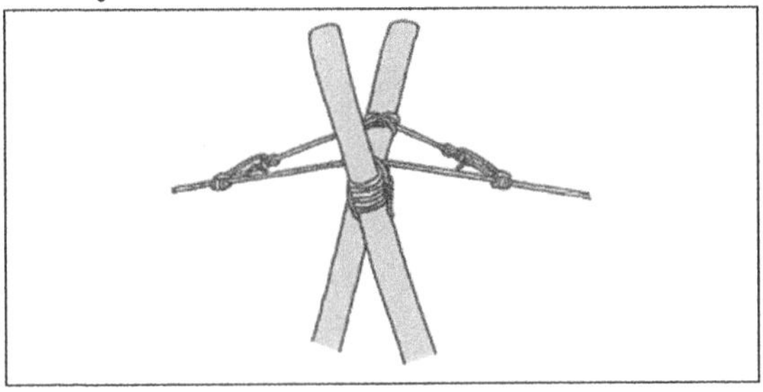

Figure 10-26. CARRYING ROPE FOR USE ON A SUSPENSION TRAVERSE

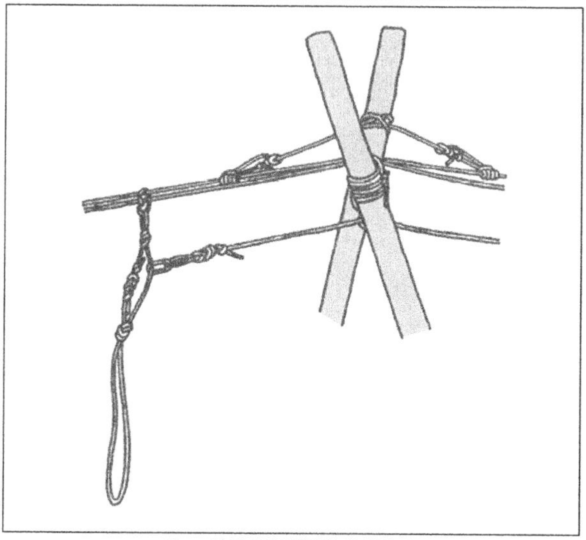

Figure 10-27. SUSPENSION TRAVERSE

10-9. **RAPPELLING**. Rappelling is a quick method of descent but can be extremely dangerous. These dangers include anchor failure, equipment failure, and individual error. Anchors in a mountainous environment should be selected carefully. Great care must be taken to load the anchor slowly and to ensure that no excessive stress is placed on the anchor. The best way to ensure this is to prohibit bounding rappels, and to use only walk-down rappels.

a. **Hasty and Body Rappels**. Hasty (Figure 10-28) and body rappels (Figure 10-29) are easier and faster than other methods, but should only be used on moderate pitches--never on vertical or overhanging terrain. Wear gloves to prevent rope burns.

**Figure 10-28. Hasty Rappel**

**Figure 10-29. Body Rappel**

b. **Seat-Hip Rappel**. The seat-hip rappel uses a mechanical rappel device that is inserted in a sling rope seat and is fastened to the rappeller to provide the necessary friction to remain in control. This method provides a faster and more controlled descent than any other method (Figure 10-30). There are several mechanical rappel devices (Figure 10-31 and 10-32) that can be used for the seat-hip rappel.

**Figure 10-30. FIGURE EIGHT DESCENDER**

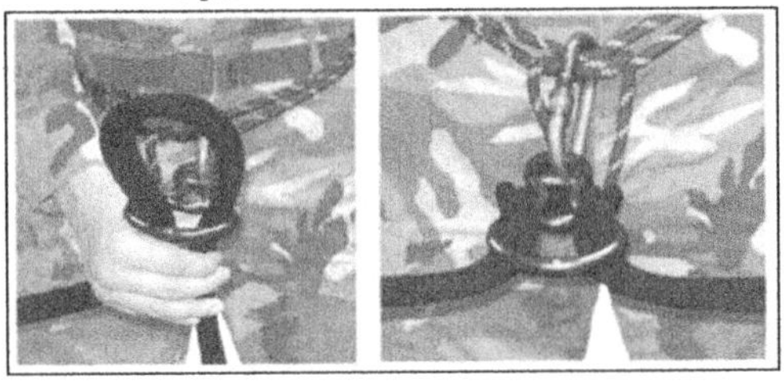

Figure 10-30. FIGURE EIGHT DESCENDER (CONTINUED)

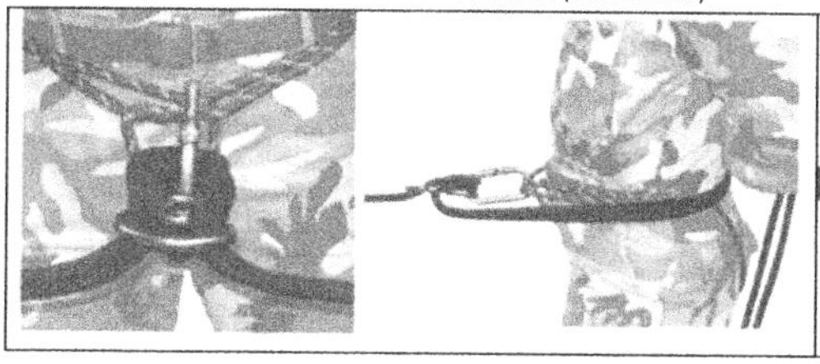

Figure 10-31. CARABINER WRAP

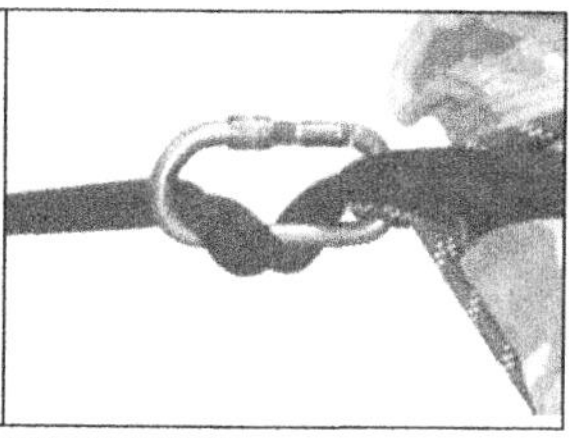

Figure 10-32. SEAT HIP RAPPEL WITH CARABINER WRAP DESCENDER

c. **Site Selection**. The selection of the rappel point depends on factors such as mission, cover, route, anchor points, and edge composition (loose or jagged rocks). There must be good anchors (primary and secondary). The anchor point should be above the rappeller's departure point. Suitable loading and off-loading platforms should be available.

- Each rappel point has primary and secondary anchors.
- Rappel point has equal tension between all anchor points.
- Double rope is used when possible.
- Ropes must reach the off loading platform.
- Site has suitable on and off loading platforms.
- Personnel working near the edge are tied in.
- Select a smooth route free of loose rock and debris.

## Chapter 11

# EVASION AND SURVIVAL

This chapter will cover basic material dealing with evading the enemy and survival techniques. Field expedient methods of acquiring food/ water, land navigation, constructing shelters, and making fires will be presented.

11-1. **EVASION**. When you become isolated or separated in a hostile area, either as an individual or as a group, your evasion and survival skills will determine whether or not you return to friendly lines.

a. When unable to continue the mission or unable to rejoin your unit, leave the immediate area, and move to your last rally point.

b. Observe activity in the area and form a plan.

c. Traveling alone offers the least possibility of detection, but traveling in groups of two to three is more desirable.

d. Plan a primary and alternate route. Consider distance, cover, food, and water. The easiest and shortest route may not be the best.

e. Food and water are daily requirements. You can do without food for several days; water, however, is essential.

f. Move at night. Use the daylight to observe, plan, and rest in a hide position.

g. Linkup only during daylight hours. Place friendly lines under observation.

h. Attempt to identify the unit you approach, note their movements and routine.

i. After carefully considering your approach route, make voice contact with the unit as soon as possible.

11-2. **SURVIVAL**.

With training, equipment, and the ***will to survive***, you can overcome *any* obstacle you may face. ***You will survive.*** Understand the emotional states associated with survival. "Knowing thyself" is extremely important in a survival situation. It bears directly on how well you cope with serious stresses, anxiety, pain, injury, illness; cold, heat, thirst, hunger, fatigue, sleep deprivation, boredom, loneliness and isolation.

a. **Memory Aid**. You can overcome and reduce the shock of being isolated behind enemy lines if you keep the key word S-U-R-V-I-V-A-L foremost in your mind. Its letters can help guide you in your actions.

(1) S - Size up the situation, the surroundings, your physical condition, and your equipment.

(2) U - Undue haste makes waste; don't be too eager to move. Plan your moves.

(3) R - Remember where you are relative to friendly and enemy units and controlled areas; water sources (most important in the desert); and good cover and concealment. This information will help you make intelligent decisions.

(4) V - Vanquish fear and panic.

(5) I – Improvise/Imagine. You can improve your situation. Learn to adapt what is available for different uses. Use your imagination.

(6) V - Value living. Remember your goal - to get out alive. Remain stubborn. Refuse to give in to problems and obstacles. This will give you the mental and physical strength to endure.

(7) A - Act like the natives; watch their daily routines and determine when, where, and how they get their food and water.

(8) L - Live by your wits. Learn basic skills.

b. **Navigation**. In a survival situation, you might find yourself without a compass. The ability to determine direction can enable you to navigate back to your unit or to a friendly sanctuary. In sunlight, there are two simple ways to determine direction: the shadow-tip and the watch.

(1) ***Shadow Tip***. Use the sun to find approximate true North. Use this in light bright enough to cast shadows. Find a fairly straight stick about 3 feet long, and follow the diagram below (Figure 11-l).

**Figure 11-1. SHADOW-TIP METHOD**

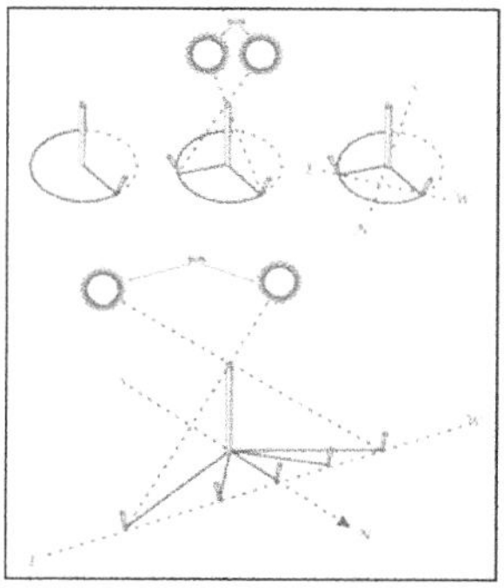

(2) ***Watch Method***. You can also determine direction using a watch (Figure 11-2). The steps you take will depend on whether you are in the northern Temperate Zone or in the southern Temperate Zone. The northern temperate zone is located between 23.4 north and 26.6 north. The southern Temperate Zone is located between 23.4 south and 66.6 south.

**Figure 11-2. WATCH METHOD**

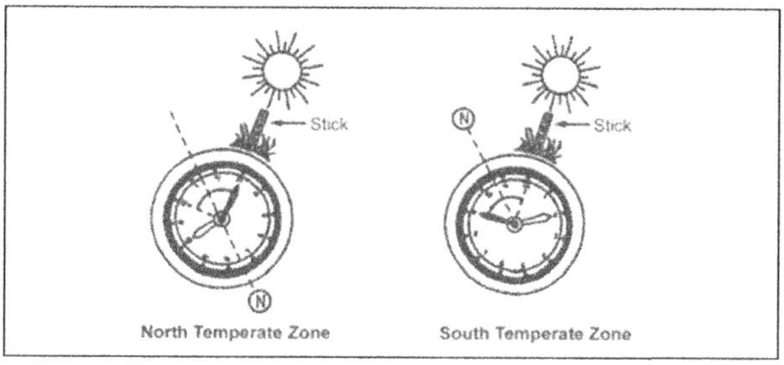


c. **Northern Temperate Zone**. Procedures in the northern temperate zone using a conventional watch are as follows:

(1) Place a small stick in the ground so that it casts a definite shadow.

(2) Place your watch on the ground so that the hour hand points toward and along the shadow of the stick.

(3) Find the point on the watch midway between the hour hand and 12 o'clock and draw an imaginary line from that point through and beyond the center of the watch. This imaginary line is a north-south line. You can then tell the other directions.

***Note:*** If your watch is set on daylight savings time, then use the midway point between the hour hand and 1 o'clock to draw your imaginary line.

d. **Southern Temperate Zone**. Procedures in the southern temperate zone using a conventional watch are as follows:

(1) Place a small stick in the ground so that it casts a definite shadow.

(2) Place your watch on the ground so that 2 o'clock points to and along the shadow.

(3) Find the midway point between the hour and 12 o'clock and draw an imaginary line from the point through and beyond the center of the watch. This is a north-south line.

e. A hasty shortcut using a conventional watch is simply to point the hour hand at the sun in the northern temperate zone (or point the 12 at the sun in the southern temperate zone) and then follow the last step of the watch method above to find your directions. This shortcut, of course, is not as accurate as the regular method but quicker. Your situation will dictate which method to use.

11-3. **WATER**. Water is one of your most urgently needed resources in a survival situation. You can't live long without it, especially in hot areas where you lose so much through sweating. Even in cold areas, you need a minimum of 2 quarts of water a day to maintain efficiency. More than three-fourths of your body is composed of fluids. Your body loses fluid as a result of heat, cold, stress, and exertion. The fluid your body loses must be replaced for you to function effectively. So, one of your first objectives is to obtain an adequate supply of water.

a. **Purification**. Purify all water before drinking. Either--

(1) Boil it for at least one minute (plus 1 more minute for each additional 1,000 feet above sea level) or for a maximum of 10 minutes anywhere.

(2) Use water purification tablets.

(3) Add eight drops of 2-1/2 percent iodine solution to a quart (canteen full) of water. Let it stand for 10 minutes before drinking.

(4) Collect rain water directly in clean containers or on plants. This is generally safe to drink without purifying. Never drink urine or sea water; the salt content is too high. Use old, bluish sea ice, but newer, grayer ice may be salty. Glacier ice is safe to melt and drink.

b. **Desert Environment**. In a desert environment, water has a huge effect on Rangers. If a unit fails to plan properly for water, and resupply is unavailable, then they can run out of water. In the desert, look for four signs of water: animal trails, vegetation, birds, and civilization. Adequate water is critical in a hot environment if a unit is to survive and maintain the physical condition necessary to accomplish the mission. Unit leaders must enforce water discipline and plan for water resupply. The leader can use the following planning considerations for water resupply:

(1) Units average water consumption.

(2) Drop sites.

(3) Aviation support.

(4) DZ and LZ parties.

(5) Caches.

(6) Targets of opportunity (enemy).

c. **Survival Water Still**. Dig a belowground still (Figure 11-3). Select a site where you believe the soil will contain moisture (such as a dry stream bed or a spot where rain water has collected). The soil should be easy to dig, and be in sunlight most of the day:

(1) Dig a bowl-shaped hole about 3 feet across and 2 feet deep.

(2) Dig a sump in center of the hole. The depth and the perimeter of the sump will depend on the size of the container that you have to set inside of it. The bottom of the sump should allow the container to stand upright.

(3) Anchor the tubing to the bottom of the container by forming a loose overhand knot in the tubing.

(4) Place the container upright in the sump.

(5) Extend the unanchored end of the tubing up, over, and beyond the lip of the hole.

(6) Place plastic sheeting over the hole covering the edge with soil to hold it in place.

(7) Place a rock in the center of the plastic.

(8) Allow the plastic to lower into the hole until it is about 15 inches below ground level. The plastic now forms an inverted cone with the rock at its apex. Make sure that the apex of the cone is directly over your container. Also, make sure the plastic cone does not touch the sides of the hole, because the earth will absorb the condensed water.

(9) Put more soil on the edges of the plastic to hold it securely in place and to prevent loss of moisture.

(10) Plug the tube when not being used so that moisture will not evaporate.

d. You can drink water without disturbing the still by using the tube as a straw. You may want to use plants in the hole as a moisture source. If so, when you dig the hole you should dig out additional soil from the sides of the hole to form a slope on which to place the plants. Then proceed as above.

Figure 11-3. SURVIVAL WATER STILL

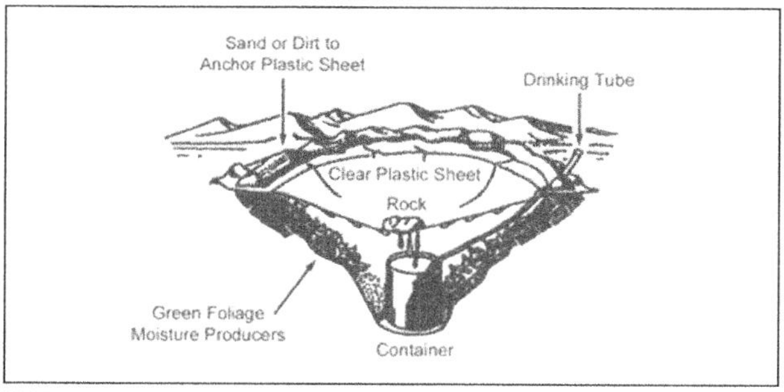


11-4. **PLANT FOOD**. There are many plants throughout the world. Tasting or swallowing even a small portion of some can cause severe discomfort, extreme internal disorders, or death. Therefore, if you have the slightest doubt as to the edibility of a plant, apply the universal edibility test described below before eating any part of it.

a. **Universal Edibility Test**. Before testing a plant for edibility, make sure there are a sufficient number of plants to make testing worth your time and effort. You need more than 24 hours to apply the edibility test outlined below:

(1) Test only one part of a potential food plant at a time.

(2) Break the plant into its basic components, leaves, stems, roots, buds, and flowers.

(3) Smell the food for strong or acid odors. Keep in mind that smell alone does not indicate a plant is edible.

(4) Do not eat for 8 hours before starting the test.

(5) During the 8 hours you are abstaining from eating, test for contact poisoning by placing a piece of the plant you are testing on the inside of your elbow or wrist. Usually 15 minutes is enough time to allow for reaction.

(6) During the test period, take nothing by mouth except purified water and the plant part being tested.

(7) Select a small portion and prepare it the way you plan to eat it.

(8) Before putting the prepared plant part in your mouth, touch a small portion (a pinch) to the outer surface of the lip to test for burning or itching.

(9) If after 3 minutes there is no reaction on your lip, place the plant part on your tongue, holding there for 15 minutes.

(10) If there is no reaction, thoroughly chew a pinch and hold it in your mouth for 15 minutes. *Do not swallow.*

(11) If no burning, itching, numbing, stinging, or other irritation occurs during the 15 minutes, swallow the food.

(12) Wait 8 hours. If any ill effects occur during this period, induce vomiting and drink a lot of water.

(13) If no ill effects occur, eat 1/2 cup of the same plant part prepared the same way. Wait another 8 hours. If no ill effects occur, the plant part as prepared is safe for eating.

b. **Poisonous Plants**. ***Do not*** eat unknown plants that have the below characteristics:

(1) Have a milky sap or a sap that turns black when exposed to air.

(2) Are mushroom like.

(3) Resemble onion or garlic.

(4) Resemble parsley, parsnip, or dill.

(5) Have carrot-like leaves, roots, or tubers.

11-5. **ANIMAL FOOD**. Animal food contains the most food value per pound. Anything that creeps, crawls, swims, or flies is a possible source of food, however you must first catch, kill and butcher it before this is possible. There are numerous methods for catching fish and animals in a survival situation. You can catch fish by using a net across a small stream, (Figure 11-4) or by making fish traps and baskets. Improvise fish hooks and spears as indicated in Figure 11-5, and use them for conventional fishing, spearing and digging.

Figure 11-4. SETTING A GILL NET IN THE STREAM

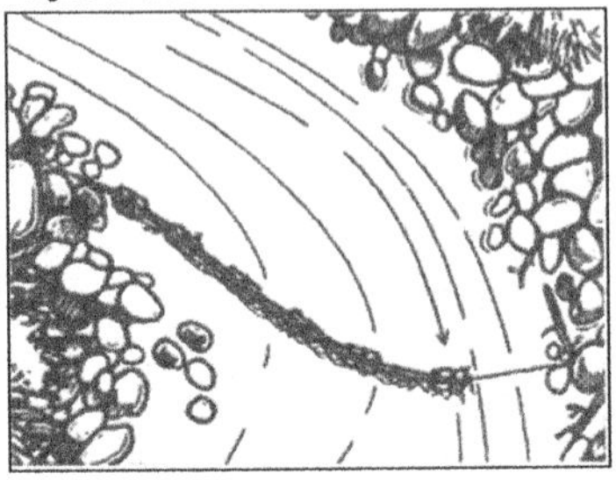

Figure 11-5. SPEARS AND FISH HOOKS

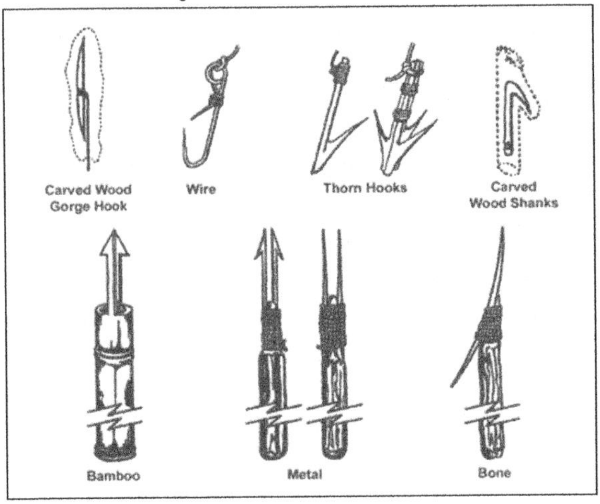


11-6. **TRAPS AND SNARES**. For an unarmed survivor or evader, or when the sound of a rifle shot could be a problem, trapping or snaring wild game is a good alternative. Several well-placed traps have the potential to catch much more game than a Ranger with a rifle is likely to shoot.

a. To be effective with any type of trap or snare, you must---

(1) Know the species of animal you intend to catch.

(2) Know how to construct a proper trap.

(3) Avoid alarming the prey with signs of your presence.

b. There are no catchall traps you can set for all animals. You must determine what species are in a given area and set your traps specifically with those animals in mind. Look for the following:

(1) Runs and trails.
(2) Tracks.
(3) Droppings.
(4) Chewed or rubbed vegetation.
(5) Nesting or roosting sites.
(6) Feeding and watering areas.

c. Position your traps and snares where there is proof that animals pass through. You must determine if it is a "run" or a "trail." A trail will show signs of use by several species and will be rather distinct. A run is usually smaller and less distinct and will only contain signs of one species. You may construct a perfect snare, but it will not catch anything if haphazardly placed in the woods. Animals have bedding areas, waterholes, and feeding areas with trails leading from one to another. You must place snares and traps around these areas to be effective.

d. An evader in a hostile environment must conceal traps and snares. It is equally important, however, to avoid making a disturbance that will alarm the animal and cause it to avoid the trap. Therefore, if you must dig, remove all fresh dirt from the area. Most animals will instinctively avoid a pitfall-type trap. Prepare the various parts of a trap or snare away from the site, carry them in, and set them up. Such actions make it easier to avoid disturbing the local vegetation, thereby alerting the prey. Do not use freshly cut, live vegetation to construct a trap or snare. Freshly cut vegetation will "bleed" sap that has an odor the prey will be able to smell. It is an alarm signal to the animal.

e. You must remove or mask the human scent on and around the trap you set. Although birds do not have a developed sense of smell, nearly all mammals depend on smell even more than on sight. Even the slightest human scent on a trap will alarm the prey and cause it to avoid the area. Removing the scent from a trap is difficult but masking it is relatively easy. Use the fluid from the gall and urine bladders of previous kills. Do not use human urine. Mud, particularly from an area with plenty of rotting vegetation, is also good. Use it to coat your hands when handling the trap and to coat the trap when setting it. In nearly all parts of the world, animals know the smell of burned vegetation and smoke. It is only when a fire is actually burning that they become alarmed. Therefore, smoking the trap parts is an effective means to mask your scent. If one of the above techniques is not practical, and if time permits, allow a trap to weather for a few days and then set it. Do not handle a trap while it is weathering. When you position the trap, camouflage it as naturally as possible to prevent detection by the enemy and to avoid alarming the prey.

f. Traps or snares placed on a trail or run should use canalization. To build a channel, construct a funnel-shaped barrier extending from the sides of the trail toward the trap, with the narrowest part nearest the trap. Canalization should be inconspicuous to avoid alerting the prey. As the animal gets to the trap, it cannot turn left or right and continues into the trap. Few wild animals will back up, preferring to face the direction of travel. Canalization does not have to be an impassable barrier. You only have to make it inconvenient for the animal to go over or through the barrier. For best effect, the canalization should reduce the trail's width to just slightly wider than the targeted animal's body. Maintain this constriction at least as far back from the trap as the animal's body length, then begin the widening toward the mouth of the funnel.

(1) Use a treadle snare against small game on a trail (Figure 11-6). Dig a shallow hole in the trail. Then drive a forked stick (fork down) into the ground on each side of the hole on the same side of the trail. Select two fairly straight sticks that span the two forks. Position these two sticks so that their ends engage the forks. Place several sticks over the hole in the trail by positioning one end over the lower horizontal stick and the other on the ground on the other side of the hole. Cover the hole with enough sticks so that the prey must step on at least one of them to set off the snare. Tie one end of a piece of cordage to a twitch-up or to a weight suspended over a tree limb. Bend the twitch-up or raise the suspended weight to determine where you will tie a 5 centimeter or so long trigger. Form a noose with the other end of the cordage.

(2) Route and spread the noose over the top of the sticks over the hole. Place the trigger stick against the horizontal sticks and route the cordage behind the sticks so that the tension of the power source will hold it in place. Adjust the bottom horizontal stick so that it will barely hold against the trigger. As the animal places its foot on a stick across the hole, the bottom horizontal stick moves down, releasing the trigger and allowing the noose to catch the animal by the foot. Because of the disturbance on the trail, an animal will be wary. You must therefore use canalization.

Figure 11-6. TREADLE SNARE

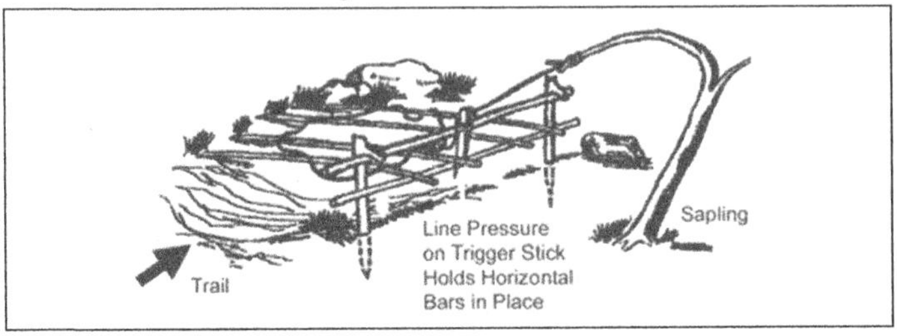


g. Trapping game can be accomplished through the use of snares, traps, or deadfalls. A snare is a noose that will slip and strangle or hold any animal caught in it. You can use inner core strands of parachute suspension lines, wire, bark of small hardwood saplings as well as hide strips from previously caught animals to make snares.

(1) The drag noose snare, Figure 11-7, is usually the most desirable in that it allows you to move away from the site, plus it is one of the easiest to make and fastest to set. It is especially suitable for catching rabbits. To make the drag noose snare, make a loop in the string using a bowline or wireman's knot. (When using wire, secure the loop by intertwining the end of the wire with the wire at the top of the loop). Pull the other end of the string (or wire) through the loop to form a noose that is large enough for the animal's head but too small for its body; tie the string (or attach the wire) to a sturdy branch. The branch should be large enough to span the trail and rest on the bush or support (two short forked sticks) you have selected. A snared animal will dislodge the drag stick, pulling it until it becomes entangled in the brush. Any attempt to escape tightens the noose, strangling or holding the animal.

Figure 11-7. DRAG NOOSE SNARE

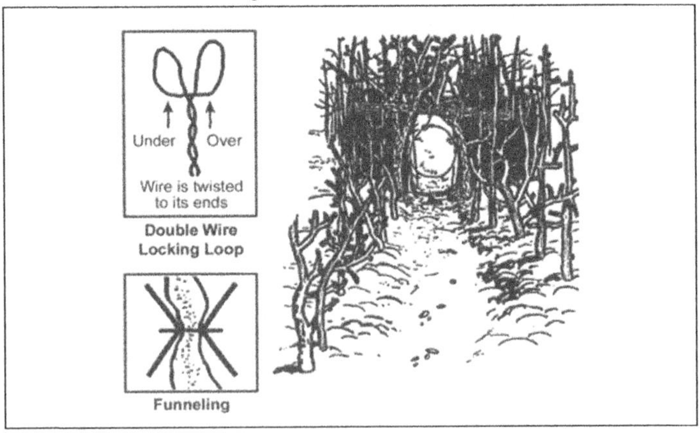

(2) Another type snare is the locking type snare loop (Figure 11-8) that will lock when pulled tight, ensuring the snared animal cannot escape. Use lightweight wire to make this snare, such as, trip wire or the wire from a vehicle or aircraft electrical system. To construct this snare, cut a piece of wire twice the length of the desired snare wire. Double the wire and attach the running ends to a securely placed object, such as the branch of a tree. Place a stick about 1/2 inch in diameter through the loop end of the wire; holding the wire taut, turn the stick in a winding motion so that the wire is twisted together. You should have four to five twists per inch. Detach the wire from the branch and then remove the loop from the stick; make a Figure Eight in the l/2-inch loop by twisting the loop over itself then fold the Figure Eight so the small loops are almost overlapping; run the loose wire ends through these loops. This forms a stiff noose that is strong. Tie the loose end to the stick (for a drag noose square) or branch you are using to complete the snare. This is an excellent snare for catching large animals.

**Figure 11-8. FORMING A LOCKING-TYPE SNARE LOOP**

(3) Another means of obtaining game is the use of the deadfall trap (Figure 11-9 and Figure 11-10).

**Figure 11-9. TRIGGER WITH DEADFALL**

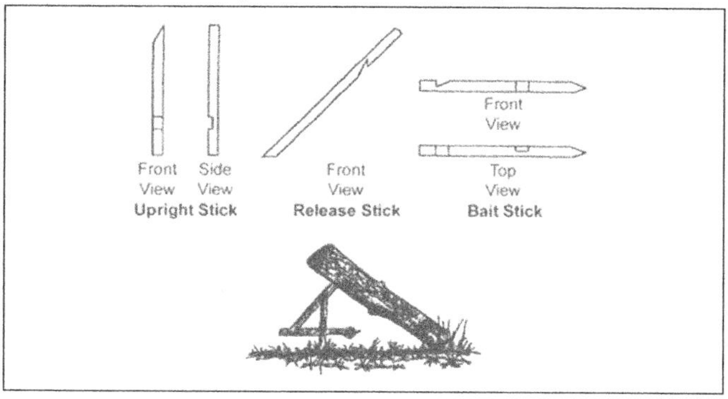

Figure 11-10. TRIP-STRING DEADFALL TRAP

h. Once you have obtained your fish or game, you must clean/butcher and cook/store it. Improper cleaning storing can result in inedible fish and game.

(1) ***Fish***. You must know how to tell if fish are free of bacterial decomposition that makes the fish dangerous to eat. Although cooking may destroy the toxin from bacterial decomposition, do not eat fish that appear spoiled.

(a) Signs of spoilage are:

- A peculiar odor.
- A suspicious color. (Gills should be red or pink. Scales should be a pronounced-not faded shade of gray).
- A dent remaining after pressing the thumb against the flesh.
- A slimy rather than moist or wet body.
- A sharp or peppery taste.

(b) Eating spoiled or poisoned fish may cause diarrhea, nausea, cramps, vomiting, itching; paralysis, or a metallic taste in the mouth. These symptoms appear suddenly 1 to 6 hours after eating. If you are near the sea, drink sea water immediately upon on set of such symptoms and force yourself to vomit.

(c) Fish spoil quickly after death, especially on a hot day, so prepare fish for eating as soon as possible after you catch them.

(d) Cut out the gills and large blood vessels that lie next to the backbone. (You can leave the head if you plan to cook the fish on a spit).

(e) Gut fish that are more than 4 inches long cut along the abdomen and scrape out the intestines.

(f) Scale or skin the fish.

(g) You can impale a whole fish on a stick and cook it over an "open fire." However, boiling the fish with the skin on is the best way to get the most food value. The fats and oil are under the skin, and by boiling the fish, you can save the juices for broth. Any of the methods used for cooking plant food can be used for cooking fish. Fish is done when the meat flakes off.

(h) To dry fish in the sun, hang them from branches or spread them on hot rocks. When the meat has dried, splash it with sea water, if available, to salt the outside. Keep seafood only if it is well dried or salted.

(2) ***Snakes***. All poisonous and nonpoisonous fresh water and land snakes are edible. To prepare snakes for eating use the following steps (Figure 11-11):

(a) Grip the snake firmly behind the head and cut off the head with a knife.

(b) Slit the belly and remove the innards. (You can use the innards for baiting traps and snares).

(c) Skin the snake. (You can use the skin for improvising, belts, straps, or similar items).

> **CAUTION**
>
> Take extreme care in securing snakes--the bite of some poisonous snakes can be fatal. Even after a snake's head is cut off, its reflex action can cause it to bite, injecting poison. The best time to capture snakes is in the early morning or late evening when temperatures are low and they move slow. Kill or use a long stick to pin down its head and capture it. To pick up a snake, place the index finger on the top rear of its head with your thumb and middle finger on either side of the head behind the jaws. Keep your index finger on top of snake's head to prevent it from turning inside its skin and biting you.

**Figure 11-11. CLEANING A SNAKE**

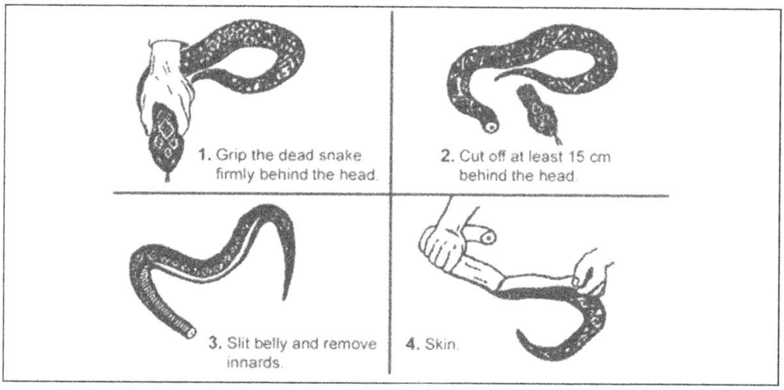


(3) ***Fowl***. Your first step after killing a fowl for eating or preserving is to pluck its feathers. If plucking is impractical, you can skin the fowl. Keep in mind, however, that a fowl cooked with the skin on retains more food value. Waterfowl are easier to pluck while dry, but other fowl are easier to pluck after scalding. After you pluck the fowl--

(a) Cut off its neck close to the body.

(b) Cut an incision in the abdominal cavity and clean out the insides. Save the neck, liver, and heart for stew. Thoroughly clean and dry the entrails to use for cordage.

(c) Wash out the abdominal cavity with fresh clean water. You can boil fowl or cook it on a spit over a fire. You should boil scavenger birds such as vultures and buzzards for at least 20 minutes to kill any parasites. Use the feathers from fowl for insulating your shoes clothing, or bedding. You can also use feathers for fish lures.

(4) **Medium-Sized Mammals**. The game you trap or snare will generally be alive when you find it and therefore dangerous. Be careful when you approach a trapped animal. Use a spear or club to kill it so you can keep a safe distance from it. After you kill an animal, immediately bleed it by cutting its throat. If you must drag the carcass any distance, do so before you cut off the hide so that the carcass is protected from dirt and debris that might contaminate it. Clean the animal near a stream if possible so that you can wash and cool the carcass and edible parts. Fleas and parasites will leave a cooled body so if the situation allows, wait until the animal cools before cleaning and dressing the carcass. To skin and dress the animal (Figure 11-12 and 11-13).

Figure 11-12. SKINNING AND BUTCHERING LARGE GAME

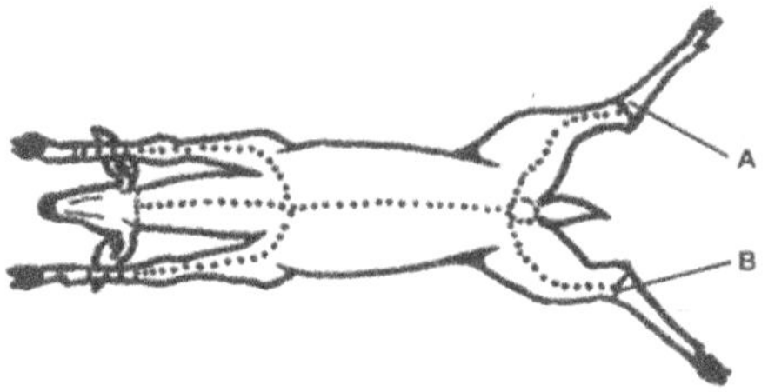


Figure 11-13. SKINNING SMALL GAME

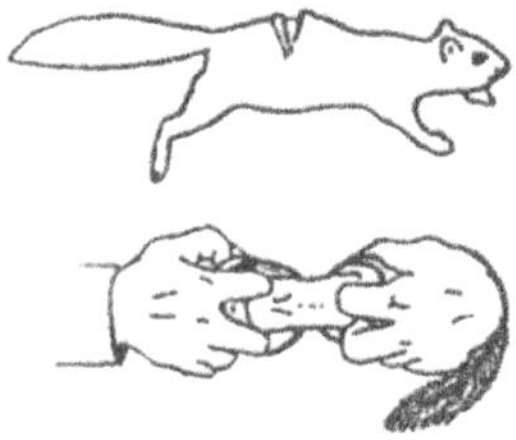


(a) Place carcass, belly up, on a slope if available. You can use rocks or brush to support it.

(b) Remove genitals or udder.

(c) Remove musk glands to avoid tainting meat.

(d) Split hide from tail to throat. Make the cut shallow so that you do not pierce the stomach.

(e) Insert your knife under the skin, taking care not to cut into the body cavity. Peel the hide back several inches on each side to keep hair out of the meat.

(f) Open the chest cavity by splitting the sternum. You can do this by cutting to one side of the sternum where the ribs join.

(g) Reach inside and cut the windpipe and gullet as close to the base of the skull as possible.

(h) With the forward end of the intestinal tract free, work your way to the rear lifting out internal organs and intestines. Cut only where necessary to free them.

(i) Carefully cut the bladder away from the carcass so that you do not puncture the bladder (urine can contaminate meat). Pinch the urethra tightly and cut it beyond the point you are pinching.

(j) Remove the bladder.

(k) From the outside of the carcass, cut a circle around the anus.

(l) Pull the anus into the body cavity and out of the carcass.

(m) Lift or roll the carcass to drain all blood.

***Note:*** Try to save as much blood as you can as it is a vital source of food and salt. Boil the blood.

(n) Remove the hide, and make cuts along the inside of the legs to just above the hoof or paw. Peel the skin back, using your knife in a slicing motion to cut the membrane between the skin and meat. Continue this until the entire skin is removed.

(o) Most of the entrails are usable. The heart, liver, and kidneys are edible. Cut open the heart and remove the blood from its chambers. Slice the kidneys and if enough water is available, soak or rinse them. In all animals except those of the deer family, the gall bladder (a small, dark-colored, clear-textured sac) is attached to the liver.

(p) Sometimes, the sac looks like a blister on the liver. To remove the sac, hold the top portion of it and cut the liver around and behind the sac. If the gall bladder breaks and gall gets on the meat, wash it off immediately so the meat will not become tainted. Dispose of the gall.

(q) Clean blood splattered on the meat will glaze over and help preserve the meat for a short time. However, if an animal is not bled properly, the blood will settle in the lowest part of its body and will spoil in a short time. Cut out any meat that becomes contaminated.

(r) When temperatures are below 40 degrees, you can leave meat hanging for several days without danger of spoilage. If maggots get on the meat, remove the maggots and cut out the discolored meat. The remaining meat is edible. Maggots, which are the larvae of insects, are also edible.

(s) Blood, which contains salts and nutrients, is a good base for soups.

(t) Thoroughly clean the intestines and use them for storing or smoking food or lashings for general use. Make sure they are completely dry to preclude rotting.

(u) The head of most animals contains a lot of meat, which is relatively easy to get. Skin the head, saving the skin for leather. Clean the mouth thoroughly and cut out the tongue. Remove the outer skin from the tongue after cooking. Cut or scrape the meat from the head. If you prefer, you can roast the head over an open fire before cutting off the meat. Eyes are edible. Cook them but discard the retina (this is a plastic like disc). The brain is also edible; in fact, some people consider it a delicacy. The brain is also used to tan leather, the theory being that the brain of an animal is adequate to tan its hide.

(v) Use the tendons and ligaments of the body of large animals for lashings.

(w) The marrow in bones is a rich-food source. Crack the bones and scrap out the marrow, and use bones to make weapons.

(x) If the situation and time allow, you should preserve the extra meat for later use. If the air is cold enough, you can freeze the meat. In warmer climates however, you will need to use a drying or smoking process to preserve it. One night of heavy smoking will make meat edible for about 1 week. Two nights will make it remain edible for 2 to 4 weeks. To prepare meat for drying or smoking, cut it with the grain in quarter inch strips. To air dry the meat, hang it in the wind and hot sun out the reach of animals; cover it so that blow flies cannot land on it.

(y) To smoke meat, you will need an enclosed area – for instance, a teepee (Figure 11-14) or a pit. You will also need wood from deciduous trees, preferably green. Do not use conifer trees such as pines, firs, spruces, or cedars, as the smoke from these trees give the meat a disagreeable taste.

(z) When using the para-teepee or other enclosed area with a vent at the top, set the fire in the center and let it burn down to coals, then stoke it with green wood. Place the strips of meat on a grate or hang them from the top of the enclosure so that they are about 2 feet above the smoking coals. To use the pit method of smoking meat dig, a hole about 3 feet (1 meter) deep and 1 1/2 feet (1/2 meter) in diameter. Make a fire at the bottom of the hole. After it starts burning well, add chipped green wood or small branches of green wood to make it smoke. Place a wooden grate about 1 1/2 feet (1/2 meter) above the fire and lay the strips of meat on the grate. Cover the pit with poles, boughs, leaves, or other material.

11-7. **SHELTERS**. A shelter can protect you from the sun, insects, wind, rain, snow, hot or cold temperatures, and enemy observation. In some areas your need for shelter may take precedence over your need for food, possibly even your need for water. After determining your shelter site, you should keep in mind the type of shelter (protection) you need. You need to know how to make different types of shelters. Only two are described in this handbook. Additional information is available in FM 21-76.

a. **Planning Considerations**.

- How much time and effort are needed to build the shelter?
- Will the shelter adequately protect you from the elements (rain, snow, wind, sun, and so on)?
- Do you have tools to build it? If not, can you improvise tools from materials in the area?
- Do you have the type and amount of manmade materials needed to build it? If not, are there sufficient natural materials in the area?

Figure 11-14. SMOKING MEAT

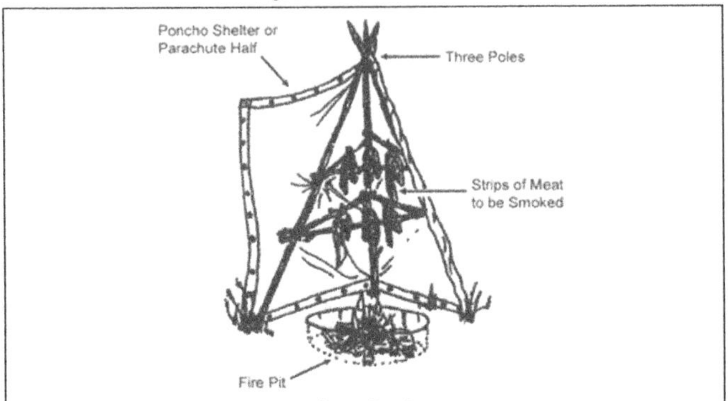


Tepee Smoker

b. **Types**.

(1) ***Poncho Lean-To***. It takes only a short time and minimal equipment to build this lean-to (Figure 11-15). You need a poncho, 6 to 10 feet of rope, three stakes about 6 inches long, and two trees (or two poles) 7 to 9 feet apart. Before you select the trees you will use (or decide where to place the poles), check the wind direction. Make sure the back of your lean-to will be into the wind. To make the lean-to:

(a) Tie off the hood of the poncho. To do this, pull the draw cord tight; roll the hood long ways, fold it into thirds, and tie it with the draw cord.

(b) Cut the rope in half. On one long side of the poncho, tie half of the rope to one corner grommet, and the other half to the other corner grommet.

(c) Attach a drip stick (about a 4-inch stick) to each rope 1/4 to 3/4 inches away from the grommet. These drip sticks will keep rainwater from running down the ropes into the lean-to. Using drip lines is another way to prevent dripping inside the shelter. Tie lines or string about 4 inches long to each grommet along the top edge of the shelter. This allows water to run to and down the line without dripping into the shelter.

(d) Tie the ropes about waist high on the trees (uprights). Use a round turn and two half hitches with quick-release knot.

(e) Spread the poncho into the wind and anchor to the ground. To do this, put three sharpened sticks through the grommets and into the ground.

Figure 11-15. PONCHO LEAN-TO

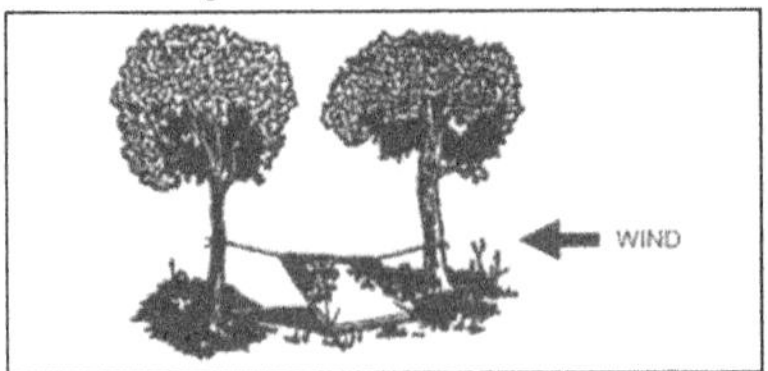

(f) If you plan to use the lean-to for more than one night, or if you expect rain, make a center support to the lean-to. You can do this by stretching a rope between two upright poles or trees that are in line with the center of the poncho.

(g) Tie another rope to the poncho hood; pull it upward so that it lifts the center of the poncho, and tie it firmly to the rope stretched between the two uprights.

(h) Another method is to cut a stick to place upright under the center of the lean-to. This method, however, will restrict your space and movements in the shelter.

(i) To give additional protection from wind and rain, place boughs, brush, your rucksack, or other equipment at the sides of the lean-to.

(j) To reduce heat loss to the ground, place some type of insulating material, such as leaves or pine needles, inside your lean-to.

***Note:*** When at rest, as much as 80 percent of your body heat can be lost to the ground.

(l) To increase your security from enemy observation, lower the silhouette of the lean-to by making two modifications.

- Secure the support lines to the trees knee-high rather than waist-high.
- Use two knee-high sticks in the two center grommets (sides of lean-to), and angle the poncho to the ground, securing it with sharpened sticks as above.

(2) ***Field Expedient Lean-To***. If you are in a wooded area and have sufficient natural materials, you can make an expedient lean-to (Figure 11-16) without the aid of tools or with only a knife. You need more time to make it than the shelter previously mentioned, but it will protect you from most environmental elements. You will need two trees, (or two upright poles), about 6 feet apart; one pole about 7 feet long and 1 inch in diameter. Five to eight poles about 10 feet long and 1 inch in diameter for beams, cord or vines for securing, the horizontal support to the trees and other poles, saplings, or vines to crisscross the beams. To make this lean-to:

**Figure 11-16. FIELD-EXPEDIENT LEAN-TO**

(a) Tie the 7-foot pole to the two trees at point about waist to chest high. This is your horizontal support. If there is a fork in the tree, you can rest the pole in it instead of tying the pole in place. If a standing tree is not available, construct a bipod using an Y-shaped sticks or two tripods.

(b) Place one end of the beams (10-foot poles) one side of the horizontal support. As with all lean-to type shelters, make sure the backside of the lean-to is placed into the wind.

(c) Crisscross sapling or vines on the beams.

(d) Cover the framework with brush, leaves, pine needles, or grass, starting at the bottom and working your way up like shingling.

(e) Place straw, leaves, pine needles, or grass inside the shelter for bedding.

(f) In cold weather, you can add to the comfort of your lean-to by building a fire-reflector wall (Figure 11-16). Drive four stakes about 4 feet long into the ground to support the wall. Stack green logs on top of one another between the support stales. Bind the top of the support stakes so the green logs will stay in place. Fill in the spaces between the logs with twigs or small branches. With just a little more effort, you can have a drying rack. Cut a few 3/4-inch diameter poles. The length depends on distance between the lean-to support and the top of the fire-reflector wall. Lay one end of the poles on the lean-to horizontal support and the other ends on top of the reflector wall. Place and tie into place smaller sticks across these poles. You now have a place to dry clothes, meat, or fish.

11-8. **FIRES**. A fire can full fill several needs. It can keep you warm, it can keep you dry: you can use it to cook food, to purify water, and to signal. It can also cause you problems when you are in enemy territory: it creates smoke, which can be smelled and seen from a long distance: It causes light which can be seen day or night and it leaves signs of your presence. Remember you should always weigh your need for a fire against your need to avoid enemy protection. When operating in remote areas you should always take a supply of matches in a waterproof case and keep them on your person.

a. **Selection**. When selecting a site to build a fire, you should consider the following:

(1) The area (terrain and climate) in which you are operating.

(2) The material and tools available.

(3) How much time you have.

(4) Why you need a fire.

(5) The nearness of the enemy.

b. **Preparation**. If you are in a wooded or brush-covered area, clear brush away, and scrape the surface soil from the spot you selected. The cleared circle should be at least 3 feet (1 meter) in diameter so that there is little chance of the fire spreading. To prepare the site for a fire, ensure that it is dry and that it--look for a dry spot that--

(1) Offers protection from the wind.

(2) Is suitably placed in relation to your shelter (if any).

(3) Concentrates the heat in the direction you desire.

(4) Has a supply of wood or other fire burning material.

c. **Dakota Fire Hole**. In some situations, an underground fireplace will best meet your needs. It conceals the fire to some extent and serves well for cooking food. To make an underground fireplace or Dakota fire hole (Figure 11-17):

**Figure 11-17. DAKOTA FIRE HOLE**

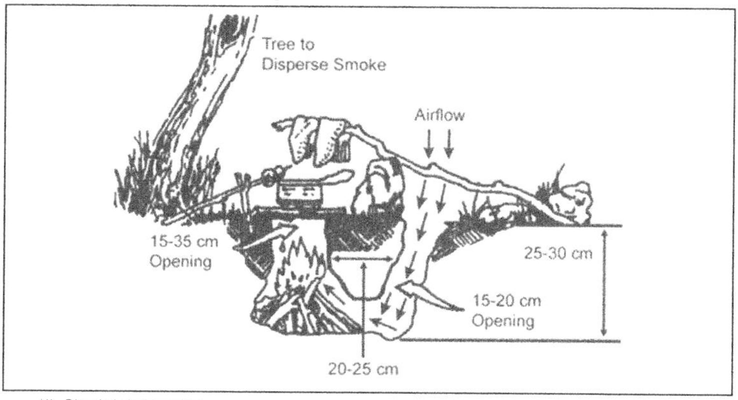


(1) Dig a hole in the ground.

(2) On the upwind side of this hole, poke one large connecting hole for ventilation.

d. **Aboveground fire**. If you are in a snow covered or wet area, you can use green logs to make a dry base for your fire (Figure 11-18). Trees with wrist-size trunks are easily broken in extreme cold. Cut or break several green logs and lay them side by side on top of the snow. Add one or two more layers, laying the top layer logs in a direction opposite those of the layer below it.

**Figure 11-18. BASE FOR FIRE IN SNOW COVERED AREA**

e. **Methods**. There are several methods for laying a fire for quick fire making. Three easy methods are Tepee, lean-to, and cross-ditch. Tepee (Figure 11-19). Arrange tinder and a few sticks of kindling in the shape of a cone. Fire the center. As the cone burns away, the outside logs will fall inward, feeding the heart of the fire. This type of fire burns well even with wet wood.

(1) ***Lean-To*** (Figure 11-19). Push a green stick into the ground at a 30 degree angle. Point the end of the stick in the direction of the wind. Place some tinder (at least a handful) deep inside this lean-to stick. Light the tinder. As the kindling catches fire from the tinder, add more kindling.

Figure 11-19. METHODS FOR LAYING A FIRE

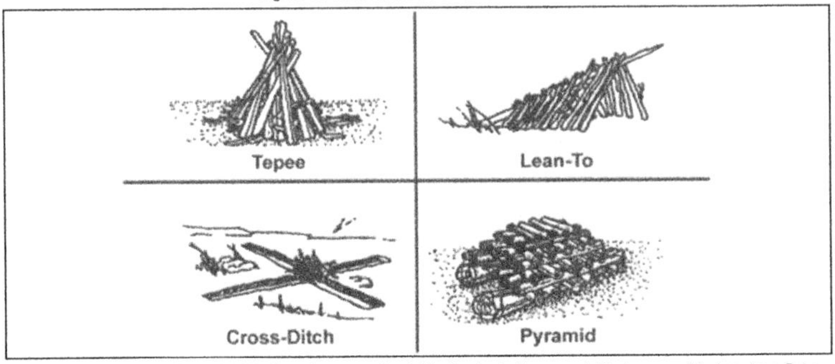


(2) ***Cross-Ditch*** (Figure 11-19). Scratch a cross about 1 foot in size in the ground. Dig the cross 3 inches deep. Put a large wad of tinder in the middle of the cross. Build a kindling pyramid above the tinder. The shallow ditch allows air to sweep under the fire to provide a draft.

NOTES

# Chapter 12
# FIRST AID

Patrolling, more than some other types of missions, puts Rangers in harms way. CASEVAC planning is vital. Also, because trained medical personnel might be unavailable at the initial point of injury, everyone must know how to diagnose and treat injuries, wounds, and common illnesses. The unit should also have a plan for handling KIAs.

12-1. **LIFESAVING STEPS**. Whatever the injury, (1) open the airway and restore breathing; (2) stop the bleeding and protect the wound; (3) check, treat and monitor for shock; and (4) MEDEVAC the casualty.

12-2. **CARE UNDER FIRE**. When still under fire, (1) maintain situational awareness; (2) return fire; (3) protect the casualty; (4) Identify and control severe bleeding with bandage or tourniquet; (5) and move the casualty to cover.

12-3. **PRIMARY SURVEY**. Use the ABC's to help you remember how to identify and deal with life-threatening injuries, airway complications, breathing problems, or uncontrolled hemorrhaging (bleeding).

| | |
|---|---|
| A | Airway. Open airway by patient position or with airway adjuncts. |
| B | Breathing. Seal open chest wounds with occlusive dressing. |
| C | Circulation. Identify uncontrolled bleeding and control with pressure or tourniquet. Start IV if needed. |
| D | Disability. Determine Level of consciousness. |
| E | Exposure. Fully expose patient. (Environment dependent) |

12-4. **AIRWAY MANAGEMENT**. The *airway* is most often obstructed at the base of the tongue.

a. If this happens, open the airway using the chin lift (nontrauma, Figure 12-1) or jaw thrust (trauma, Figure 12-2).

Figure 12-1. CHIN LIFT

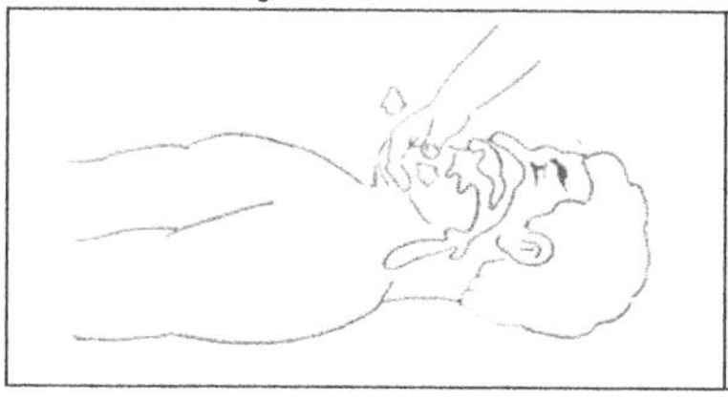

Figure 12-2. JAW THRUST

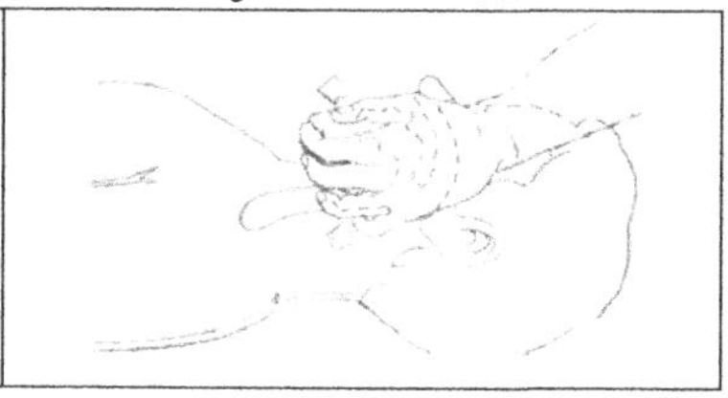

b. **Remove** debris (teeth, blood clots, bone) from the oral cavity; use suction if you have it; and place airway adjuncts to allow the victim to breathe through their nose (Figure 12-3) or mouth (Figure 12-4).

Figure 12-3. NASAL AIRWAY

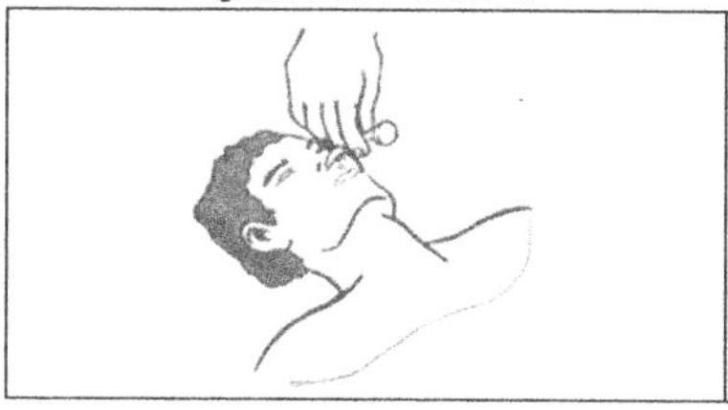

Figure 12-4. MOUTH AIRWAY

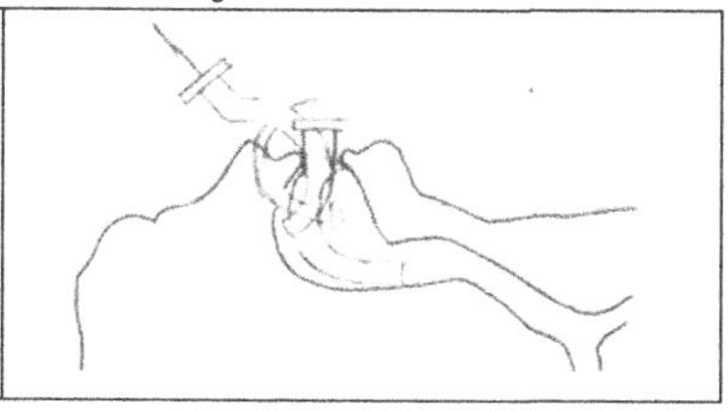

12-5. **BREATHING**. If the patient is having difficulty breathing, (1) expose the chest and identify open chest injuries; (2) place occlusive dressing (plastic and tape, or a COTS chest dressing) over open entry and exit chest wounds; and (3) place the patient on his injured side, or position him where he can breathe most comfortably.

12-6. **BLEEDING**. Quickly identify and control bleeding. (1) Apply a standard or pressure dressing. (2) If this does not control the bleeding, immediately apply a tourniquet. (3) Reassess dressings frequently to ensure bleeding has been controlled.

12-7. **SHOCK**. Shock causes an inadequate flow of oxygen to the tissues. Shock can cause bleeding (hemorrhagic) or not (nonhemorrhagic). Signs and symptoms of shock can include increased pulse, increased respiration; and decreased level of consciousness. (1) Open the victim's airway; (2) restore his breathing; (3) control the bleeding; (4) initiate an IV or a saline lock; and (5) monitor his condition.

12-8. **EXTREMITY INJURIES**. Identify and control bleeding. If you suspect a fracture, splint it as it lies. Do not reposition injured extremity.

12-9. **ABDOMINAL INJURIES**. Identify and control bleeding. Treat for shock. (1) If internal organs are exposed, cover them with dry, sterile dressing. Do not place organs back in wound area. Do not handle internal organs. (2) Place patient in position of comfort. Flex knees to relax abdomen. (3) Do not give anything by mouth to the patient.

12-10. **BURNS**. Remove patient from burn source. Then (1) remove all clothing and jewelry from the area; (2) cover with dry, sterile dressings, and ensure fingers and toes have dressings between them before covering entire area; and (3) Evacuate immediately any casualties with burns of the face, neck, hands, genitalia, or over 20 percent of his body surface.

12-11. **WEATHER (HEAT AND COLD) INJURIES**. Tables 12-1, 12-2, and 12-3 show first aid for (respectively) heat injuries, cold injuries, and environmental injuries.

**Table 12-1. HEAT INJURIES**

| *INJURY* | *SIGNS/SYMPTOMS* | *FIRST AID* |
|---|---|---|
| Heat Cramps | Casualty experiences muscle cramps in arms, legs and/or stomach, may also have wet skin and extreme thirst. | 1. Move the casualty to a shaded area and loosen clothing.<br>2. Allow casualty to drink 1 quart of cool water slowly per hour.<br>3. Monitor casualty and provide water as needed.<br>4. Seek medical attention if cramps persist. |
| Heat Exhaustion | Casualty experiences loss of appetite, headache, excessive sweating, weakness or faintness, dizziness, nausea, muscle cramps. The skin is moist, pale, and clammy. | 1. Move the casualty to a cool, shaded area and loosen clothing.<br>2. Pour water on casualty and fan to increase cooling effect of evaporation.<br>3. Provide at least one quart of water to replace lost fluids.<br>4. Elevate legs.<br>5. Seek medical aid if symptoms continue. |
| Heatstroke (Sunstroke) | Casualty stops sweating (hot, dry skin), may experience headache, dizziness, nausea, vomiting, rapid pulse and respiration, seizures, mental confusion. Casualty may suddenly collapse and lose consciousness. THIS IS A MEDICAL EMERGENCY! | 1. Move casualty to a cool, shaded area, loosen clothing, and remove outer clothing if the situation permits.<br>2. Immerse in cool water. If cool bath is not available, massage arms and legs with cool water. Fan casualty to increase the cooling effect of evaporation.<br>3. If conscious, slowly consume one quart of water.<br>4. SEEK MEDICAL AID AND EVACUATE AS SOON AS POSSIBLE. Perform any lifesaving measures. |

Table 12-2. COLD INJURIES

| *INJURY* | *SIGNS/SYMPTOMS* | *FIRST AID* |
|---|---|---|
| Chilblain | Red, swollen, hot, tender, itching skin. Continued exposure may lead to infected (ulcerated bleeding) skin lesions. | 1. Area usually responds to locally applied warming (body heat).<br>2. Do Not rub or massage area.<br>3. Seek medical treatment. |
| Immersion (trench) foot | Affected parts are cold, numb, and painless. As parts warm, they may be hot, with burning and shooting pains. Advanced stage: skin pale with bluish cast: pulse decreases, blistering, swelling, heat hemorrhages, and gangrene may follow. | 1. Gradual warming by exposure to warm air.<br>2. DO NOT massage or moisten skin.<br>3. Protect affected parts from trauma.<br>4. Dry feet thoroughly: avoid walking.<br>5. Seek medical treatment. |
| Frostbite | ***SUPERFICIAL***<br>Redness, blisters in 24-36 hours and sloughing of the skin.<br>***DEEP***<br>Preceded by superficial frostbite; skin is painless, pale- yellowish, waxy, "wooden" or solid to touch, blisters form in 12-36 hours | ***SUPERFICIAL***<br>1. Keep casualty warm; gently warm affected parts.<br>2. Decrease constricting clothing, increase exercise and insulation.<br>***DEEP***<br>1. Protect the part from additional injury.<br>2. Seek medical treatment as fast as possible. |
| Snow Blindness | Red, scratchy, or watery eyes; headache; increased pain in eyes with exposure to light. | 1. Cover the eyes with a dark cloth.<br>2. Seek medical treatment. |
| Dehydration | Similar to heat exhaustion. | 1. Keep warm, loosen clothes.<br>2. Replace lost fluids, rest, and additional medical treatment. |
| Hypothermia | Casualty is cold, shivers uncontrollably until shivering stops. Rectal (core) temp less 95° F can affect consciousness. Uncoordinated movements, shock, and coma may occur as body temperature drops. | ***MILD HYPOTHERMIA***<br>1. Warm body evenly and without delay. (Heat source must be provided.)<br>2. Keep dry, protect from elements.<br>3. Warm liquids may be given to conscious casualty only.<br>4. Be prepared to start CPR.<br>5. Seek medical treatment immediately.<br>***SEVERE HYPOTHERMIA***<br>1. Quickly stabilize body temperature.<br>2. Attempt to prevent further heat loss.<br>3. Handle the casualty gently.<br>4. Evacuate to nearest medical treatment facility as soon as possible. |

Table 12-3. ENVIRONMENTAL INJURIES

| *TYPE* | *FIRST AID* |
|---|---|
| Snake bite | 1. Get the casualty away from the snake.<br>2. Remove all rings and bracelets from the affected extremity.<br>3. Reassure the casualty and keep him quiet.<br>4. Apply constricting band(s) 1 to 2 finger widths close to the bite. You should be able to slip 1 finger between the band and skin.<br>***ARM OR LEG BITE*** - Place one band above and one band below the bite site.<br>***HAND OR FOOT BITE*** - Place one band above the wrist or ankle.<br>5. Immobilize the affected limb below the level of the heart.<br>6. Kill the snake, if possible, (without damaging its head or endangering yourself) and send it with the casualty.<br>7. Seek medical treatment immediately. |
| Brown recluse or black widow, spider bite | 1. Keep the casualty calm.<br>2. Wash the area.<br>3. Apply ice or a freeze pack, if available.<br>4. Seek medical treatment. |
| Tarantula bite, scorpion sting, ant bite | 1. Wash the area.<br>2. Apply ice or a freeze pack, if available.<br>3. Apply baking soda, calamine lotion, or meat tenderizer to the bite site to relieve pain and itching.<br>4. If site of bite(s) or sting(s) is on the face, neck (possible airway blockage), or genital area, or if reaction is severe, or if the sting is by the dangerous Southwestern scorpion, keep the casualty as quiet as possible, administer the epinephrine auto-injector (EAI) if needed, and then seek immediate medical aid. |
| Wasp or bee sting | 1. If the stinger is present, remove by scraping with a knife or finger nail. DO NOT squeeze venom sack on stinger, more venom may be injected.<br>2. Wash the area.<br>3. Apply ice or freeze pack, if available.<br>4. If allergic signs or symptoms appear, be prepared to administer the EAI, and then seek medical assistance. |
| Human and other animal bites | 1. Cleanse the wound thoroughly with soap or detergent solution.<br>2. Flush bite well with water.<br>3. Cover bite with a sterile dressing.<br>4. Immobilize injured extremity.<br>5. Transport casualty to a medical treatment facility.<br>6. For animal bites, without endangering yourself or damaging the animal's head, kill the animal and send its head with the casualty.<br>7. For human bites, try to extract some of the attacker's saliva from the wound and send that in a sealed, identified container with the casualty. |

Table 12-3. ENVIRONMENTAL INJURIES (CONTINUED)

| | |
|---|---|
| Poison Ivy, Oak, Sumac | 1. Gently clean affected area. Clean areas 2-3 times daily. Wash clothing.<br>2. Apply topical anti-itch lotion or ointment as needed, and cover.<br>3. Avoid scratching or itching the area.<br>4. Observe for signs of infection (increasing redness, tenderness, warm to touch).<br>5. Seek medical attention if rash persists or signs of infection develop. |

12-12. **POISONOUS PLANT IDENTIFICATION**

Poison plants include, among others, poison ivy, oak, and sumac, as well as a few more such as stinging nettles which are not discussed here.

a. **Poison Ivy**. Poison ivy is grows in two forms, a vine or shrub. The compound leaves of poison ivy consist of three pointed leaflets; the middle leaflet has a much longer stalk than the two side ones. The leaflet edges can be smooth or toothed but are rarely lobed. The leaves vary greatly in size, from 8 to 55 mm (0.31" to 2.16") in length. They are reddish when they emerge in the spring, turn green during the summer, and become various shades of yellow, orange, or red in the autumn. Small greenish flowers grow in bunches attached to the main stem close to where each leaf joins it. Later in the season, clusters of poisonous, berrylike drupes form. They are whitish, with a waxy look.

b. **Poison Oak**. Poison oak is a widespread deciduous shrub throughout mountains and valleys of North America, generally below 5,000 feet elevation. It commonly grows as a climbing vine with aerial (adventitious) roots that adhere to the trunks of oaks and sycamores. Poison oak also forms dense thickets in chaparral and coastal sage scrub. Poison oak leave typically have three leaflets (sometimes five), the terminal one on a slender rachis (also called a stalk or petiolule). Eastern poison ivy often has a longer rachis and the leaflet margins tend to be less lobed and serrated (less "oak-like"). Like many members of the sumac family (Anacardiaceae) new foliage and autumn leaves often turn brilliant shades of pink and red.

c. **Poison Sumac**. Poison Sumac is a woody perennial shrub or small tree growing from 5 - 25 feet tall that grows in peat bogs and swamps. To identify Poison Sumac, look for the fruit that grows between the leaf and the branch. Leaves are arranged in an alternate pattern on the vine. There are about 7-13 leaflets forming a feather-like appearance. The foliage has brilliant orange or scarlet coloring in the fall.

Figure 12-5. POISONOUS PLANTS

12-13. **FOOT CARE.** Use moleskin to prevent blisters prior to movement or foot march. Drain large blisters. Clean area, puncture with needle, drain blister. Place moleskin over area. Observe for signs of infection. Keep feet as clean and dry as possible. Use foot powder and change socks. Let feet air dry as mission permits.

12-14. **CASUALTY RESCUE AND TRANSPORT SYSTEM.** Remove the hoistable CRTS litter from its pack, and place the litter on the ground. Unfasten the retainer strap, step on the foot end of the litter, and unroll the litter completely to the opposite end. Then--

a. Bend the CRTS litter in half and backroll it. Repeat with the opposite end of the litter. The litter will now lay flat. Lift the sides of the litter, and fasten the four cross-straps to the buckles directly opposite the straps.

b. Place the CRTS litter next to the casualty. Insure the head end of the stretcher is adjacent to the head of casualty. Place the cross-straps under the litter. Logroll the casualty and slide the litter as far under the casualty as possible. Gently roll the casualty down on the litter. Slide the casualty to the center of the litter. Be sure to keep his spinal column as straight as possible.

c. Pull the straps out from under the CRTS litter and fasten it to the buckles. Position the foot end of the litter at the head of the casualty. Have one rescuer straddle the litter and support the casualty's head, neck, and shoulders.

d. Grasp the foot straps of the litter, and slide the litter under the casualty. Center the casualty on the litter, and feed foot straps thru the unused grommets at the foot end of the litter. Fasten the straps to the buckles.

12-15. **HYDRATION AND ACCLIMATIZATION**

a. Table 12-4 shows strategies for minimizing dehydration and increasing acclimatization and good hydration practices.

**Table 12-4. HYDRATION MANAGEMENT AND ACCLIMATIZATION**

| ***STRATEGY*** | ***SUGGESTIONS FOR IMPLEMENTATION*** |
|---|---|
| Start early | 1. Start at least 1 month prior to School<br>2. Be flexible and patient: performance benefits take longer than the physiological benefits |
| Mimic the training environment climate | 1. In warm climates, acclimatize in the heat of day.<br>2. In temperate climates workout in a warm room wearing sweats. |
| Ensure adequate heat stress | 1. Induce sweating.<br>2. Work up to 100 minutes of continuous physical exercise in the heat. Be patient. The first few days, you may not be able to go 100 minutes without resting.<br>3. Once you can comfortably exercise for 100 minutes in the heat, then continue for at least 7-14 days with added exercise intensity (loads or training runs). |
| Teach yourself to drink and eat | 1. Your thirst mechanism will improve as you become heat acclimatized, but you will still under-drink if relying on thirst sensation.<br>2. Heat acclimatization will increase your water requirements.<br>3. Dehydration will negate most benefits of physical fitness and heat acclimatization.<br>4. You will sweat out more electrolytes when not acclimatized, so add salt to your food, or drink electrolyte solutions during the first week of heat acclimatization.<br>5. A convenient way to learn how much water your body needs to replace is to weigh yourself before and after the 100 minutes of exercise in the heat. For each pound lost, you should drink about one-half quart of fluid.<br>6. Do not skip meals, as this is when your body replaces most of its water and salt losses. |

12-16. **WORK, REST, AND WATER CONSUMPTION.** Table 12-5 shows a work, rest, and water consumption table. The guidance applies to the average size, heat acclimated Soldier wearing ACU, not hot weather gear, except as specified otherwise. The work and rest times and fluid replacement volumes shown will help the Soldier sustain his performance and hydration for at least 4 hours of work in the specified heat category. Fluid needs can vary based on individual differences (±1 quart per hour).

a. "NL" means that there is no limit to work time per hour. "Rest" means minimal physical activity such as sitting or standing, preferably in the shade.

b. Consume no more than 1.5 quarts of fluid per hour, and no more than 12 quarts per day.

c. If you are wearing body armor in a humid climate, then add 5° F to the WBGT. if wearing MOPP 4 clothing, add 10° F to the WBGT.

d. Work categories include easy, moderate, and hard.

1) ***Easy Work***. This includes, for example, maintaining weapons; walking on hard surfaces at 2.5 mph with a load weighing no more than 30 pounds; participating in marksmanship training; and participating in drills or ceremonies.

2) ***Moderate Work***. This includes, for example, walking in loose sand at 2.5 mph (maximum) or under with no load; walking on a hard surface at 3.5 mph (maximum) with a load weighing no more than 40 pounds; performing calisthenics; patrolling; or conducting individual movement techniques such as the low or high crawl.

3) ***Hard Work***. This includes, for example, walking on a hard surface at 3.5 mph with a load weighing 40 or more pounds; walking in loose sand at 2.5 mph while carrying a load; and conducting field assaults.

**Table 12-5. WORK, REST, AND WATER CONSUMPTION TABLE**

| HEAT CATEGORY | WBGT INDEX, F° | *EASY WORK* | | *MODERATE WORK* | | *HARD WORK* | |
|---|---|---|---|---|---|---|---|
| | | WORK/ REST | WATER INTAKE (QT/H) | WORK/ REST | WATER INTAKE (QT/H) | WORK/ REST | WATER INTAKE (QT/H) |
| 1 | 78° to 81.9° | NL | 0.50 | NL | 0.75 | 40/20 | 0.75 |
| 2 (Green) | 82° to 84.9° | NL | 0.50 | 50/10 | 0.75 | 30/30 | 1.00 |
| 3 (Yellow) | 95° to 87.9° | NL | 0.75 | 40/20 | 0.75 | 30/30 | 1.00 |
| 4 (Red) | 88° to 89.9° | NL | 0.75 | 30/30 | 0.75 | 20/40 | 1.00 |
| 5 (Black) | 90° or more | 50/10 min | 1.00 | 20/40 | 1.00 | 10/50 | 1.00 |

NOTES

## Chapter 13
# DEMOLITIONS

This chapter introduces Rangers to the characteristics of explosives, to initiation systems, MDI components, detonation systems, safety considerations, expedient explosives, breaching charges, and timber cutting charges (FM 5-250). The two categories of explosives are low and high (Table 13-1).

- *Low explosives* have a detonating velocity up to 1,300 feet per second, which produces a pushing or shoving effect.
- *High explosives* have a detonating velocity of 3,280 to 27,888 feet per second, which produces a shattering effect.

**Table 13-1. CHARACTERISTICS OF US DEMOLITIONS EXPLOSIVES**

| NAME | APPLICATIONS | DETONATION VELOCITY | | RE FACTOR* | FUME TOXICITY | WATER RESISTANCE |
|---|---|---|---|---|---|---|
| | | MIN/SEC | FT/SEC | | | |
| AMMONIUM NITRATE | CRATERING CHG | 2,700 | 8.800 | 0.42 | DANGEROUS | POOR |
| PETN | DET CORD<br>BLASTING CAPS<br>DEMOLITION CHGS | 8,300 | 27,200 | 1.66 | SLIGHT | EXCELLENT |
| RDX | BLASTING CAPS<br>COMPOSITION EXPLOSIVE | 8,350 | 27,400 | 1.60 | DANGEROUS | EXCELLENT |
| TRINITROTOLUENE (TNT) | DEMOLITION CHG<br>COMPOSITION EXPLOSIVE | 6,900 | 22,600 | 1.00 | DANGEROUS | EXCELLENT |
| TETRYL | BOOSTER CHG<br>COMPOSITION EXPLOSIVE | 7,100 | 23,300 | 1.25 | DANGEROUS | EXCELLENT |
| NITROGLYCERIN | COMMERCIAL DYNAMITE | 7,700 | 25,200 | 1.50 | DANGEROUS | GOOD |
| BLACK POWDER | TIME FUSE | 400 | 1,300 | 0.55 | DANGEROUS | POOR |
| AMATOL 80/20 | BURSTING CHG | 4,900 | 16,000 | 1.17 | DANGEROUS | POOR |
| COMPOSITION A3 | BOOSTER CHG<br>BURSTING CHG | 8,100 | 26,500 | --- | DANGEROUS | GOOD |
| COMPOSITION B | BURSTING CHG | 7,800 | 25,600 | 1.35 | DANGEROUS | EXCELLENT |
| COMPOSITION C4 (M112) | CUTTING CHG<br>BREACHING CHG | 8.040 | 26,400 | 1.34 | SLIGHT | EXCELLENT |
| COMPOSITION H6 | CRATERING CHG | 7,190 | 23,600 | 1.33 | DANGEROUS | EXCELLENT |
| TETRYTOL 75/25 | DEMOLITION CHG | 7,000 | 23,000 | 1.20 | DANGEROUS | EXCELLENT |
| PENTOLITE 50/50 | BOOSTER CHG<br>BURSTING CHG | 7,450 | 24,400 | --- | DANGEROUS | EXCELLENT |

| NAME | APPLICATIONS | DETONATION VELOCITY MIN/SEC | FT/SEC | RE FACTOR* | FUME TOXICITY | WATER RESISTANCE |
|---|---|---|---|---|---|---|
| M1 DYNAMITE | DEMOLITION CHG | 6,100 | 20,000 | 0.92 | DANGEROUS | FAIR |
| DET CORD | PRIMING DEMOLITION CHG | 6,100 TO 7,300 | 20,000 TO 24,000 | --- | SLIGHT | EXCELLENT |
| SHEET EXPLOSIVE M118 AND M186 | CUTTING CHG | 7,300 | 24,000 | 1.14 | DANGEROUS | EXCELLENT |
| BANGALORE TORPEDO M1A2 | DEMOLITION CHG | 7,800 | 25,600 | 1.17 | DANGEROUS | EXCELLENT |
| SHAPED CHARGES M2A3, M2A4, AND M3A1 | CUTTING CHG | 7,800 | 25,600 | 1.17 | DANGEROUS | EXCELLENT |

** TNT = 1.00 RELATIVE EFFECTIVENESS*

13-1. **INITIATING (PRIMING) SYSTEMS**. The best way to prime demolition systems is with modernized demolition initiators (MDI). These consist of blasting caps attached to various lengths of time fuse or shock tube. MDI can be used with a fuse igniter and detonating cord to create many firing systems. In the absence of MDI, field expedient methods may be used.

a. **Shock Tube**.

(1) Thin, plastic tube of extruded polymer with a layer of special explosive material on the interior surface.

(2) Explosive material propagates a detonation wave which moves along the shock tube to a factory crimped and sealed blasting cap.

(3) Detonation is normally contained within the plastic tubing. However, burns may occur if the shock tube is held.

(4) Advantages of shock tube.

(a) The shock tube offers the instantaneous action of electric initiation without the risk of accidental initiation caused by radio transmitters, static electricity, and so on.

(b) Extremely reliable.

(c) May be extended using left over pieces from previous operations.

b. **Blasting Caps**. Five MDI blasting caps are available which replace the M6 electric and M7 Non-electric blasting cap. Three of these are high-strength caps and two are low-strength. High-strength blasting caps can be used to prime all standard military explosives (including detonating cord) or to initiate the shock tube of other MDI blasting caps.

(1) ***M11***.

- Factory crimped to 30 feet of shock tube.
- A movable "J" hook is attached for quick and easy attachment to detonating cord.
- A red flag is attached 1 meter from the blasting cap and a yellow flag 2 meters from the blasting cap.

(2) ***M14***.

- Factory crimped to 7.5 feet of time fuse.
- May be initiated using a fuse igniter or match.
- Burn-time for total length is about five minutes.
- Yellow bands indicate calibrated one-minute time intervals.

(3) ***M15***.

- Two blasting caps factory crimped to 70 feet of shock tube.
- Each blasting cap has delay elements to allow for staged detonations.
- Low-strength blasting caps. Used as a relay device to transmit a shock tube detonation impulse from an initiator to a high strength-blasting cap.

(4) ***M12***. Factory crimped to 500 feet of shock tube on a cardboard spool.

(5) ***M13***. Factory crimped to 1,000 feet of shock tube.

c. **Matches**. If fuse igniter is unavailable, light the time blasting fuse with a match. Split the fuse at the end (Figure 13-1), and place the head of an unlit match in the powder train. Light the inserted match head with a flaming match, or rub the abrasive on the match box against it. It may be necessary to use two match heads during windy conditions. Burn time will increase with altitude and colder temperatures.

d. **M81 Fuse Igniter**. This is used to ignite time blasting fuse or to initiate the shock tube of MDI blasting caps.

**Figure 13-1. LIGHTING TIME FUSE WITH A MATCH**

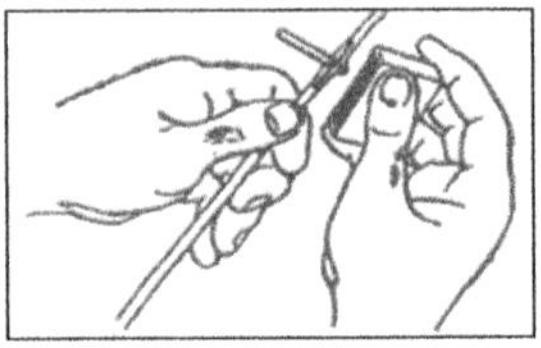

***Notes:*** Low strength blasting caps cannot reliably set off explosives. They should only be used to set off additional shock tubes. The M60 fuse igniter may still be used to ignite time blasting fuse. However, it will not reliably initiate the shock tube.

13-2. **DETONATION (FIRING) SYSTEMS**. The two types of firing systems are MDI alone, or MDI plus detonating cord.

a. **MDI Alone**. An MDI firing system is one in which the initiation set, transmission and branch lines are constructed using MDI components and the explosive charges are primed with MDI blasting caps. Construct the charge in the following manner.

(1) Emplace and secure explosive charge (C4, TNT, cratering charge, and so on.) on target.

(2) Place a sandbag or other easily identifiable marker over the M11, M14, or M15 blasting cap to be used.

(3) Connect to an M12 or M13 transmission line if desired.

(4) Connect blasting cap with shock tube to an M14 cap with time fuse. Cut the time blasting fuse to desired delay time.

(5) Prime the explosive charge by inserting the blasting cap into the charge.

(6) Visually inspect firing system for possible misfire indicators such as cracks, bulges, or corrosion.

(7) Return to the firing point and secure a fuse igniter to the cut end of the time fuse.

(8) Remove the safety cotter pin from the igniter's body.

(9) Actuate the charge by grasping the igniter body with one hand while sharply pulling the pull ring.

b. **MDI and Detonating Cord**. Construct the charge using the above steps for MDI stand-alone system. Incorporate detonating cord branch lines into the system using the "J" hooks of the M11 shock tube. Taping the ends of the Detonation Cord reduces the effect of moisture on the system.

13-3. **SAFETY**. MDI is not recommended for below ground use, except in quarry operations with water-gel or slurry explosives. Use detonating cord when it is necessary to bury primed charges.

a. Do not handle misfires downrange until the required 30 minute waiting period for both primary and secondary initiation systems has elapsed and other safety precautions have been accomplished.

b. Never yank or pull hard on the shock tube. This may actuate the blasting cap.

c. Do not dispose of used shock tubes by burning because of potentially toxic fumes given off from the burning plastic.

d. Do not use M1 dynamite with the M15 blasting cap. The M15 delay-blasting cap should be used only with water-gel or slurry explosives.

e. Always use protective equipment when handling demolitions. Minimum protection consists of leather gloves, ballistic eye protection, and helmet.

13-4. **EXPEDIENT EXPLOSIVES--IMPROVISED SHAPED CHARGE**. An improvised shaped charge (Figure 13-2) concentrates the energy of the explosion released on a small area, making a tubular or linear fracture in the target.

a. The versatility and simplicity of these charges make them effective against targets, especially those made of concrete or those with armor plating.

(1) Bowls, funnels, cone shaped glasses, (champagne glasses with stem removed) used as cones. Champagne or cognac bottles are excellent.

(2) Charge characteristics.

(a) *Cavity Liners*. These are made of copper, tin, or zinc. If none is available, cut a cavity out of the plastic explosive.

(b) *Cavity Angle*. This will work with 30 to 60 degree angles. The cavity angle in most high-explosive antitank (HEAT) ammunition is 42 to 45 degrees.

(c) *Explosive Height (In Container)*. 2 x the height of the cone measured from the base of the cone to the top of the explosive.

(d) *Standoff*. Normal standoff is one and one half the cone's diameter.

(e) *Detonation Point*. The exact top center of the charge is the detonation point. Cover the blasting cap with a small amount of C4 if any part of the blasting cap is exposed.

b. Remove the narrow neck of a bottle or the stem of a glass by wrapping it with a piece of soft, absorbent twine or by soaking the string in gasoline and lighting it. Place two bands of adhesive tape, one on each side of the twine, to hold the twine firmly in place. The bottle or stem must be turned continuously with the neck up, to heat the glass uniformly.

c. A narrow band of plastic explosive placed around the neck and burned gives the same result. After the twine or plastic has burned, submerge the neck of the bottle in water and tap it against some object to break it off. Tape the sharp edge of the bottle to prevent cutting hands while tamping the explosive in place.

d. Do not immerse the bottle in water before the plastic has been completely burned, or it could detonate.

**Figure 13-2. IMPROVISED SHAPED CHARGE**

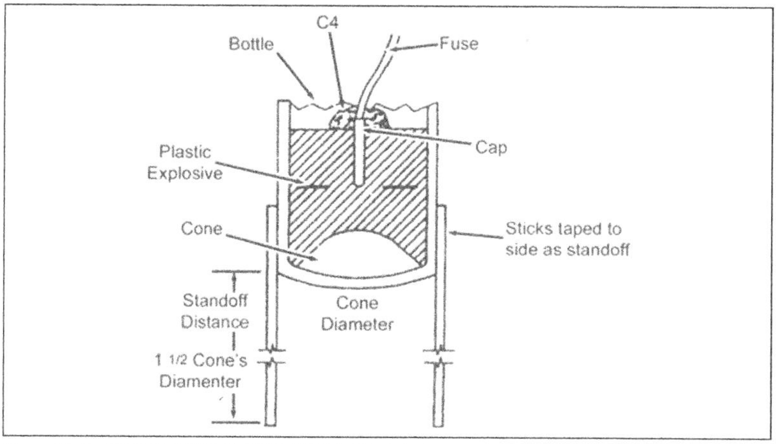


13-5. **EXPEDIENT EXPLOSIVES--PLATTER CHARGE**. This device turns a metal plate into a powerful blunt-nosed projectile (Figure 13-3). The plate should be steel, preferably round but square will work, and should weigh from 2 to 6 pounds.

a. The weight of the explosive should equal the weight of the platter.

b. Uniformly pack the explosive behind the platter. You will only need a container if the explosives fail to remain firmly against the platter. You can use tape to anchor the explosives, if needed.

c. Prime the charge at the exact, rear center of the charge. If any part of the blasting cap is exposed, cover the blasting cap with a small quantity of C4.

d. Aim charge at the direct center of the target, and ensure that the charge is on the opposite side of the platter from the target. Effective range is 35 yards for a small target. With practice, you might hit a 55-gallon drum at 25 yards 90 percent of the time. A gutted fuse igniter can serve as an expedient aiming device.

**Figure 13-3. PLATTER CHARGE**

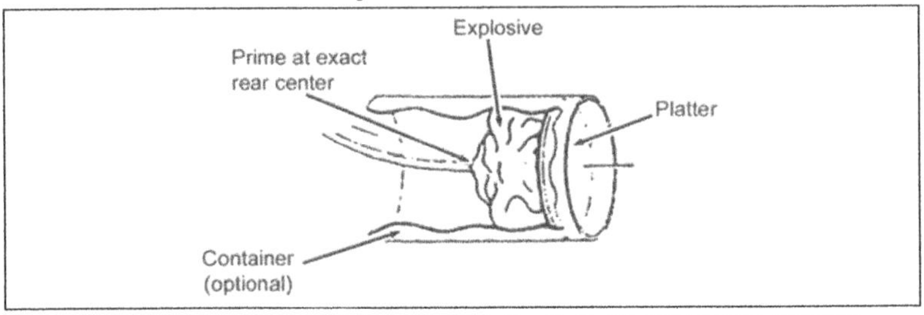


13-6. **EXPEDIENT EXPLOSIVES--GRAPESHOT CHARGE**. To use this antipersonnel fragmentation mine (Figure 13-4)--

a. **Hole**. Create a hole in the center, bottom of the container, for the blasting cap.

b. **Explosives**. Place explosives evenly on the bottom of the container. Remove all voids and air pockets by pressing the C4 into place using a non-sparking instrument.

c. **Buffer**. Place buffer material directly over top of the explosives.

d. **Projectiles**. Place projectiles over top of the buffer materials, then cover to prevent spilling from movement.

e. **Aim**. Aim at target from about 100 feet. Use a small amount of C4 on any exposed portion of the blasting cap.

**Figure 13-4. GRAPESHOT CHARGE**

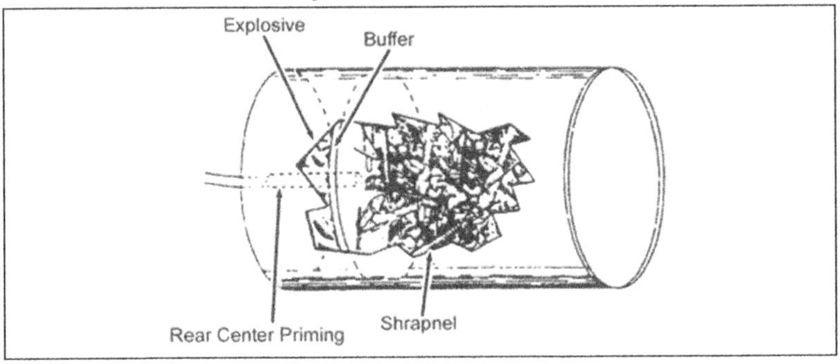

13-7. **DEMOLITION KNOTS.** Several knots are used in demolitions. Figures 13-5 and 13-6 show a few simple knots that can join demolitions to detonation cord.

**Figure 13-5. VARIOUS JOINING KNOTS USED IN DEMOLITIONS**

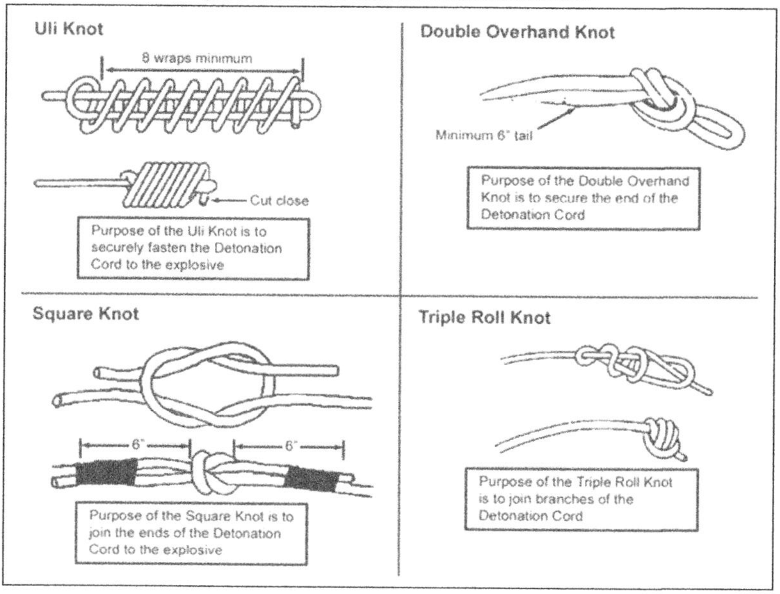

Figure 13-6. BRITISH JUNCTION

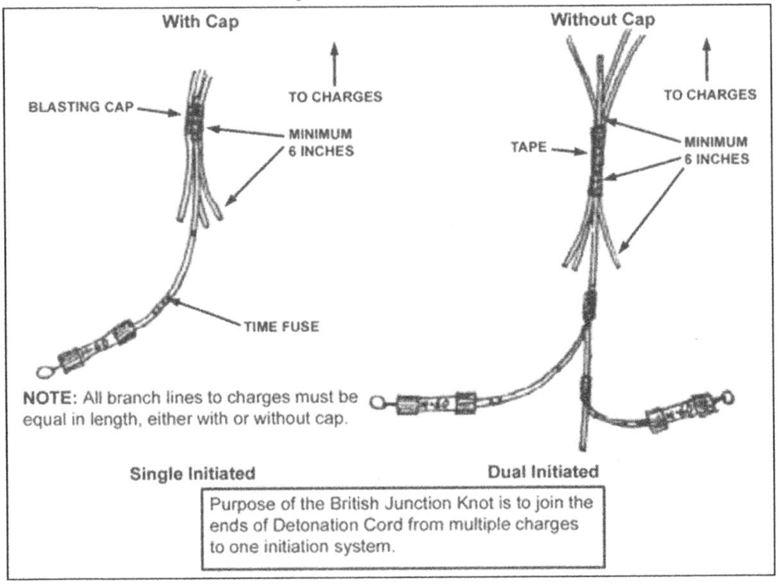

13-8. **MINIMUM SAFE DISTANCES**. Rangers must remain especially aware of their situation when using demolitions. (Table 13-2 shows minimum safe distances for employing demolitions.) For charges over 500 pounds, use the formulas shown in Figure 13-7.

**Table 13-2. MINIMUM SAFE DISTANCE FOR PERSONNEL IN OPEN (BARE CHARGE)**

| EXPLOSIVE WEIGHT (LB) | SAFE DISTANCE | | EXPLOSIVE WEIGHT (LB) | SAFE DISTANCE | |
|---|---|---|---|---|---|
| | FEET | METERS | | FEET | METERS |
| 27 OR LESS | 985 | 300 | 175 | 1,838 | 560 |
| 30 | 1,021 | 311 | 200 | 1,920 | 585 |
| 35 | 1,073 | 327 | 225 | 1,999 | 609 |
| 40 | 1,123 | 342 | 250 | 2,067 | 630 |
| 45 | 1,168 | 356 | 275 | 2,136 | 651 |
| 50 | 1,211 | 369 | 300 | 2,199 | 670 |
| 60 | 1,287 | 392 | 325 | 2,258 | 688 |
| 70 | 1,355 | 413 | 350 | 2,313 | 705 |
| 80 | 1,415 | 431 | 375 | 2,369 | 722 |
| 90 | 1,474 | 449 | 400 | 2,418 | 737 |
| 100 | 1,526 | 465 | 425 | 2,461 | 750 |
| 125 | 1,641 | 500 | 500 | 2,625 | 800 |
| 150 | 1,752 | 534 | --- | --- | --- |

**Figure 13-7. MINIMUM SAFE DISTANCES FOR CHARGES OVER 500 POUNDS**

$$\text{safe distance (meters)} = 100\sqrt[3]{\text{pounds of explosive}}$$

$$\text{safe distance (feet)} = 300\sqrt[3]{\text{pounds of explosive}}$$

***Notes:*** For charges on targets, the minimum radius of danger is 1,000 meters. Minimum safe distance when in a missile-proof shelter from the point of detonation is 100 meters.

13-9. **BREACHING CHARGES**. For Table 13-3, the left column represents the thickness of reinforced concrete wall. The remaining 7 columns show the number of packages of C4 required to breach the wall using the charge placements shown in the drawings above the columns.

a. Use Table 13-3, 13-4 and 13-5 for breaching charges.

b. Use the formula in Figure 13-8 to calculate the charges.

c. Multiply number of packages of C4 from Table 13-3 by conversion factor from Table 13-4 for materials other than reinforced concrete.

**Table 13-3. BREACHING CHARGES FOR REINFORCED CONCRETE**

| Rein-forced-Concrete Thick-ness (ft) | Placement Methods | | | | | | |
|---|---|---|---|---|---|---|---|
| | Placed in center of mass | Tamped or stemmed (Fill) | Deep water | Elevated untamped | Shallow water | Earth tamping | Ground placed, untamped |
| | C = 1.0 | C = 1.0 | C = 1.0 | C = 1.8 | C = 2.0 | C = 2.0 | C = 3.6 |
| | Packages of M112 (C4) | | | | | | |
| 2.0 | 1 | 5 | 5 | 9 | 10 | 10 | 17 |
| 2.5 | 2 | 9 | 9 | 17 | 18 | 18 | 33 |
| 3.0 | 2 | 13 | 13 | 24 | 26 | 26 | 47 |
| 3.5 | 4 | 21 | 21 | 37 | 41 | 41 | 74 |
| 4.0 | 5 | 31 | 31 | 56 | 62 | 62 | 111 |
| 4.5 | 7 | 44 | 44 | 79 | 88 | 88 | 157 |
| 5.0 | 9 | 48 | 48 | 85 | 95 | 95 | 170 |
| 5.5 | 12 | 63 | 63 | 113 | 126 | 126 | 226 |
| 6.0 | 13 | 82 | 82 | 147 | 163 | 163 | 293 |
| 6.5 | 17 | 104 | 104 | 186 | 207 | 207 | 372 |
| 7.0 | 21 | 111 | 111 | 200 | 222 | 222 | 399 |
| 7.5 | 26 | 137 | 137 | 245 | 273 | 273 | 490 |
| 8.0 | 31 | 166 | 166 | 298 | 331 | 331 | 595 |

NOTE: The results of all calculations for this table have been rounded UP to the next whole package.

Table 13-4. CONVERSION FACTORS FOR MATERIALS OTHER THAN REINFORCED CONCRETE

| ***MATERIAL*** | ***CONVERSION FACTOR*** |
|---|---|
| • EARTH | 0.1 |
| • ORDINARY MASONRY<br>• HARDPAN<br>• SHALE<br>• ORDINARY CONCRETE<br>• ROCK<br>• GOOD TIMBER<br>• EARTH CONSTRUCTION | 0.5 |
| • DENSE CONCRETE<br>• FIRST-CLASS MASONRY | 0.7 |

Table 13-5. MATERIAL FACTOR (K) FOR BREACHING CHARGES

| Material | R | K |
|---|---|---|
| Earth | All values | 0.07 |
| Poor masonry, shale, hardpan, good timber, and earth construction | Less than 1.5 m (5 ft) | 0.32 |
| | 1.5 m (5 ft) or more | 0.29 |
| Good masonry, concrete block, and rock | 0.3 m (1 ft) or less | 0.88 |
| | Over 0.3 m (1 ft) to less than 0.9 m (3 ft) | 0.48 |
| | 0.9 m (3 ft) to less than 1.5 m (5 ft) | 0.40 |
| | 1.5 m (5 ft) to less than 2.1 m (7 ft) | 0.32 |
| | 2.1 m (7 ft) or more | 0.27 |
| Dense concrete and first-class masonry | 0.3 m (1 ft) or less | 1.14 |
| | Over 0.3 m (1 ft) to less than 0.9 m (3 ft) | 0.62 |
| | 0.9 m (3 ft) to less than 1.5 m (5 ft) | 0.52 |
| | 1.5 m (5 ft) to less than 2.1 m (7 ft) | 0.41 |
| | 2.1 m (7 ft) or more | 0.35 |
| Reinforced concrete (factor does not consider cutting steel) | 0.3 m (1 ft) or less | 1.76 |
| | Over 0.3 m (1 ft) to less than 0.9 m (3 ft) | 0.96 |
| | 0.9 m (3 ft) to less than 1.5 m (5 ft) | 0.80 |
| | 1.5 m (5 ft) to less than 2.1 m (7 ft) | 0.63 |
| | 2.1 m (7 ft) or more | 0.54 |

Figure 13-8. FORMULA FOR COMPUTING SIZES OF CHARGES TO BREACH CONCRETE, MASONRY, AND ROCK

$$P = R^3KC$$

where—

*P* = *TNT required, in pounds*

*R* = *breaching radius, in feet*

*K* = *material factor, which reflects the strength, hardness, and mass of the material to be demolished (Table 3-5)*

*C* = *tamping factor, which depends on the location and tamping of the charge (Figure 3-15)*

13-10. **TIMBER CUTTING CHARGES**. Table 13-6 shows timber-cutting charge sizes. Figures 13-9 through 13-15 show the types of charges and the formulas to use with each.

Table 13-6. TIMBER-CUTTING CHARGE SIZE

| | Packages of C4 Required (1.25-lb Packages) by Timber Diameter (in) | | | | | | | | | | | |
|---|---|---|---|---|---|---|---|---|---|---|---|---|
| Charge Type | 6 | 8 | 10 | 12 | 15 | 18 | 21 | 24 | 27 | 30 | 33 | 36 |
| Internal | 1 | 1 | 1 | 1 | 1 | 1 | 2 | 2 | 2 | 3 | 3 | 4 |
| External | 1 | 1 | 2 | 3 | 4 | 5 | 7 | 9 | 11 | 14 | 17 | 20 |
| Abatis | --- | --- | --- | --- | --- | --- | --- | 7 | 9 | 11 | 14 | 16 |
| NOTE: Packages required are rounded UP to the next whole package. | | | | | | | | | | | | |

Figure 13-9. ABATIS

Figure 13-10. FORMULA FOR FALLEN TREE OBSTACLES OR TEST SHOT

$$P = \frac{D^2}{50} = P = 0.02D^2$$

where—

$P$ = TNT required per tree, in pounds

$D$ = diameter or least dimension of dimensioned timber, in inches

Figure 13-11. TIMBER-CUTTING RING CHARGE

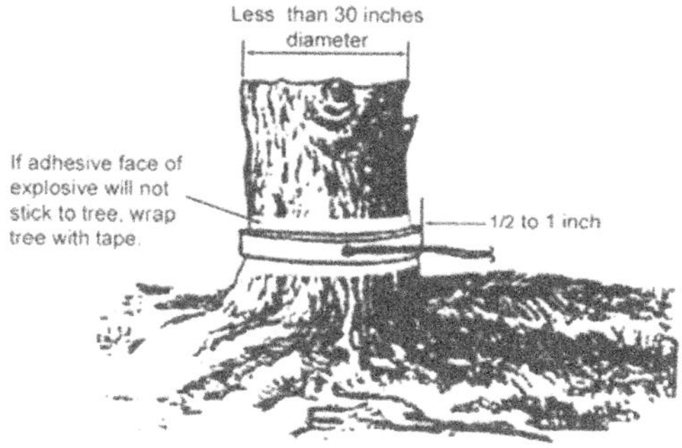

Figure 13-12. TIMBER-CUTTING CHARGE (EXTERNAL)

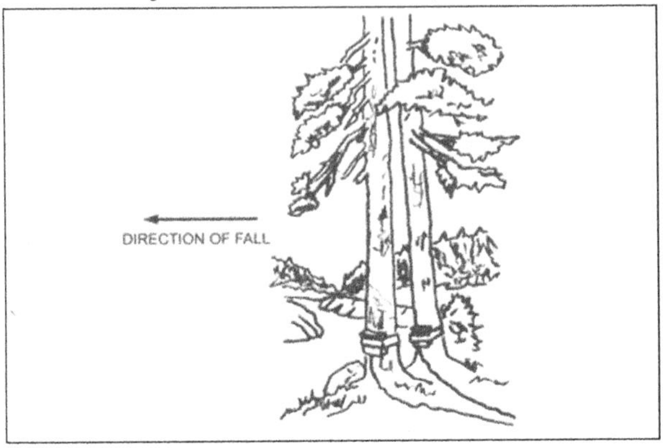


Figure 13-13. FORMULA FOR EXTERNAL TIMBER-CUTTING CHARGE

$$P = \frac{D^2}{40} \text{ or } P = 0.025D^2$$

where—

*P = TNT required per target, in pounds*

*D = diameter or least dimension of dimensioned timber, in inches*

Figure 13-14. TIMBER-CUTTING CHARGE (INTERNAL)

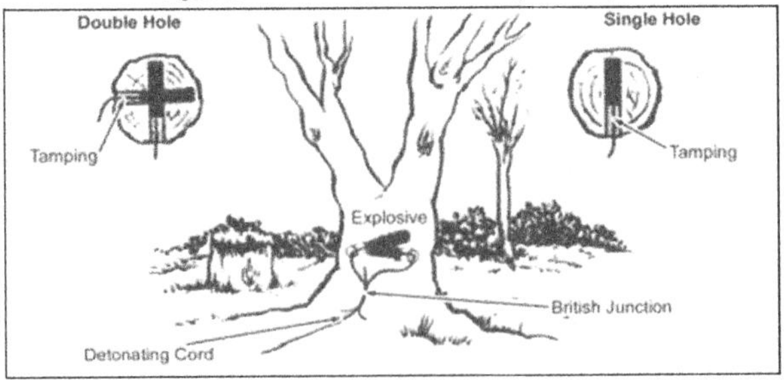

Figure 13-15. FORMULA FOR INTERNAL TIMBER-CUTTING CHARGE

$$P = \frac{D^2}{250} \text{ or } P = 0.004D^2$$

where—

$P$ = TNT required per tree, in pounds

$D$ = diameter or least dimension of dressed timber, in inches

NOTES

## Chapter 14
# RANGER URBAN OPERATIONS

Urban operations are planned and conducted in an area of operations (AO) that includes one or more urban areas. An urban area is a topographical complex where man-made construction or high population density is the dominant feature (FM 30-06.11).
Urban terrain is likely to be one of the most significant future areas of operations for American forces throughout the world. Expanding urban development affects military operations as the terrain is altered. The increasing focus on stability and support operations, urban terrorism, and civil disorder emphasizes that combat in urbanized areas is unavoidable. Urban areas are the power centers, the centers of gravity, and thus the future battlefield. For further study, see FM 30-06.11, FM 90-10, FM 90-1, FM 7-8, 75th Ranger Regiment Advanced UO SOP, and Ranger Training Circular 350-1-2.

14-1. **TYPES**. Infantry units must be trained to conduct urban combat under high-intensity conditions. High-intensity urban combat requires the employment of combat power of the joint combined arms team. An Infantry unit's mission is normally to recon, isolate, penetrate, systematically clear, defend the urban area, and engage and defeat the enemy with decisive combat power. Although the changing world situation may have made urban combat under high-intensity conditions less likely for US forces, it represents the high end of the combat spectrum, and units must be trained for it. High-intensity urban operations can be casualty-intensive for both sides. With the integrated firepower of the joint, combined arms team, leaders must make every attempt to limit unnecessary destruction of critical infrastructure and casualties among noncombatants.

a. **Precision Conditions**. Infantry units train to defeat an enemy that is mixed with non-combatants in precision urban combat. Leaders plan to limit civilian casualties and collateral damage through the establishment of strict rules of engagement (ROE) and the employment of precision weapons and munitions. The ROE provides the focus for the use and restraint of combat power. The ROE may be significantly more restrictive than under high-intensity conditions.

b. **Surgical Conditions**. Operations conducted under surgical conditions include special-purpose raids, small precision strikes, or small-scale personnel seizure or recovery operations in an urban environment (for example, hostage rescue). Joint special operation forces usually conduct these operations. They may closely resemble US police operations performed by Special Weapons and Tactics (SWAT) teams. They may even involve cooperation between US forces and host nation police. Though regular units may not usually be involved in the actual surgical operation, they may support it by isolating the area, by providing security or crowd control, or providing search and rescue teams.

c. **Transitions**. Leaders must always be prepared to transition rapidly from one type of urban combat to another, and back. Real-world combat shows us that urban operations can deteriorate rapidly and without warning. A force involved in stability and support operations can suddenly find itself in a high-intensity combat situation.

14-2. **PRINCIPLES**

a. **Surprise**. Strike the enemy at a time or place or in a manner for which he is unprepared. Key to success: gives the assaulting element the advantage.

b. **Security**. Never permit the enemy to acquire unexpected advantage.

(1) Maintain during all phases of the operation.

(2) Four-dimensional battlefield (height, depth, width, subterranean).

(3) Always maintain 360 degree security (include elevated and subterranean areas).

(4) Mission is never complete as long as you remain in the urban environment. The status of actors in the urban environment does not afford the sense of security offered by "open" terrain. The key to survivability is a constant state of situational awareness.

c. **Simplicity**. Prepare clear, uncomplicated plans, and provide subordinates with concise orders to ensure thorough understanding.

(1) Always keep plans simple.

(2) Ensure everyone understands the mission and the commander's intent.

(3) Plan and prepare for the worst.

d. **Speed**. Rate of military action.

(1) Act as security.

(2) Move in a careful hurry.

(3) Smooth is fast and fast is smooth.

(4) Never move faster than you can accurately engage targets.
(5) Exercise tactical patience.

e. **Violence of Action**. Eliminate the enemy with sudden, explosive force.
(1) Combined with speed gives surprise.
(2) Prevents enemy reaction.
(3) Both physical and mental.

14-3. **METT-TC**. To effectively plan combat operations in urban environments, leaders must use Troop Leading Procedures and conduct a thorough analysis utilizing METT-TC factors. The following lists specific planning guidance that must be incorporated when planning for urban operations. For more specifics on mission planning, refer to Chapter 2, Ranger Handbook.

a. **Mission**. Know correct Task Organization to accomplish the mission (Offense, Defense, or Stability and Support Operations).

b. **Enemy**.

(1) ***Disposition***. Analyze the arrayal of enemy forces in and around your objective, known and suspected. Example: Known or suspected locations of minefields, obstacles, and strong points.

(2) ***Composition and Strength***. Analyze the enemy's task organization, troops available, suspected strength, and amount of support from local civilian populace based on intelligence estimates. Is the enemy a conventional or unconventional force?

(3) ***Morale***. Analyze the enemy's current operational status based on friendly intelligence estimates. For example, is the enemy well supplied; have they had recent success against friendly forces, taken many casualties; and what is the current weather?

(4) ***Capabilities***. Determine what the enemy can employ against your forces. Example: Enemy's weapons, artillery assets, engineer assets, air defense assets, NBC threats, thermal/NVG capabilities, close air support, armor threat, and so on.

(5) ***Probable Course(s) of Action***. Based on friendly intelligence estimates, determine how the enemy will fight within his area of operation (in and around your area of operation).

c. **Terrain**. Leaders conduct a detailed terrain analysis of each urban setting, considering the types of built-up areas and composition of existing structures. They use OCOKA when analyzing terrain, in and around the area of operation.

(1) ***Observation and Fields of Fire***. Always be prepared to conduct urban operations under limited visibility conditions. This includes the effects of reduced illumination, as well as natural and manmade obscuration. Leaders should ensure that Rangers are equipped with adequate resources, which allow them to successfully operate in the urban environment under these types of conditions. (2) ***Cover and Concealment***. Leaders must perform a thorough analysis of peripheral as well as intra-urban areas. Leaders should identify routes to objectives, which afford assault forces with the best possible cover and concealment. Additionally, leaders should take advantage of limited visibility conditions, which would allow forces to move undetected to their final assault / breaching positions. When in the final assault position, forces should move as rapidly as tactically possible to access structures, which afford additional cover and concealment. Leaders must learn to properly employ obscurants and exercise "tactical patience" to fully take advantage of these effects. Finally, all members of the urban force must practice noise and light discipline. Soldiers must avoid unnecessary voice communications, learn the proper use of white light, and limit contact with surfaces that may alert the enemy of their presence.

(3) ***Obstacles***. There are many manmade and natural obstacles on the periphery, as well as within the urban environment. Leaders should conduct a detailed reconnaissance of routes and objectives (this must include subterranean complexes), taking into consideration route adjustments and special equipment needs.

(4) ***Key Terrain***. Analyze which buildings, intersections, bridges, LZ/PZ, airports, and elevated areas that provide a tactical advantage to you or the enemy. Additionally, the leader must identify critical infrastructure within his area of operations, which would provide the enemy with a tactical advantage on the battlefield. These may include, but are not limited to, communication centers, medical facilities, governmental facilities, and facilities that are of psychological significance.

(5) ***Avenues of Approach***. Consider roads, intersections, inland waterways, and subterranean constructions (subways, sewers, and basements). Leaders should classify areas as go, slow go, or no-go based on the navigability of the approach.

***Note:*** Military maps may not provide enough detail for urban terrain analysis or reflect the underground sewer systems, subways, water systems, or mass transit routes.

(6) ***Troops***. Analyze your forces utilizing their disposition, composition, strength, morale, capabilities, and so on. Leaders must also consider the type and size of the objective to plan effective use of troops available.

(7) ***Time***. Operations in an urban environment have a slower pace and tempo. Leaders must consider the amount of time required to secure, clear, or seize the urban objective and stress and fatigue Rangers will encounter. Additional time must also be allowed for area analysis efforts; these may include, but are not limited to--

- Maps and urban plans Recon and analysis.
- Hydrological data analysis.
- Line-of-sight surveys.
- Long Range Surveillance and Scout reconnaissance.
- Artillery.
- Armor.
- Aviation.
- Armor.
- Engineers.

(8) ***Civilians***. Authorities such as the National Command will establish the Rules of Engagement. Commanders at all levels may provide further guidance regarding civilians occupying the area of operations (AO). Leaders must daily reiterate the ROE to subordinates, and immediately inform them of any changes to the ROE. Rangers must have the discipline to identify the enemy from noncombatants and ensure civilians understand and follow all directed commands.

***Note:*** Civilians may not speak English, may be hiding (especially small children), or dazed from a breach. Civilians must not be given the means to resist. Rehearse how clearing/search teams will react to these variables. Never compromise the safety of your Rangers.

14-4. **CLOSE QUARTERS COMBAT.** Due to the very nature of a CQC encounter, engagements will be very close (within 10 meters) and very fast (targets exposed for only a few seconds). Most close quarter's engagements are won by who hits first and puts the enemy down. It is more important to knock a Ranger down as soon as possible than it is to kill him. In order to win a close quarters engagement, Rangers must make quick, accurate shots by mere reflex. This is accomplished by reflexive fire training. Remember, no matter how proficient you are, always fire until the enemy goes down. All reflexive fire training is conducted with the eyes open.

***Note:*** Research has determined that, on average, only three out of ten people actually fire their weapons when confronted by an enemy during room clearing operations. Close quarters combat success for the Ranger begins with the Ranger being psychologically prepared for the close quarter's battle. The foundation for this preparedness begins with the Ranger's proficiency in basic rifle marksmanship. Survival in the urban environment does not depend on advanced skills and technologies. Rangers must be proficient in the basics.

14-5. **REHEARSALS**. Similar to the conduct of other military operations, leaders need to designate time for rehearsals. Urban operations require a variety of individual, collective, and special tasks, which are not associated with operations on less complex terrain. These task require additional rehearsal time for clearing, breaching, obstacle reduction, casualty evacuation, and support teams. Additionally, rehearsal time must be identified for rehearsals with combined arms elements. These may include, but are not limited to--

a. **Stance**. Feet are shoulder width apart, toes pointed straight to the front (direction of movement). The firing side foot is slightly staggered to the rear of the non-firing foot. Knees are slightly bent and the upper body leans slightly forward. Shoulders are not rolled or slouched. Weapon is held with the butt stock in the pocket of the shoulder. The firing side elbow is kept in against the body. The stance should be modified to ensure the Ranger maintains a comfortable boxer stance.

(1) ***Low-Carry Technique***. The butt stock of the weapon is placed in the pocket of the shoulder. The barrel is pointed down so the front sight post and day optic is just out of the field of vision. The head is always up identifying targets. This technique is safest and is recommended for use by the clearing team once inside the room.

(2) ***High-Carry Technique***. The butt stock of the weapon is held in the armpit. The barrel is pointed slightly up with the front sight post in the peripheral vision of the individual. Push out on the pistol grip and thrust the weapon forward and pull straight back into the pocket of the shoulder to assume the proper firing position. This technique is best suited for the line-up outside the door. Exercise caution with this technique always maintaining situational awareness, particularly in a multi-floored building.

***Note:*** Muzzle awareness is critical to the successful execution of close quarter's operations. Rangers must never point their weapons or cross the bodies of their fellow Rangers at any time. Additionally Rangers should always avoid exposing the muzzle of their weapons around corners; this is referred to as "flagging."

b. **Malfunction**. If a Ranger has a malfunction with his weapon during any CQC training, he will take a knee to conduct immediate action. Once the malfunction is cleared, there is no need to immediately stand up to engage targets. Rangers can save precious seconds by continuing to engage from one knee. Whenever other members of the team see a Ranger down, they must automatically clear his sector of fire. Before rising to his feet, the Ranger warns his team members of his movement and only rises after they acknowledge him. If a malfunction occurs once committed to a doorway, the Ranger must enter the room far enough to allow those following him to enter and move away from the door. This drill must be continually practiced until it is second nature.

c. **Approach to a Building or Breach Point**. One of the trademarks of ranger operations is the use of limited visibility conditions. Whenever possible, breaching and entry operations should be executed during hours and conditions of limited visibility. Rangers should always take advantage of all available cover and concealment when approaching breach and entry points. When natural or manmade cover and concealment is not available, Rangers should employ obscurants to conceal their approach. There are times when Rangers will want to employ obscurants to enhance existing cover and concealment. Members of the breach / entry team should be numbered for identification, communication, and control purposes.

(1) The # 1 Ranger should always be the most experienced and mature member of the team, usually the team leader. The # 1 Ranger is responsible for frontal and entry and breach point security. The team leader is responsible for initiating all voice and physical commands. The team leader must exercise situational awareness at all time with respect to the task, friendly force, and enemy activity.

(2) The # 2 Ranger is directly behind the # 1 Ranger in the order of movement, and he moves through the breach point in the opposite direction from the #1 Ranger.

(3) The # 3 Ranger will simply go opposite the # 2 Ranger inside the room at least 1 meter from the door.

(4) The # 4 Ranger moves the opposite the # 3 Ranger and is responsible for rear security (and is normally the last Ranger into the room). An additional duty of the # 4 Ranger is breaching.

***Note:*** Consider how much firepower each Ranger delivers. Where do you put the SAW gunner in the order? You must weigh the benefits of firepower against those of quick, accurate shots. If the # 4 Ranger has breaching responsibilities, it should not be the SAW gunner, because this would reduce your firepower.

d. **Actions Outside the Point of Entry**. Entry point position and individual weapon positions are important. The clearing team members should stand as close to the entry point as possible, ready to enter. Weapons are oriented in such a manner that the team provides itself with 360 degree security at all times. Team members must signal to one another that they are ready at the point of entry. This is best accomplished by sending up a "squeeze." If a tap method is used, an inadvertent bump may be misunderstood as a tap.

e. **Enter Building/Clear Room**. See Battle Drill 5, Chapter 6.

f. **Locking Down the Room**.

(1) Control the situation within the room.

(2) Use clear, concise arm and had signals. Voice commands should be kept to a minimum to reduce the amount of confusion and to prevent the enemy--who might be in the next room--from discerning what is going on. This enhances the opportunity for surprise and allows the assault force the opportunity to detect any approaching force.

(3) Physically and psychologically dominate.

(4) Establish security and report status.

(5) Cursory search of the room to include the ceiling (3 Dimensional Fight).

(6) Identify the dead using reflexive response techniques (Eye thump method).

(7) Search the room for PIR, precious cargo as per the mission and time available.

(8) Evacuate personnel.

(9) Mark room clear using chemical lights, engineer tape, chalk, paint, VS-17 panels, and so on.

14-6. **TTPS FOR MARKING BUILDINGS AND ROOMS**. Units have long identified a need to mark specific buildings and rooms during UO. Sometimes rooms need to be marked as having been cleared, or buildings need to be marked as containing friendly forces. The US Army Infantry School is currently testing a remote marking device that can be used to mark doors from as far away as across a wide street. In the past, units have tried several different field-expedient marking devices; some with more success than others. Chalk has been the most common. It is light and easily obtained but not as visible as other markings. Some of the other techniques have been to use spray paint, and paintball guns.

a. **NATO Standard Marking SOP**. The North Atlantic Treaty Organization (NATO) has developed a standard marking SOP for use during urban combat. It uses a combination of colors, shapes, and symbols. These markings can be fabricated from any material available. (Figure 14-1 shows examples.)

**Figure 14-1. EXAMPLE NATO STANDARD MARKINGS**

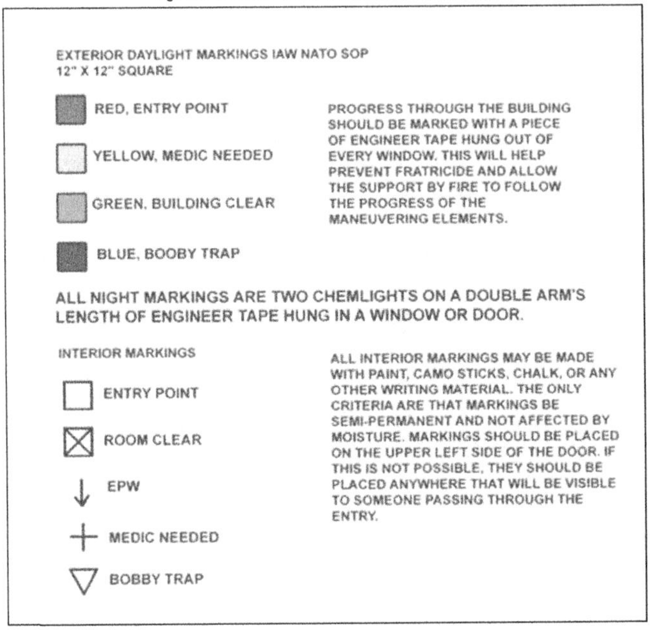


b. **Spray Paint**. Canned spray paint is easily obtained and comes in a wide assortment of colors including florescent shades that are highly visible in daylight. It cannot be removed once used. Cans of spray paint are bulky and hard to carry with other combat equipment. Paint is not visible during darkness nor does it show up well through thermal sights.

c. **Paintball Guns**. Commercial paintball guns have been purchased by some units and issued to small unit leaders. Some models can be carried in standard military holsters. They can mark a building or door from about 30 meters. The ammunition and propellant gas is not easily obtainable. The ammunition is fragile and often jams the gun if it gets wet. The available colors are not very bright, and just like spray paint, cannot be seen at night or through thermal sights.

d. **Wolf Tail**. A simple, effective, easy-to-make, lightweight device called a "wolf tail" can be fabricated to mark buildings, doorways, and windows (Figure 14-1). A unit has changed its tactical SOP to require that each Infantryman carry one of these devices in his ACU cargo pocket. Wolf tails, when used IAW a simple signaling plan understood by all members of the unit, can aid in command and control, reduce the chances of fratricide, and speed up casualty collection during urban combat.

(1) The wolf tail marking device is simple to make and versatile. It can be used together with the NATO marking scheme. Rolled up, it makes a small, easily accessible package that can be carried in the cargo pocket of the ACUs. It can be recovered easily and used again if the situation changes. All its components can be easily obtained through unit supply. It combines a variety of visual signals (colored strapping and one or more chemical lights of varying colors) with a distinctive heat signature that is easily identified through a thermal weapon sight. An infrared chemlight can be used either as a substitute for the colored chemlight(s) or in addition to them.

(2) Constructing the wolf-tail marking device requires the following material:

- A 2-foot length of nylon strap (the type used for cargo tie-downs) (engineer tape can be substituted).
- About 5 feet of 550 cord.
- A small weight such as a bolt or similar object.
- Duct tape.
- Chemical lights (colored or IR).
- Two 9-volt batteries.

(3) Assemble the items by tying or taping the cord to the small weight. Attach the other end of the cord to the nylon strapping, securing it with duct tape. Attach the 9-volt batteries in pairs to the lower end of the strapping with several wraps of duct tape, making sure that the negative terminals are opposite the positive, but not actually touching. Use more duct tape to attach the chemical lights, about 2 inches above the batteries, to the strapping.

(4) When you want to mark your position, push the batteries together firmly until the male and female plugs lock. This shorts out the battery, causing it to heat up rapidly. The hot battery is easily identified through the thermal sights of tanks or BFVs. The batteries stay visible for about 45 minutes. Activating the chemical lights provides an easily identified light source visible to the naked eye. You can use infrared chemical lights if you want them to be seen through night vision devices but not with the naked eye.

**Figure 14-2. EXAMPLE OF A WOLF-TAIL MARKING DEVICE**

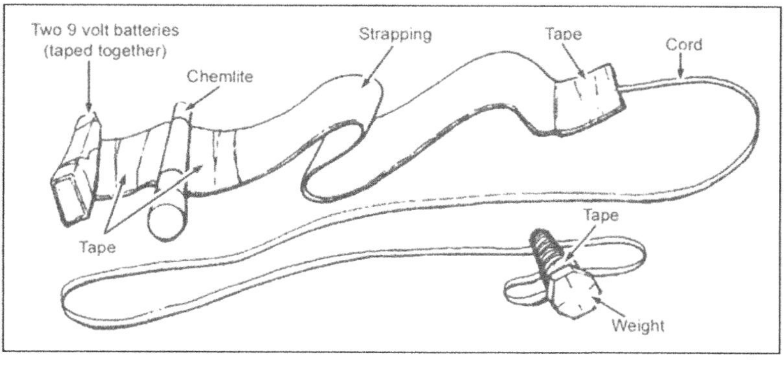


***Note:*** One option is to place chemical lights and batteries at both ends of the wolf tail to mark the inside and outside of a building or room.

(5) Use the cord and the small weight to hold the wolf tail in position by tying or draping it out a window or hanging it on a door, wherever it is best seen by other friendly troops. Squads or platoons can vary the numbers and colors of chemical lights, or use multiple battery sets to identify precisely what unit is in which building.

(6) Medics and combat lifesavers can carry a standardized variation that can be used to clearly identify a building as containing wounded personnel needing evacuation. This could be a white strap with multiple red chemical lights, or any other easily identified combination.

14-7. **URBAN ASSAULT BREACHES**. Understanding how to employ and incorporate breaching as part of a leader's planning process is an important part of urban operations. It is imperative that elements of your patrol be skilled in the art of breaching. Whether infiltrating or exfiltrating from an objective, leaders must plan for either option. One constant disadvantage with the employment of explosives is that placement requires Rangers to expose themselves to possible enemy fires. Breach teams need to be supported by fires or obscurants and breaching operations should be performed during hours of limited visibility whenever possible. Breaching classes follow:

a. **Mechanical**. Mechanical breaching should be an important part of a leader's breaching training program because it is usually an option. Mechanical breaching is best described as gaining access by the use of tools or saws. Although most tools and saws used are recognizable and self explanatory to the individual Ranger, one must practice on various techniques to increase speed an effectiveness. This reduces fatigue and expedites the actual assault.

(1) **Tools**.

(a) Hooligan tools..........................Doors/windows of all types.
(b) Sledge hammer.......................Heavy duty doors, locks, and window frames.
(c) Picket pounder.........................Doors of all types, light walls.
(d) Bolt Cutters..............................Chain link fence, locks, and wire obstacles.
(e) Pick Ax.....................................Lightweight doors and locks.
(f) Saws........................................Fences, light doors, locks.

b. **Ballistic**. Ballistic breaching is defined as a forced entry or exit by the use of weapons. Whether using shotguns, M16A2/M4, M249 SAW, specific considerations must be addressed:

(a) Type of round and ricochet factor.
(b) Composition of the breaching point.
(c) Composition of the floor beyond the door.
(d) Personnel behind the door (friendly/enemy).
(e) Always shoot at a 45-degree angle.

c. **Explosive**. Explosive breaching is the most viable because it is the most effective. When employing explosives during breaching operations, leaders must consider three major factors.

(1) ***Overpressure***. The amount of PSI released from the concussion of the blast.

(2) ***Missile Hazard***. Fragmentation or projectiles sent at tremendous speed from the explosion area. This occurs from either the charge or target being breached.

(3) ***Minimum Safe Distance Requirements***. Use of explosives in the urban environment must consider the presence of noncombatants and friendly forces. Additionally, the are many hazardous materials located in the urban environment, these may include chemicals as well as construction materials. There is always a risk of secondary explosions and fires, when employing explosive breaching techniques.

(4) ***Charges***. Various charges can be used for explosive breaching. Leaders must conduct extensive training on the use of the charges to get proper target feedback. Figure 14-3 shows examples of charges used for explosive breaching.

**Figure 14-3. EXAMPLE CHARGES FOR EXPLOSIVE BREACHING**

| Charge | Target |
|---|---|
| Water Impulse................ | Steel/wood doors |
| Flexible linear................. | Wood doors |
| Ranger wall breach........ | Masonry/brick walls |
| Chain link ladder............ | Chain link fence |
| E-Type silhouette........... | Wood doors |
| Brashier breacher.......... | Concertina wire |

# Chapter 15
# VEHICLE CONVOY OPERATIONS

This chapter outlines a technique for conducting vehicle convoy operations. Convoy operations present a challenge to the Ranger leader. Trucks and other combat vehicles produce a large signature on the battlefield and increase your unit's value as a target. Vehicle movement is restricted to roads and terrain that they can traverse; therefore, a sound plan must be implemented to minimize the possibility of compromise.

15-1. **PLANNING CONSIDERATIONS**. This paragraph explains how to plan a vehicle convoy IAW the eight steps of the TLP.

a. **Receive the Mission.** (PL will extract the following information from Company OPORD).

(1) Truck Support (number and type of Truck, ACL).
(2) Weather. Road conditions.
(3) Truck pickup and drop-off location / marking.
(4) Truck movement Timeline (Pick-up time, roll time, H-hour).
(5) Truck Routes (primary and alternate, Check points).
(6) Drop-off locations (primary and alternate).
(7) IED/Contact compromise and contingencies.

b. **Issue the Warning Order**. During the WARNO the PL gives a basic overview of the Truck Movement and gives specific Task to Maneuver Units.

(1) Manifest Personnel (PSG).
(2) Pick up point security and set-up (PSG).
(3) Primary and alternate routes (Security SL).

c. **Make a tentative Plan**. PL considers METT-TC.

(1) ***Mission***. Develop your Ground Tactical Plan and a then develop a truck plan that best supports it.
(2) ***Enemy***. Consider enemy patrols and activity in AO.
(3) ***Terrain***. Evaluate routes and dropoff and pickup points using OCOKA.
(4) ***Troops***.
(a) Truck ACL.
(b) Chalks and chalk leaders identified.
(c) Chalks planed with tactical integrity & self sufficiency.
(d) Key leaders and weapons are cross-loaded.
(e) Pickup point marking team identified.
(f) Pickup point security teams identified.
(5) ***Time***. PL considers the following when evaluating time available.
(a) Movement to Pickup point.
(b) Recon of Pickup point.
(c) Emplacement of security.
(d) Marking the Pickup point.
(e) Pickup point Posture time.

d. **Initiate Necessary Movement**. PL decides whether to move the platoon to the pickup location or to conduct a map reconnaissance and complete the plan.

e. **Conduct Reconnaissance**. (Is not necessary if pickup point is secured and established by Higher HQ's).

(1) Leader's Recon consists of PL, RTO, Chalk Leaders and Pickup point Security element.
(2) PL determines suitability of Pickup point.
(3) PL determines security plan (either overwatch or strong point).

f. **Complete the Plan**. Complete Truck Movement Annex on Page ??? the Ranger Handbook.

g. **Issue the OPORD**.

h. **Supervise**.

(1) Back brief from chalk leaders.
(2) Rock drill of truck movement and contingency plans.

15-2. **FIVE PHASES OF TRUCK MOVEMENT**. Each phase must support the ground tactical plan, which specifies actions in the objective area to accomplish the commander's intent for the assigned mission, be it a raid, ambush, recon or other follow-on missions.

a. **Staging Plan**.
(1) Secure the pickup point (strongpoint or overwatch).
(2) Mark pickup point (day/night).

b. **Loading Plan**. Each Ranger is assigned a truck.

c. **Movement Plan**.
(1) Troops awake and alert during movement.
(2) Platoon leader and chalk leaders tracking route progress.
(3) Compromise and contingency plan.
(a) React to IED.
(b) React to ambush.
(c) Truck breakdown.

d. **Dropoff Plan**.
(1) Establish security of dropoff point (overwatch or strongpoint).
(2) Dismount vehicles.
(3) Recon and secure assembly area.
(4) Adjust perimeter as chalks arrive.
(5) PSG clears the trucks when last chalk departs.

e. **Ground Tactical Plan**. Execute after platoon is reconsolidated or minimum force required is assembled.

15-3. **CONVOY TECHNIQUES**. The following convoy techniques have been included for your convenience:

- Convoy warning order
- Convoy brief
- React to Ambush Near/Far
- Convoy forced to stop (Methods 1 and 2)
- Break contact
- Recovery/CASEVAC Operations

## Convoy Warning Order

1. SITUATION: A brief statement of the enemy and friendly situation. Who, What, Where
2. MISSION: Who, What, When, Where, Why
3. TASK ORGANIZATION: Based on tasking from higher WARNO.
4. INITIAL TIME SCHEDULE:

| When | Who | What |
|---|---|---|
| | | |
| | | |
| | | |
| | | |

5. SPECIAL INSTRUCTIONS: PCC/PCI Guidance, Rehearsals, additional tasks to be accomplished.
6. SERVICE SUPPORT:

Class I: (Rations & Water)

| | | |
|---|---|---|
| | | |
| | | |
| | | |

Class III: (POL)

| | | |
|---|---|---|
| | | |
| | | |
| | | |

Class V: (Ammunition /Pyrotechnics)

| Weapon System | Rounds | Type |
|---|---|---|
| | | |
| | | |
| | | |

| Pyrotechnic Device | Number | Location |
|---|---|---|
| | | |
| | | |
| | | |

Class VIII: (Medical)

| | | |
|---|---|---|
| | | |
| | | |
| | | |

7. UNIFORM AND EQUIPMENT COMMON TO ALL.

**CONVOY BRIEF** (Modified from FM 55-30)

Movement Order No. ____

**References:** A. ______________________ *(Maps, tables and relevant documents)*

B. ______________________

**TASK ORGANIZATION:** (Internal organization for convoy – Manifest)

1. **SITUATION:**

a. **Enemy Forces:**

(1) Weather. General forecast.

(2) Light Data (EENT, % Illumination, MR, MS, BMNT)

(3) Discuss Enemy.

Identification of enemy (If known).

Composition / capabilities / strength / equipment

Location

(Hot Spots highlighted on map)

b. **Friendly Forces:**

(1) Operational support provided by higher headquarters.

Helicopter / Gunships

Quick Reaction Forces (QRFs)

MP Escorts / Rat Patrols

Fire Support elements

c. **Attachments:** (From outside the organization)

2. **MISSION: (WHO, WHAT, WHEN, WHERE, & WHY)**

3. **Employment Tactics, Techniques and Procedures**

a. Readiness Level
- i. Ride / Scanning (Observation)
- ii. At the Ready

b. Scanning / Sector of Fire
- i. Driver
  1. Sector of Scan is 9 – 1 clock position
  2. Observation with Mirrors
  3. Sector of Fire is 9 – 11 o'clock
- ii. TC
  4. Sector of Scan is 11 – 3 clock position
  5. Sector of Fire is 1 – 3 clock position
- iii. Other Systems are based off vehicle type and load

b. Target ID
- i. Communicate / Signal

c. Body Positioning
- i. Engage as you Train (Right or Left handed Firing)
- ii. Firing-side Shoulder Down
- iii. Maintain your Body Position
- iv. Weapon to Head

d. Rules of Engagement Concerns – Employ the appropriate systems based off the threat.

e. Point of Aim
- i. Moving Platform – Stationary Target requires aim to the rear (trail) and low dependant upon speed.

## CONVOY BRIEF (Modified from FM 55-30) (Continued)

**3. Employment Tactics, Techniques and Procedures**

e. Point of Aim
   ii. Moving Platform – Moving Target requires aim directly on and low.
   iii. Stationary Platform – Moving Target requires aim to the front (lead) and low dependent upon speed.

f. Rate of Fire
   i. Burst Mode
   ii. Steady Suppression (ROE)

g. Magazine Awareness
   i. Serviceability / maintenance
   ii. Tracer Mix
   iii. Magazine Storage / Placement
   iv. Mounted, Reload when time is available

**Dismounted, Seek cover prior to the need to Reload**

## REACT TO AMBUSH (NEAR)

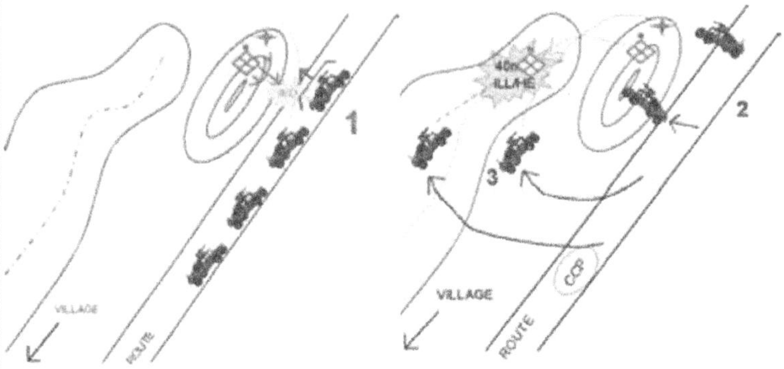


### (Enemy Initiates Weapons Fire / RPG / IED)

1. Lead element of convoy reacts to contact and immediately returns fire. Lead fire team fires 40mm HE (at night also white star if NVGs drown out by vehicle headlights – leader calls it. FYI, most travel at night will be with headlights on).
2. Lead elements assualt through ambush – fire team dismounts while gun trucks use M249, MK-19, and .50 cal in close support.
3. Trail fire team and trail gun truck move to block most likely enemy escape route. Fire team uses 40mm HE/Illumination if needed. (Use of 40mm becomes less likely as units get closer to urban areas due to collateral damage risk.) Establish CCP upon action completion or enemy neutralized.

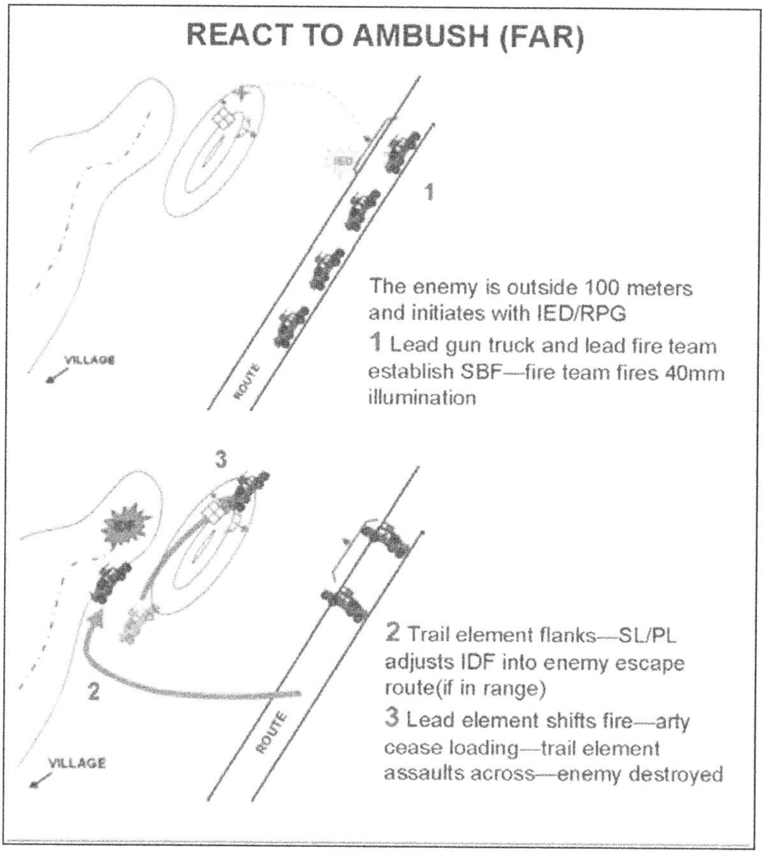
REACT TO AMBUSH (FAR)
IED
1
VILLAGE
ROUTE
The enemy is outside 100 meters and initiates with IED/RPG
1 Lead gun truck and lead fire team establish SBF—fire team fires 40mm illumination
3
2
VILLAGE
ROUTE
2 Trail element flanks—SL/PL adjusts IDF into enemy escape route(if in range)
3 Lead element shifts fire—arty cease loading—trail element assaults across—enemy destroyed

# CONVOY FORCED TO STOP
## (Weapons Fire / RPG / IED / Indirect)

1. Vehicles forced to stop. Activate vehicle turn signal to indicate direction of contact.
2. Vehicle(s) / Personnel not in direct contact, report on internal communication, identifying truck number, type of contact, clock direction and eight digit grid if available.

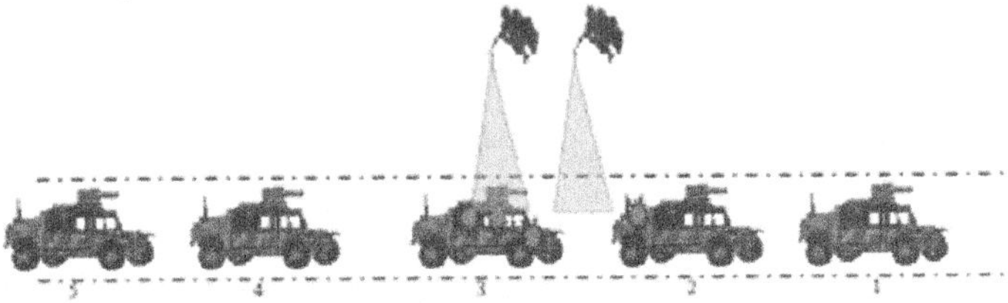


3. Personnel on vehicle(s) forced to stop dismount on the non-contact side, assume covered position(s) and provide initial base of fire.
4. The <u>entire</u> convoy halts; personnel will dismount vehicle(s) on the non-contact side and provide additional base of fire on the enemy if in range. Vehicle(s) not in contact will reposition to attack the attackers. M16's should now be on <u>Semi-Automatic</u> in order to conserve ammunition.
5. PL/CC/SC will move to better assess the situation and position the Gun Truck(s) in order to best suppress the enemy while maintaining <u>standoff</u>. Gun trucks will close with and engage the enemy.
6. Once the PL/CC/SC determines the convoy has either gained fire superiority or defeated the enemy contact, Recovery / CASEVAC operations will begin.
7. If the PL/CC/SC determines the convoy cannot gain fire superiority, leader will then conduct Break Contact procedures

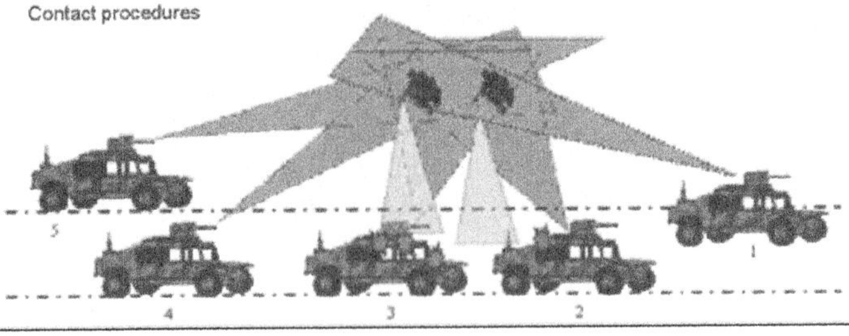

## CONVOY FORCED TO STOP - METHOD 2

### (Weapons Fire /RPG / IED / Indirect)

1. Vehicle(s) forced to stop. Activate vehicle turn signal to indicate direction of contact.

*** ALL PERSONNEL STAY IN VEHICLES ***

2. All vehicles immediately drive forward out of the kill zone.

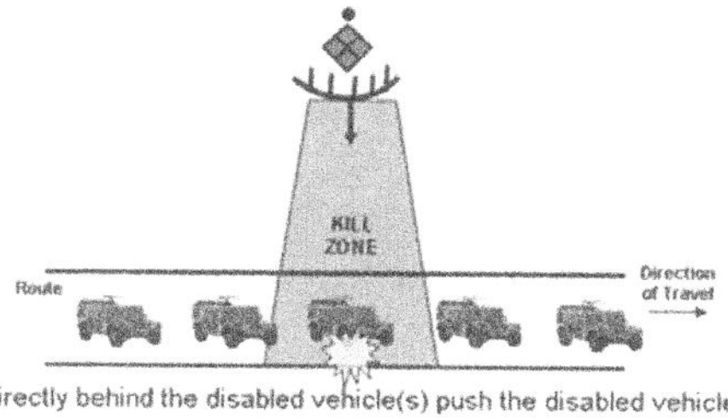


3. The vehicle(s) directly behind the disabled vehicle(s) push the disabled vehicle(s) out of the kill zone.

4. The vehicle(s) that are not disabled will establish a base of fire towards the suspected/known contact.

5. If fire superiority can be gained the PL will use the minimum force necessary to destroy the enemy.

6. If the PL determines the convoy cannot gain fire superiority, the leader will conduct break contact.

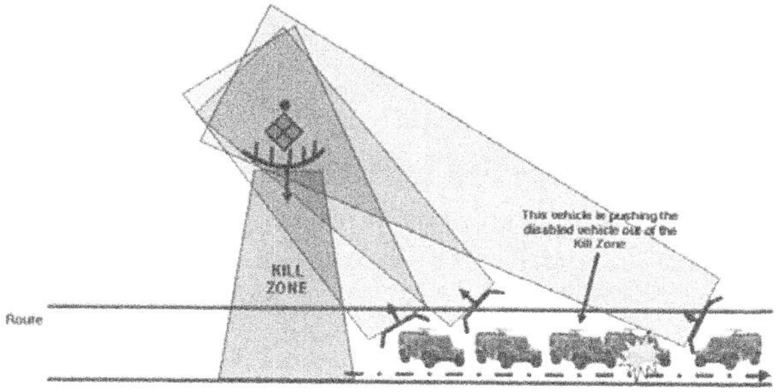

# BREAK CONTACT

**Always try to close with and destroy the enemy first. This way he cannot come back later on to attack you or another convoy again. However – if you must...**

1. The PL/CC/SC determines the convoy cannot gain fire superiority and the decision has been made to Break Contact.
2. The PL/CC/SC designates that either Rally Point "Rear" or "Forward" will be used. If necessary, both Rally Points may be used. Communication systems and appropriate pyrotechnic signals will be used to communicate Break Contact and Rally Point.
3. Personnel will deploy obscuration measures if available. Utilizing cover and concealment, Aid & Litter team(s) will evacuate all casualty(s) under support of Gun Truck and other protective fire(s).
4. Personnel will maintain position and suppression in contact zone and assist Aid & Litter team(s) as necessary.
5. Disabled vehicle(s) will be hauled back or destroyed as directed by leaders. (Thermite or explosives)

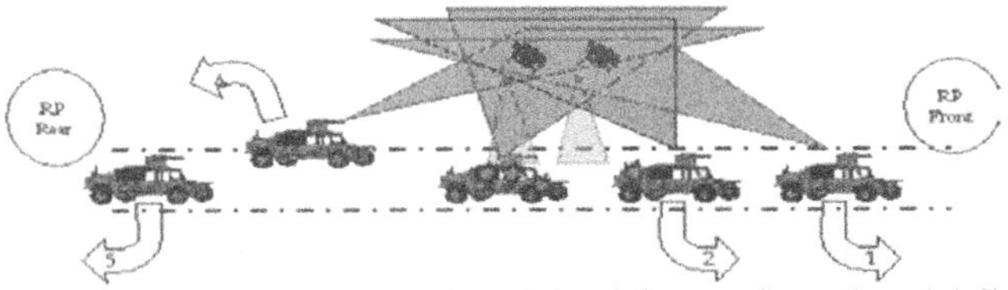


6. Vehicles will displace either backwards or forwards through the convoy lines under control of leaders. The most forward vehicle in the contact zone moves first, then the next most forward vehicle moves second. Vehicles will continue to displace. As vehicles displace, Gun Truck repositions as necessary until contact is broken.

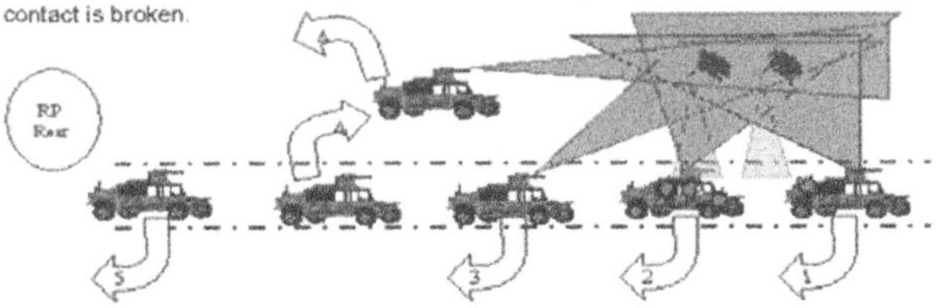


7. If Break Contact occurs with vehicles on both sides of the contact zone, displacement of vehicles will occur using an alternating displacement technique.
8. Upon occupation of the Rally Point, leaders will immediately position vehicles, setting 360 degree security, and conduct Consolidation and Reorganization.
9. If the convoy vehicles get separated when not in contact with the enemy, personnel and vehicles stay together and move to the closest Rally Point or Check Point.

# RECOVERY / CASEVAC OPERATIONS

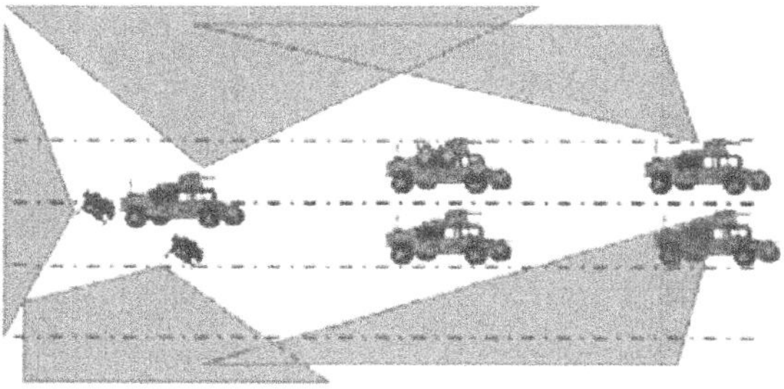

1. Once the leader assesses the enemy threat to be destroyed, neutralized, defeated and the area secured – Recovery / CASEVAC operations will begin. This keeps soldiers focused first on defeating and destroying the threat.
2. CASEVAC:

   a. Aid & Litter team will position on the safe side of the vehicle and extract casualty(s) and personnel.

   b. Treatment of casualty(s) will occur once they are safely removed from the contact area.
3. Vehicle Recovery Procedures:

   a. Recovery team will position on the safe side of the disabled vehicle.

   b. TC will dismount and assess the disabled vehicle.

   c. If determined the vehicle can be safely recovered, TC guide the recovery vehicle into position and conduct a hasty hook-up. TC will operate the disabled vehicle is necessary.

   d. Upon exiting the contact area, complete and correct hook-up procedures will occur.

   e. If assessment results in outside support necessary for recovery, leader will contact higher for guidance.
4. Once recovery operations are complete, the team will displace and conduct link-up with the convoy at the Rally Point.
5. Disabled vehicle(s) will be abandoned or destroyed as directed by leaders. (Thermite or explosives)

# GLOSSARY

| | |
|---|---|
| 1SG | first sergeant |
| 5 W's | Who, What, Where, When, and Why? |
| 5-S Rule | Search, Silence, Segregate, Safeguard, and Speed to rear |
| AA | avenue(s) of approach |
| AAR | after-action review |
| AATF | air assault task force |
| ABCDE | method of identification and response to life-threatening conditions: Airway, Breathing, Circulation, Disability, Exposure |
| ACE | ammunition, casualties, and equipment |
| ACL | allowable combat load |
| ACP | aerial checkpoint |
| ACU | Army combat uniform |
| ADA | air defense artillery |
| AG | assistant gunner |
| ATC | air traffic controller; a mechanical belay device that locks down on itself when tension is applied in opposite directions |
| ALT | alternate |
| AMC | air mission commander |
| ammo | ammunition |
| ANCD | automated net-control device |
| AO | area of operations |
| AOO | actions on the objective |
| AR | automatic rifleman |
| ABF | attack by fire |
| ATL | Alpha team leader |
| ATM | Alpha Team |
| bangalore torpedo | A manually emplaced, 1.5-meter long explosive-filled tube used to breach wire and detonate simple, pressure-activated antipersonnel mines. Ten tubes clear a 1- by 15-meter lane. |
| belay | any action taken to stop a climber's fall or to control the rate a load descends |
| binos | binoculars |
| BMNT | begin morning nautical twilight |
| BN | battalion |
| body belay | belay that uses the belayer's body to apply friction by routing the rope around the his body |
| bowline on a coil | knot used to secure a climber to the end of a climbing rope |
| BP | battle position |
| British junction knot | knot used to join the ends of detonation cords from multiple charges to one initiation system |
| BTC | bridge team commander |
| BTL | Bravo Team leader |
| BTM | Bravo Team |

| | |
|---|---|
| C2 | command and control |
| CAS | close air support |
| CASEVAC | casualty evacuation |
| CCA | close combat attack |
| CCIR | commander's critical information requirements |
| CCP | casualty collection point |
| CDR | commander |
| CDS | Camp Darby Special (map) |
| CLS | combat lifesaver |
| CO | company |
| COA | course of action |
| commo | communications |
| COMSEC | communications security |
| cordelette | short section of static rope or static cord. Also called "sling rope" |
| COTS | commercial off the shelf |
| CP | command post |
| CPR | cardiopulmonary resuscitation |
| CQC | close quarters combat |
| CS | combat support |
| CSS | *obsolete: now referred to as* sustainment |
| CTT | common task test |
| DAR | designated area of recovery |
| DOL | direction of landing |
| double figure eight knot | knot used to form a fixed loop in the end of the rope; loops are large enough to insert a carabiner |
| double overhand knot | knot used to secure the end of detonation cord |
| DP | duty position; decision point (depending on context) |
| DST | distance |
| DTG | date-time group |
| dynamic ropes | one of two classifications of kernmantle rope; used for climbing; *see also* static rope |
| DZ | drop zone |
| EA | engagement area |
| EEFI | essential elements of friendly information |
| EENT | end evening nautical twilight |
| end-of-the-rope clove hitch | intermediate anchor knot that requires constant tension |
| end-of-the-rope Prusik | knot used to attach a movable rope to a fixed rope; *see also* middle-of-the-rope Prusik |
| ENY | enemy |
| EPW | enemy prisoner of war |
| FA | field artillery |
| FDC | fire direction center |
| FFIR | friendly force information requirements |

| | |
|---|---|
| figure 8 slip knot | knot used to form an adjustable bight in the middle of a rope |
| FLIR | forward looking infrared |
| FLOT | forward line of own troops |
| FO | forward observer |
| FPF | final protective fires |
| FRAGO | fragmentary order |
| FSC | fire support coordinator |
| FSO | fire support officer |
| GOTWA | G - Where leader is GOING<br>O - Others he is taking with him<br>T - Time he plans to go<br>W - WHAT to do if the leader does not return in time<br>A - The unit's and the leader's ACTIONS on chance contact while the leader is gone |
| GPS | global positioning system |
| GRN | grenadier |
| HDG | heading |
| HE | high explosive |
| H-Hour | hit hour (the time the unit plans to accomplish the mission) |
| HI | high temperature (weather) |
| HPT | high-payoff target |
| HQ | headquarters |
| IAW | in accordance with |
| ICM | improved conventional munitions |
| ID | identification |
| IP | initial point |
| IR | intelligence requirements |
| ERRP | en route to release point |
| JAAT | joint air attack team |
| JD | Julian date |
| KIA | killed in action |
| LAW | light antiarmor weapon |
| LBV | load-bearing vest |
| LD | line of departure |
| LDA | linear danger area |
| LO | low temperature (weather) |
| LOGSTAT | logistical status |
| LP | listening post |
| LZ | landing zone |
| material factor | the strength, hardness, and mass of the material to be demolished |
| mb | millibar (a metric unit used to measure air pressure) |

| | |
|---|---|
| MDI | modernized demolition initiator |
| ME | main effort |
| mechanical belay | a belay that uses mechanical devices to help the belayer control the rope, as in rappelling |
| MEDEVAC | medical evacuation |
| METL | mission-essential task list |
| METT-TC | mission, enemy, terrain (and weather), troops (and support) available, time available, and civil considerations |
| middle of the rope clove hitch | knot that secures the middle of a rope to an anchor |
| middle-of-the-rope Prusik | knot that attaches a movable rope to a fixed rope, anywhere along the length of the fixed rope; *see also* end-of-the-rope Prusik |
| MG | machine gun/ner |
| MOPP | mission-oriented protective posture |
| MR | moonrise |
| MRE | meal, ready to eat |
| MS | moonset |
| MSD | minimum safe distance |
| MSL | mean sea level |
| munter hitch | commonly used belay that requires little equipment |
| NATO | North Atlantic Treaty Organization |
| NAV | navigation |
| NFA | no-fire area |
| NLT | no later than |
| NVD | night-vision device |
| NVG | night vision goggles |
| NVS | night vision system |
| OAKOC | observation and fields of fire, avenues of approach, key terrain, observation, and cover and concealment |
| OBJ | objective |
| OD | olive drab |
| OOM | order of movement |
| OP | observation post |
| OPORD | operations order |
| OPSKED | operational schedule |
| ORP | objective rally point |
| OT | observer-target |
| PB | patrol base |
| PCC | precombat checks |
| PCI | precombat inspection |
| PDF | principal direction of fire |

| | |
|---|---|
| PI | probability of incapacitation |
| PIR | priority intelligence requirements |
| PL | platoon leader |
| PLD | probable line of deployment |
| PLOT-CR | purpose, location, observer, trigger, communication method, resources (a format for planning fire support) |
| PLT | platoon |
| POL | petroleum, oils, and lubricants |
| PRI | primary |
| protection | a piece of equipment, natural or artificial, that is used to construct an anchor |
| PSG | platoon sergeant |
| PSI | pounds per square inch |
| PZ | pickup zone |
| R | rifleman |
| R&S | reconnaissance and surveillance |
| RACO | rear area combat operations |
| RAP | rocket-assisted projectile |
| rappel seat | a rope harness used in rappelling and climbing |
| RED | risk-estimate distance |
| REQ | required |
| rerouted figure 8 knot | anchor knot that also attaches a climber to a climbing rope |
| RFA | restrictive fire area |
| RFL | restrictive fire line |
| RFLM | rifleman |
| RHB | Ranger Handbook |
| ROE | rules of engagement |
| round-turn with two half hitches | a constant tension anchor knot |
| RP | release point |
| RTO | radio operator |
| S-2 | intelligence staff officer |
| S-3 | operations staff (and training) officer |
| SALUTE | Size, Activity, Location, Unit/Uniform, Time, and Equipment |
| SAW | squad automatic weapon |
| SBF | support by fire (position) |
| SDT | self-development test |
| SE | supporting effort |
| SEAD | suppression of enemy air defenses |
| SITREP | Situation Report |
| SITTEMP | situational template |

| | |
|---|---|
| SL | squad leader |
| sling rope | short section of static rope or static cord. Also called "cordelette" |
| SLLS | Stop, Look, Listen, Smell |
| SOC | succession of command |
| SOI | signal operating instructions |
| SOP | standing operating procedures |
| SP | start point |
| square knot | knot used to join two ropes of equal diameter; used to join the ends of the detonation cord to the explosive |
| SR | sunrise |
| SS | sunset |
| STANO | surveillance, target acquisition, and night observation |
| static ropes | one of two classifications of kernmantle rope; used where rope stretch is undesired, and when the rope is subjected to heavy static weight. *See also* dynamic rope |
| SURVIVAL | S Size up the situation, your surroundings, your physical condition, and your equipment.<br>U Undue haste makes waste; don't be too eager to move. Plan your moves.<br>R Remember where you are in relation to important friendly and locations and critical resources<br>V Vanquish fear and panic.<br>I Improvise. You can improve your situation. Use what you have. Use your Imagination.<br>V Value your life. Remember your goal: to get out alive. Remain stubborn. Refuse to give in to problems and obstacles that face you. This will give you the mental and physical strength to endure.<br>A Act like the natives; watch their daily routines. When, where, and how do they get food? Where they get water?<br>L Live by your wits. Learn basic skills. |
| suspension traverse | used to move personnel and equipment over rivers, ravines, chasms, and up or down a vertical obstacle |
| TAC | tactical air controller |
| tamping factor | depends on the location and tamping of the charge |
| technical climbing | using safe and proper equipment and techniques to climb on a rock formation in parties of two or more |
| tensionless anchor | used to anchor rope on high-load installations such as bridging |
| TL | team leader |
| TLP | troop-leading procedures |
| TL | team leader |
| TOC | tactical operations center |
| triple roll knot | knot used to join branches of detonation cord |
| TTP | tactics, techniques, and procedures |
| uli knot | knot used to securely fasten detonation cord to explosive |
| VIXL | video image crosslink |
| WARNO | warning order |
| WBGT | wet bulb globe temperature |

| | |
|---|---|
| WFFs | warfighting functions (fire support, movement and maneuver, protection, command and control, and sustainment) |
| WIA | wounded in action |
| XO | executive officer |
| WSL | weapons squad leader |

NOTES

# INDEX

Appendix A

# RESOURCES

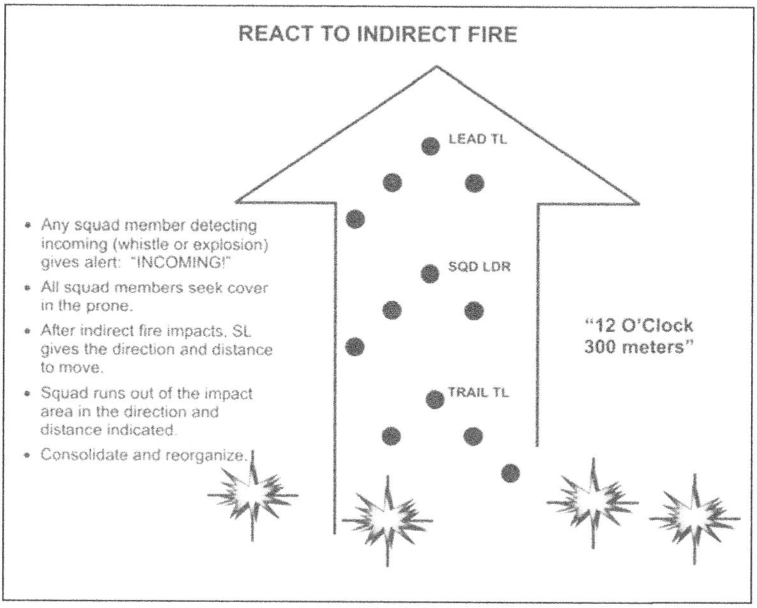

## REACT TO CONTACT

- Seek nearest cover.
- Return fire (known or suspected enemy position).
- TM LDRs control fire by using fire commands.
- Report enemy situation (number/position).
- Maintain contact (visual/oral) with team members.
- SQD LDR moves to team in contact and makes an assessment of the situation.
- Factors of his assessment:
  - Can squad move out to engagement area.
  - Can squad gain and maintain suppressive fire.
  - Location of enemy.
  - Size of enemy.
  - Vulnerable flanks.
  - Covered/concealed flanking routes.
- SQD LDR determines COA (break contact, squad attack, etc.)
- Report situation to PL

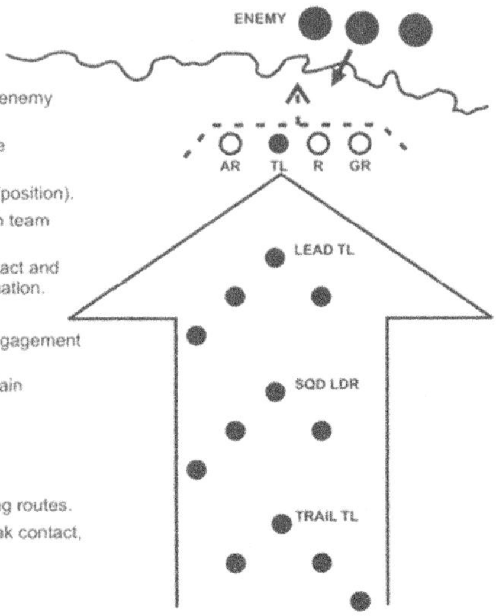

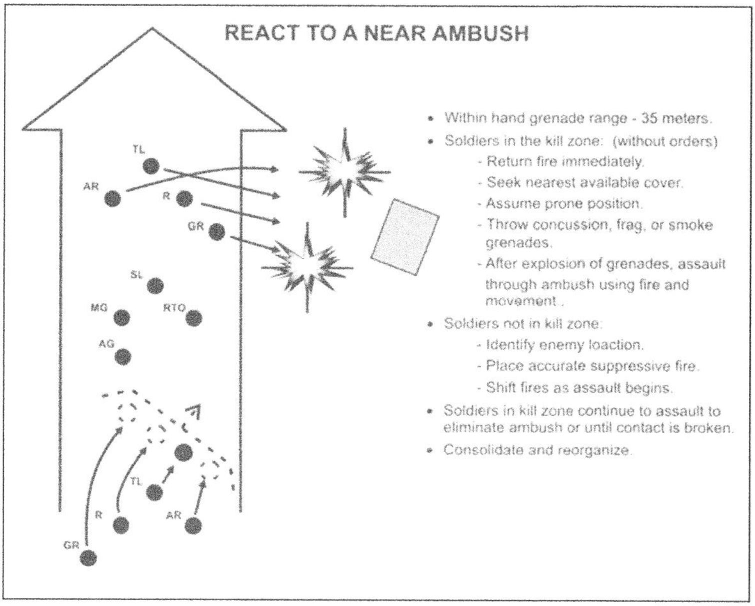
REACT TO A NEAR AMBUSH
TL
AR
R
GR
SL
MG
RTO
AG
TL
R
AR
GR
• Within hand grenade range - 35 meters.
• Soldiers in the kill zone: (without orders)
- Return fire immediately.
- Seek nearest available cover.
- Assume prone position.
- Throw concussion, frag, or smoke grenades.
- After explosion of grenades, assault through ambush using fire and movement.
• Soldiers not in kill zone:
- Identify enemy loaction.
- Place accurate suppressive fire.
- Shift fires as assault begins.
• Soldiers in kill zone continue to assault to eliminate ambush or until contact is broken.
• Consolidate and reorganize.

## REACT TO A FAR AMBUSH

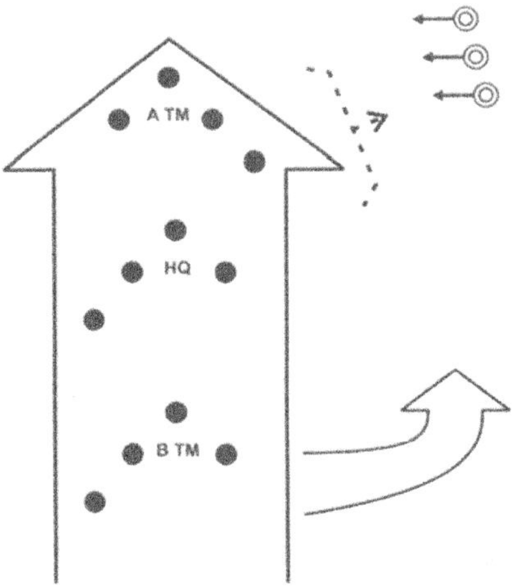


- More than 35 meters.
- TM in the kill zone: (without orders)
  - Return fire immediately.
  - Seek cover and concealment.
  - Suppress enemy (overwatch).
- SL assess situation.
  - Determines COA (flank).
- TM not in contact:
  - Moves along covered and concealed route.
  - Assaults enemy on weak flank.
  - Suppress enemy (overwatch).
- Overwatch TM continues to suppress shifts/cease fire as bounding team enters sector.
- Bounding team continues to assault through enemy.
- SL may request indirect fire.
- Consolidate and reorganize.

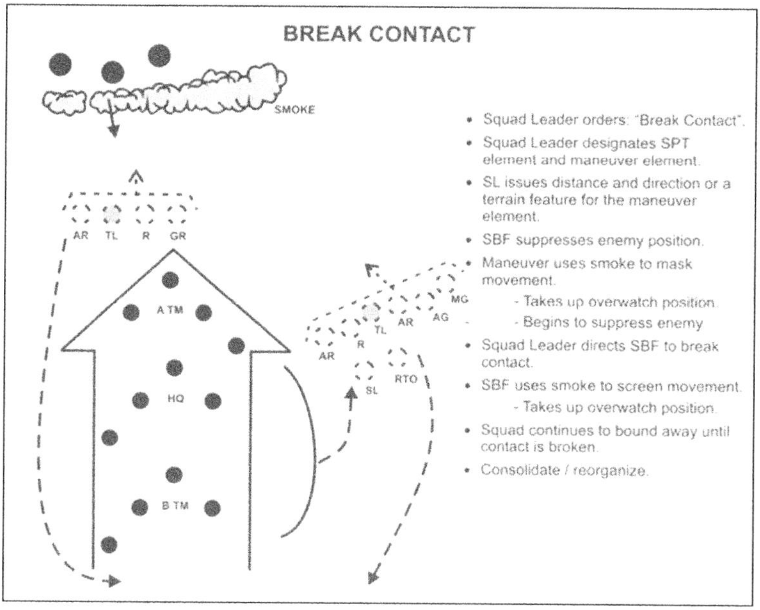
BREAK CONTACT
SMOKE
AR TL R GR
A TM
HQ
B TM
AR R TL AR AG MG
SL RTO
• Squad Leader orders: "Break Contact".
• Squad Leader designates SPT element and maneuver element.
• SL issues distance and direction or a terrain feature for the maneuver element.
• SBF suppresses enemy position.
• Maneuver uses smoke to mask movement.
- Takes up overwatch position.
- Begins to suppress enemy
• Squad Leader directs SBF to break contact.
• SBF uses smoke to screen movement.
- Takes up overwatch position.
• Squad continues to bound away until contact is broken.
• Consolidate / reorganize.

## FORMATIONS AND ORDERS OF MOVEMENT

I. Movement Formation: Fire Team Wedge; MG Team attached.

II. Three Movement Techniques utilized:

- A. Traveling technique used behind FFL when contact is not likely.
- B. Traveling Overwatch forward of the FFL when enemy contact is possible.
- C. Bounding Overwatch used forward of the FFL when enemy contact is expected.

III. Distances are based on but not dictated by visibility, terrain, and vegetation.

IV. Actions at Night: Modified Wedge

V. Actions at the Halt: Short and Long Halt (GV/LV)

VI. Leader Location: Fixed/Unfixed

**FIRE TEAM WEDGE**

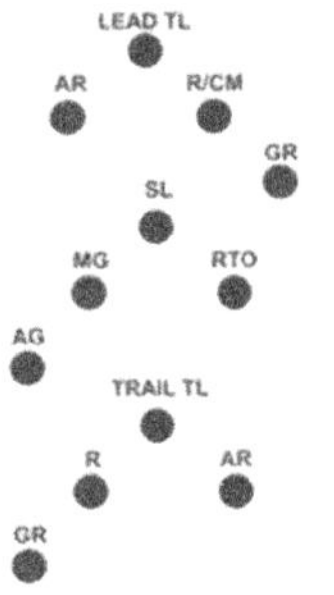


**MODIFIED WEDGE**

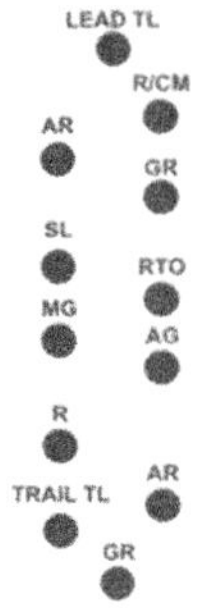

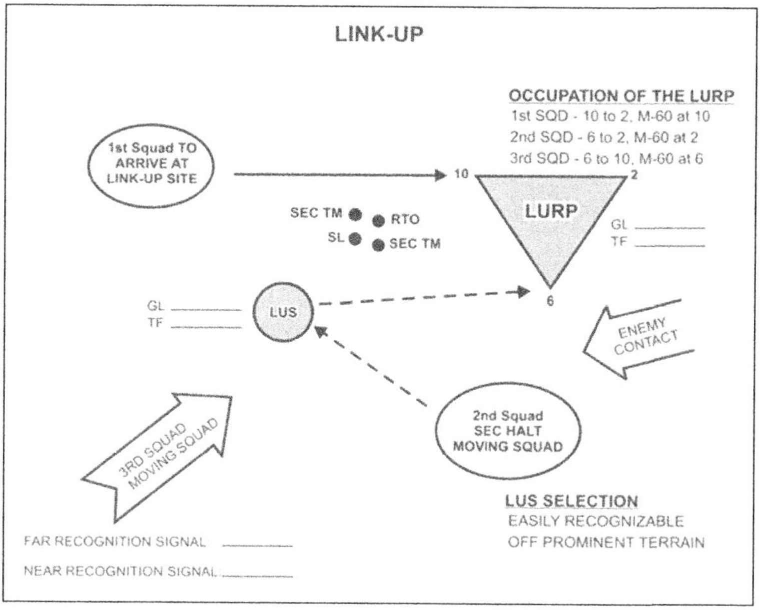
LINK-UP
OCCUPATION OF THE LURP
1st SQD - 10 to 2, M-60 at 10
2nd SQD - 6 to 2, M-60 at 2
3rd SQD - 6 to 10, M-60 at 6
1st Squad TO ARRIVE AT LINK-UP SITE
10
2
LURP
6
GL
TF
SEC TM
RTO
SL
SEC TM
GL
TF
LUS
ENEMY CONTACT
3RD SQUAD MOVING SQUAD
2nd Squad SEC HALT MOVING SQUAD
LUS SELECTION
EASILY RECOGNIZABLE
OFF PROMINENT TERRAIN
FAR RECOGNITION SIGNAL
NEAR RECOGNITION SIGNAL

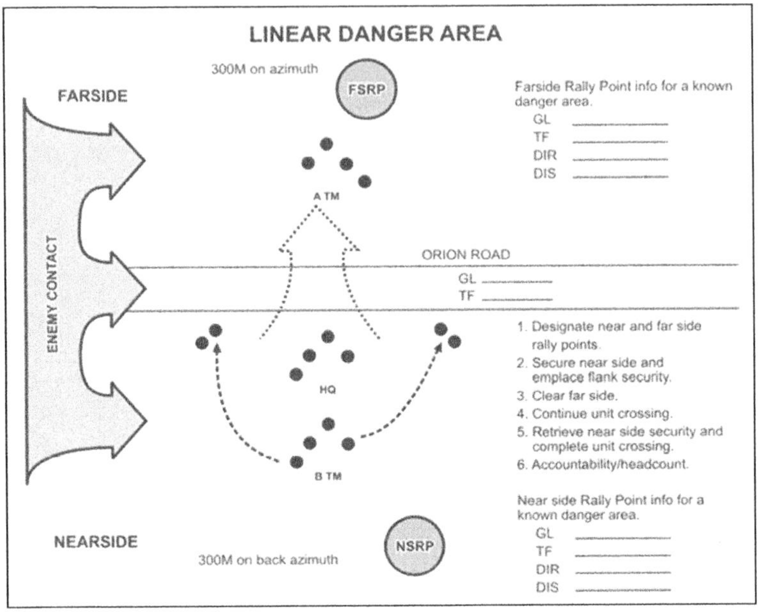
LINEAR DANGER AREA
300M on azimuth
FSRP
FARSIDE
Farside Rally Point info for a known danger area.
GL
TF
DIR
DIS
A TM
ENEMY CONTACT
ORION ROAD
GL
TF
1. Designate near and far side rally points.
2. Secure near side and emplace flank security.
3. Clear far side.
4. Continue unit crossing.
5. Retrieve near side security and complete unit crossing.
6. Accountability/headcount.
HQ
B TM
Near side Rally Point info for a known danger area.
NEARSIDE
NSRP
300M on back azimuth
GL
TF
DIR
DIS

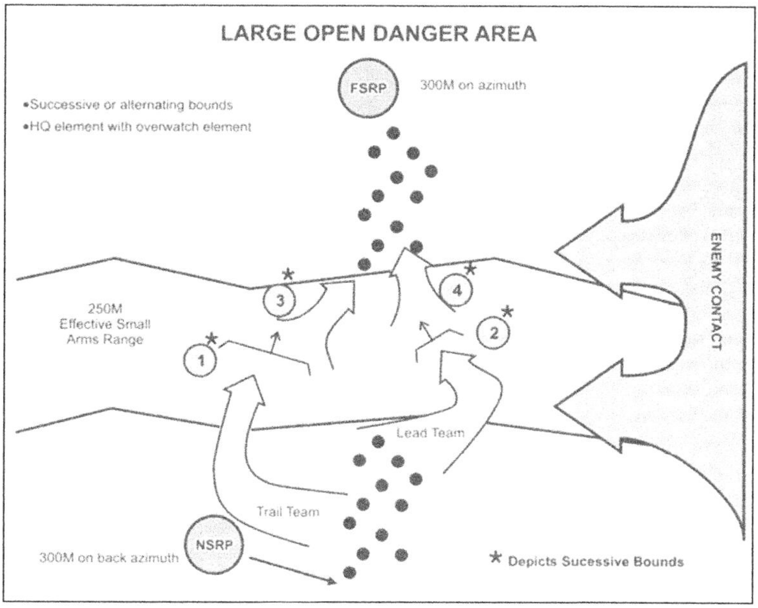
LARGE OPEN DANGER AREA
FSRP
300M on azimuth
•Successive or alternating bounds
•HQ element with overwatch element
ENEMY CONTACT
250M
Effective Small
Arms Range
1
2
3
4
Lead Team
Trail Team
NSRP
300M on back azimuth
* Depicts Sucessive Bounds

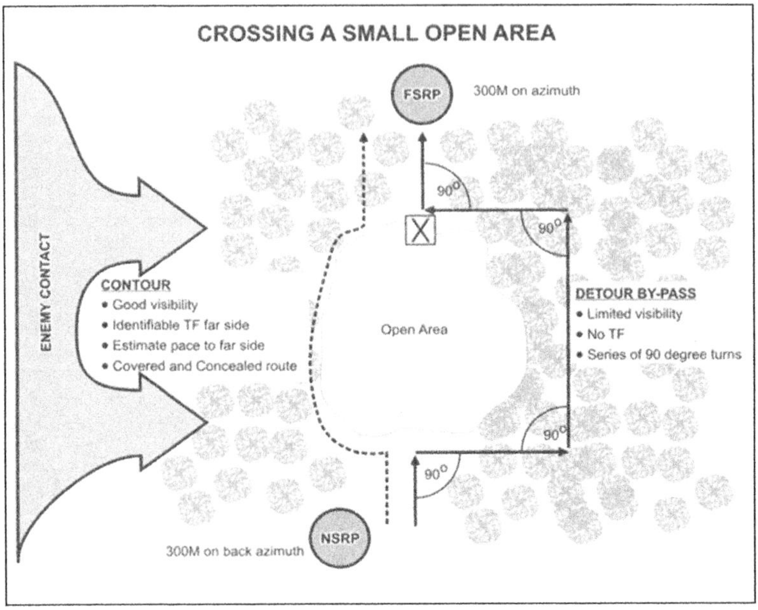
CROSSING A SMALL OPEN AREA
FSRP
300M on azimuth
ENEMY CONTACT
CONTOUR
• Good visibility
• Identifiable TF far side
• Estimate pace to far side
• Covered and Concealed route
Open Area
DETOUR BY-PASS
• Limited visibility
• No TF
• Series of 90 degree turns
90°
90°
90°
90°
NSRP
300M on back azimuth

## SQUAD ATTACK

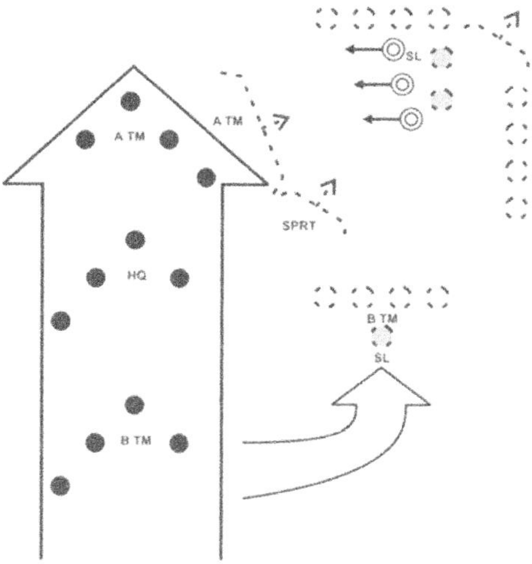


- React to contact.
- TM in the kill zone: (without orders)
  - Return fire.
  - Seek cover and concealment.
  - Suppress enemy calls out 3 D's becomes overwatch.
- SL assess situation.
  - Determines COA (flank/attack).
- TM not in contact: (with SL)
  - Moves along covered and concealed route.
  - Assaults enemy on weak flank.
- Overwatch TM continues to suppress shifts/cease fire as bounding team enters sector.
- Bounding team continues to assault through enemy.
- SL may request indirect fire.
- Consolidate and reorganize.

## RAID BOARDS (LEFT)
## GENERAL INFO

**Raid References:**
- RHB Chapter 5
- FM 3-21.8

**Purposes:**
1. Destroy
2. Liberate
3. Collect Intel

**Planning Considerations:**
- Mission
- Enemy
- Troops
- Terrain / OAKOC
- Time
- Civilians

STUDENT INPUT WRITTEN HERE

**Definition:**
A surprise attack on a fixed position or installation ending in a planned withdrawal.

**Characteristics:**
1. Surprise - Gain
2. Coordinated fires - Maintain
3. Violence of action - Retain
4. Planned withdrawal

INITIATIVE

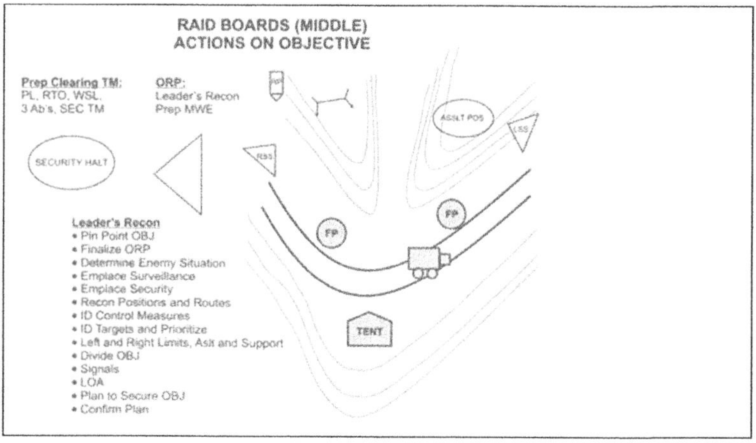
RAID BOARDS (MIDDLE)
ACTIONS ON OBJECTIVE
Prep Clearing TM:
PL, RTO, WSL,
3 Ab's, SEC TM
ORP:
Leader's Recon
Prep MWE
ASSLT POS
LSS
SECURITY HALT
RSS
FP
FP
Leader's Recon
• Pin Point OBJ
• Finalize ORP
• Determine Enemy Situation
• Emplace Surveillance
• Emplace Security
• Recon Positions and Routes
• ID Control Measures
• ID Targets and Prioritize
• Left and Right Limits, Aslt and Support
• Divide OBJ
• Signals
• LOA
• Plan to Secure OBJ
• Confirm Plan
TENT

## RAID BOARDS (MIDDLE)
## TASK ORGANIZATION

| ELEMENT | WHO | WHY | WHAT |
|---|---|---|---|
| HQ | PL, APL<br>RTO, FO, MEDIC | C2/FACILITATE<br>SIGNALS, PEQ-2, STANO | 1x119, ICOM, SKEDCO, PLUGGER |
| ASSLT 1 | SQUAD<br>(+)(-) | DESTROY | ICOM, STANO, M18A1, AT4,<br>SPECIAL TEAM KITS |
| ASSLT 2 | SQUAD<br>(+)(-) | DESTROY | ICOM, STANO, M18A1, AT4,<br>SPECIAL TEAM KITS |
| SUPPORT | WPN S SQD<br>(+)(-) | SUPPRESS | 3xM240B COMPLETE, ICOM, PEQ-2,<br>PVS-4, AT, STANO, SIGNALS, MAP |
| SECURITY | SQUAD<br>(+)(-) | DELAY | ICOMs, M18A1, AT, STANO, MAP<br>SIGNALS |

## RAID BOARDS (RIGHT)
## SOP'S

**CONTINGENCIES**

1. COMPROMISE
   a. ORP
   b. LEADERS RECON
   c. OCCUPATION

2. MASS CAL

3. COUNTERATTACK

**SOP's**

1. EPW SEARCH

2. AID LITTER

3. MED EVAC

4. CCP

5. WITHDRAW PLAN

## AMBUSH SOPS (LEFT)

**References:**
- RHB Chapter 5
- FM 3-21.8

**Ambush:**
A surprise attack from a covered and concealed position on a moving or temporarily halted target.

**Types:**
- Point
- Area

**Fundamentals:**
1. Surprise - Gain
2. Fire superiority - Maintain
3. Violence of action - Retain

**Purposes:**
1. Disrupt/Destroy
2. Collect Intelligence
3. Block or Deny Access
4. Canalize

**Categories:**
*Deliberate*: has specific target at a predetermined time and location.

*Hasty*: Platoon makes visual contact with the enemy and has time to establish an ambush without being detected.

**Planning Considerations:**
- Mission - Task/Purpose
- Enemy - MPCOA, MDCOA
- Terrain - Map Recon/Ldrs Recon
- Time - 1/3 2/3 Rule
- Troops - Task Org
- Civilians -

## AMBUSH BOARDS (MIDDLE)

| | | | |
|---|---|---|---|
| HQ | PL, PL RTO, PSG, MEDIC, FO | COMMAND AND CONTROL | PRC-119s, ICOM, NVGs, BINO's, SKEDCO, PEQ-2, SIGNALS, MAP |
| SECURITY | MINIMUM OF 3-2 MAN TEAMS | EARLY WARNING, SEALS OFF OBJ, LFS, RFS, ORP SECURITY | COMMO, NVGs, CLAYMORES, AT WPNS, BINO's, SIGNALS, MAP |
| SUPPORT | WSL, 3 GUN TEAMS COMPLETE +/- | SUPPRESSIVE FIRE ON KILL ZONE, SEC ON OBJ DURING RECON/REORG | MG's, COMMO, NVGs, AT WPNS, BINO's, SKEDCO, PEQ-2, SIGNALS., MAP |
| ASSAULT | 2 SQUADS +/- | MAIN EFFORT, BLOCK, DESTROY, CANALIZE, CAPTURE PIR | NVGs, BINO's, CLAYMORE, AT, FLEX CUFFS, POLELESS LITTER, MARKINGS, SIGNALS |

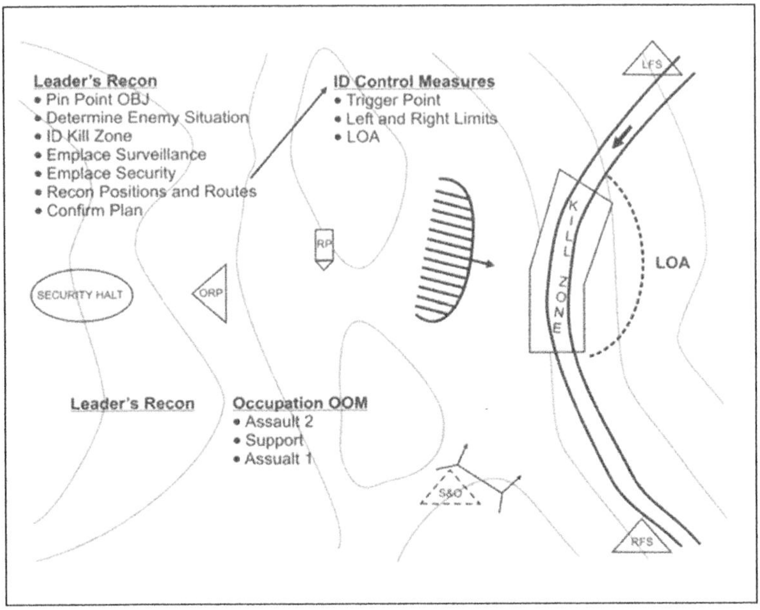
Leader's Recon
• Pin Point OBJ
• Determine Enemy Situation
• ID Kill Zone
• Emplace Surveillance
• Emplace Security
• Recon Positions and Routes
• Confirm Plan
ID Control Measures
• Trigger Point
• Left and Right Limits
• LOA
LFS
RP
KILL ZONE
LOA
SECURITY HALT
ORP
Leader's Recon
Occupation OOM
• Assault 2
• Support
• Assualt 1
S&O
RFS

## AMBUSH BOARDS (RIGHT)
## SOP'S

**CONTINGENCIES**

1. COMPROMISE
   a. ORP
   b. LEADERS RECON
   c. OCCUPATION
2. MASS CAL
3. COUNTERATTACK

**SOP's**

1. EPW SEARCH
2. AID LITTER
3. MASS CAL
4. MED EVAC
5. CCP

NOTES

Appendix B

# QUICK REFERENCE CARDS

## CASUALTY FEEDER CARD

NAME ____________________ ROS # __________

COMPANY ________ ALLEGERIES ______________

LOCATION/EVENT ______________________

TIME ______________ TIME ______________

TEMP ______________ TEMP ______________

PULSE ______________ PULSE ______________

RESP ______________ RESP ______________

B/P ______________ B/P ______________

MENTAL STATUS ______________________

A & O x 1 2 3

NPA OPA ET Tube

SAO² ______________ GLUC ______________

IV _____ AC FOREARM HAND OTHER

1000-1000-500-500-500- NS RL $D_5W$

NOTES: ______________________

______________________________

# GTA 07-01-038
## INFANTRY LEADERS' REFERENCE CARD
## January 1995

HEADQUARTERS, DEPARTMENT OF THE ARMY
References: FM 3-3, FM 6-30, FM 6-121, FM 7-8, FM 7-10, FM 7-90
(Supersedes GTA 7-1-31)

DISTRIBUTION RESTRICTION: Distribution authorized to US Government agencies only to protect technical or operational information from automatic dissemination under the International Exchange Program or by other means. Other requests for this document will be referred to Commander, US Army Training Support Center, ATTN: ATIC-DMF, Ft Eustis, VA 23604-5166.
DESTRUCTION NOTICE: Destroy by any method that will prevent disclosure of contents or reconstruction of the document.

DISTRIBUTION: US Army Training Support Centers (TSCs).

### TROOP-LEADING PROCEDURES
1. Receive mission.
2. Issue warning order.
3. Make a tentative plan.
4. Start movement.
5. Reconnoiter.
6. Complete plan.
7. Issue plan.
8. Supervise.

### FIRE REQUEST (FM 6-30)
1. Identification.
2. Warning order.
3. Target location.
4. Target description.
5. Method of engagement.
6. Method of control.

### ESTIMATE OF SITUATION (FM 7-10)
1. Conduct a detailed mission analysis.
2. Analyze the situation and develop courses of action.
3. Analyze courses of action (wargame).
4. Compare courses of action.
5. Make a decision.

### SPOT REPORT
1. Size.
2. Activity.
3. Location.
4. Unit/uniform.
5. Time.
6. Equipment.

### SHELL REPORT (FM 6-121)
1. Observer identification.
2. Location (coded).
3. Azimuth to flash or sound.
4. Time (from and to).
5. Area shelled.
6. Nature of fire.
7. Type rounds received (artillery, mortars, etc.).
8. Damage (coded).

### OPERATION ORDER (FM 7-10)
Task Organization
1. Situation
   a. Enemy
   b. Friendly
   c. Attachments/detachments
2. Mission
   - Who, what, where, when, and why
   - Task and purpose
3. Execution
   a. Concept of operation
      (1) Maneuver
      (2) Fire support
      (3) Engineer
   b. Tasks to maneuver units
   c. Tasks to combat support units
   d. Coordinating instructions
4. Service support
   a. General
   b. Materiel and services
   c. Casualty evacuation
   d. Miscellaneous
5. Command and signal
   a. Command
   b. Signal

### WEAPONS (FM 7-8)

| TYPE | MAX EFF RANGE (m) |
|---|---|
| M16A2 | 580 (pt) 800 (area) 200 (mov) |
| M203 | 150(pt) 350 (area) |
| M249 | 600 (pt) 800 (area) |
| M136 (AT4) | 300 |
| M47 (Dragon) | 1,000 (sta) 100 (mov) |
| MK19 | 1500 (pt) 2,212 (area) |
| M3 RAAWS | 700 (sta) 60 (mov) |
| M60 MG | 1,100 (600 grazing) |
| .50 Caliber MG | 1,800 (1,000 grazing) |
| TOW | 3,000 (plng purposes) |
| TOW 2 | 3,750 |
| 105-mm | 11,500 |
| 105-mm Tank | *2 to 2.5 km |
| 120-mm Tank | *2 to 2.5 km |
| 25-mm BFV | 2,200 |
| 155-mm M109A3 | 18,100 |
| M198 | 24,000 |
| 8-in Howitzer | 22,900 |

WEAPONS (MORTAR) HE ONLY (FM 7-90)

| | MIN | MAX |
|---|---|---|
| 60-mm | 70m | 3,500m |
| 81-mm (M252) | 80m | 5,800m |
| 81-mm (M29A1) | 73m | 4,790m |
| 4.2-inch | 770m | 6,840m |
| 120-mm | 200m | 7,200m |

FPFs (FM 7-90)

| GUNS | MORT | WIDTH | | DEPTH |
|---|---|---|---|---|
| 2 | 60-mm | 60m | x | 30m |
| 4 | 81 (M252) | 150m | x | 50m |
| 4 | 81 (M29A1) | 140m | x | 40m |
| 6 | 4.2-in | 240m | x | 40m |
| 6 | 120-mm | 380m | x | 60m |
| Plt | 155-mm | 200m | x | 50m |
| Btry | 155-mm | 400m | x | 50m |

*Optimum engagement ranges

### MEDEVAC REQUEST

1. Requesting unit identification.
2. Location.
3. Number of patients by type (litter or ambulatory).
4. Type of injuries.
5. Special equipment needed.
6. Tactical situation.

### AIRCRAFT REQUEST

1. Identification.
2. Precedence/priority.
3. Target description.
4. Target location.
5. Target time/date.
6. Desired ordnance/results.
7. Final control.

### DELIBERATE ATTACK CONSIDERATIONS

1. Reconnoiter, pinpoint objective/enemy positions/obstacles.
2. Determine weak points; designate supporting positions.
3. Assign platoon/squad objectives—identify the decisive point.
4. Determine main attack, supporting attack, reserve.
5. Assign breach, support, assault missions.
6. Designate fire control measures.
7. Coordinate indirect/direct fires and CAS to time of attack.
8. Control measures during attack.
9. Secure (ground and air).
10. Consolidate and reorganize.

### NBC 1 OBSERVER'S INITIAL OR FOLLOW-UP REPORT

**Instructions**

1. Line items D and H are mandatory for NBC 1 reports.
2. Line items A, E, G, I, K, L, M, S, Y, and ZA are optional for NBC 1 reports.
3. Line Items B, C, F, PAR, and PBR are reported if data is available.

**Section I. Chemical or Biological Only**

A. Strike serial number, if known (assigned by NBCE).
B. Position of observer.
C. Azimuth of attack from observer (state degrees or mils).
D. Date and time attack started (Zulu, local, or letter zone).
E. Time attack ended, if known.
F. Location of attack (UTM or place) (state actual or estimated).
G. Means of delivery, if known.
H. Type of agent and height of burst, if known.
I. Type and number of munitions or aircraft (state which).
K. Description of terrain (bare, scrubby vegetation, wooded, urban, or unknown).
S. Date and time contamination detected (Zulu, local, or letter zone).
Y. Representative downwind direction—4 digits (state degrees or mils), wind speed—3 digits (data kmph or knots).
ZA. Temperature (centigrade)—2 digits, cloud cover—1 digit, significant weather phenomena—1 digit, air stability—1 digit.
ZB. Remarks.

**Section II. Nuclear Only**

A. Strike serial number, if known (assigned by NBCE).
B. Position of observer.
C. Azimuth of attack from observer (state degrees or mils and grid or magnetic).
D. Date and time attack started (Zulu, local, or letter zone).
F. Location of attack (UTM or place) (state actual or estimated).
G. Means of delivery, if known.
H. Type of burst (state air, surface, or unknown).
J. Flash-to-bang time (seconds).
K. Crater diameter (meters), if known.
L. Cloud width at H +5 minutes (degrees or mils).
M. Cloud angle(top or bottom) or cloud height(top or bottom) at H +10 minutes (state degrees, mils, meters, or feet).
PAR. Location of radioactive cloud outline (UTM).
PAB. Downwind direction of radioactive cloud (state degrees or mils).
ZB. Remarks.

### DEFENSE PLANNING CONSIDERATIONS

1. Establish security (OP/patrols, PWs, M8).
2. Position key weapons.
   a. Coordinate with units on left and right.
   b. Establish FPF or PDF for machine gun.
   c. Ensure mutual support between machine guns.
   d. Cover armor approaches with antiarmor systems.
   e. Establish fire control measures.
3. Prepare positions.
   a. Check sectors of fire.
   b. Check overhead cover and view positions from enemy vantage.
   c. Position in depth and achieve mutual support between positions.
   d. Select/prepare alternate and supplementary positions.
4. Integrate indirect fires, CAS, and obstacles with direct and indirect fire.
5. Check communications and establish emergency signals.
6. Designate ammunition, supply, PW, and casualty points.

## IED / UXO

### Procedures when IEDs are found

**Security -** Maintain 360 degrees. Scan close in and far out, up high and down low.

**Always -** Scan you immediate surroundings for more IEDs.

**Move -** Move away. Vary safe distances, but plan for 300m minimum safe distance and adapt to your METT-TC.

**Attempt -** To confirm suspected IEDs using optics while staying back as far as possible.

**Cordon -** Off the area. Direct people out of danger area. Do not allow anyone to enter except for EOD. Question, search, and detain suspects as defined by your Existing ROE.

**Report -** Your situation using the 9 line IED / UXO spot report format.

---

This could be your hand if you try to dispose of UXOs or IEDs. The enemy has developed anti-handling to catch you when you try defusing. Leave it to experts!

**Call EOD - Don't be a Hero!**

### IED / UXO Report

**LINE 1.** DATE-TIME-GROUP: When the item was discovered.

**LINE 2.** REPORT ACTIVITY AND LOCATION: Unit and grid location of the IED/UXO.

**LINE 3.** CONTACT METHOD: Radio frequency, call sign, POC, and telephone number.

**LINE 4.** TYPE OF ORDNANCE: Dropped, projected, placed, or thrown. Give the number of items, if more than one.

**LINE 5.** NBC CONTAMINATIONS: Be as specific as possible.

**LINE 6.** RESOURCES THREATENED: Equipment, facilities, or other assets that are threatened.

**LINE 7.** IMPACT ON MISSION: Short description of current tactical situation and how the IED/UXO affects the status of the mission.

**LINE 8.** PROTECTIVE MEASURES: Any measures taken to protect personnel and equipment.

**LINE 9.** RECOMMENDED PRIORITY: Immediate, Indirect, Minor, No Threat.

---

### Priority

**Immediate:** Stops unit's maneuver and mission capability or threatens critical assets vital to the mission.

**Indirect:** Stops the unit's maneuver and mission capability or threatens critical assets important to the mission.

**Minor:** Reduces the unit's maneuver and mission capability or threatens non-critical assets of value.

**No Threat:** Has little or no effect on the unit's capabilities or assets.

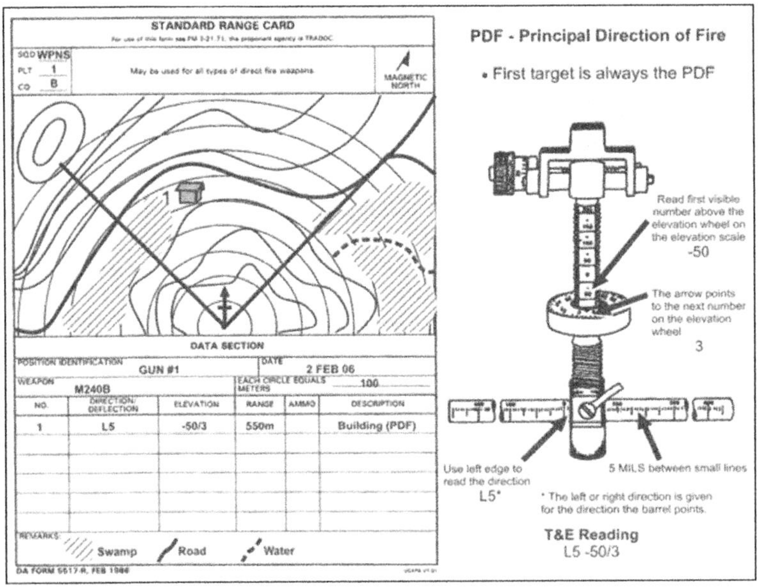
STANDARD RANGE CARD
For use of this form see FM 3-21.71, the proponent agency is TRADOC
SQD WPNS
PLT 1
CO B
May be used for all types of direct fire weapons
MAGNETIC NORTH
1
DATA SECTION
POSITION IDENTIFICATION GUN #1
DATE 2 FEB 06
WEAPON M240B
EACH CIRCLE EQUALS 100 METERS
NO. | DIRECTION/DEFLECTION | ELEVATION | RANGE | AMMO | DESCRIPTION
1 | L5 | -50/3 | 550m | | Building (PDF)
REMARKS: Swamp Road Water
DA FORM 5517-R, FEB 1986
PDF - Principal Direction of Fire
• First target is always the PDF
Read first visible number above the elevation wheel on the elevation scale
-50
The arrow points to the next number on the elevation wheel
3
Use left edge to read the direction
L5*
5 MILS between small lines
* The left or right direction is given for the direction the barrel points.
T&E Reading
L5 -50/3

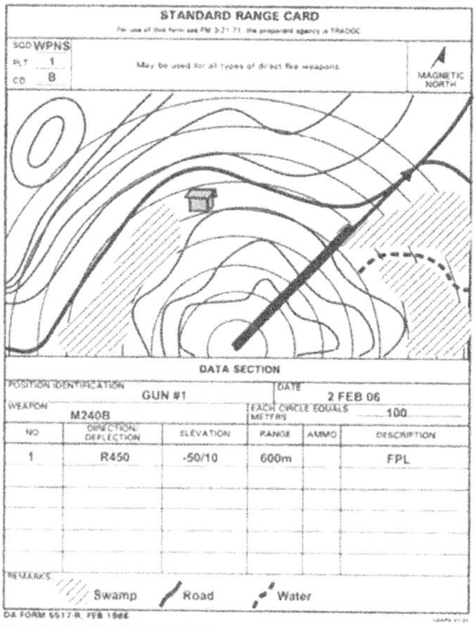


## FPL - Final Protective Line

- FPL is always target #1
- FPL will always be metal-to-metal
- Represented by thick line
  - Break in thick line for dead space out to 600 meters
  - The gap is equal to the width of the dead space

FPL - Final Protective Line

## STANDARD RANGE CARD

For use of this form see FM 3-21.71; the proponent agency is TRADOC.

| | | |
|---|---|---|
| SQD ____<br>PLT ____<br>CO ____ | May be used for all types of direct fire weapons. | MAGNETIC NORTH |

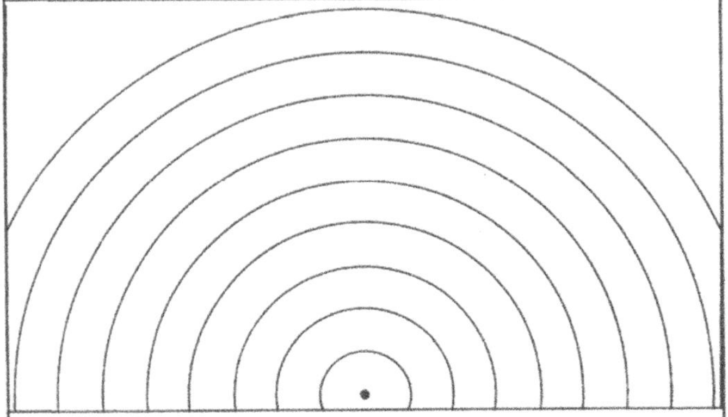

**DATA SECTION**

| POSITION IDENTIFICATION | DATE |
|---|---|
| WEAPON | EACH CIRCLE EQUALS ________ METERS |

| NO. | DIRECTION/ DEFLECTION | ELEVATION | RANGE | AMMO | DESCRIPTION |
|---|---|---|---|---|---|
| | | | | | |
| | | | | | |
| | | | | | |
| | | | | | |
| | | | | | |
| | | | | | |

REMARKS:

DA FORM 5517-R, FEB 1986 USAPA V1.01

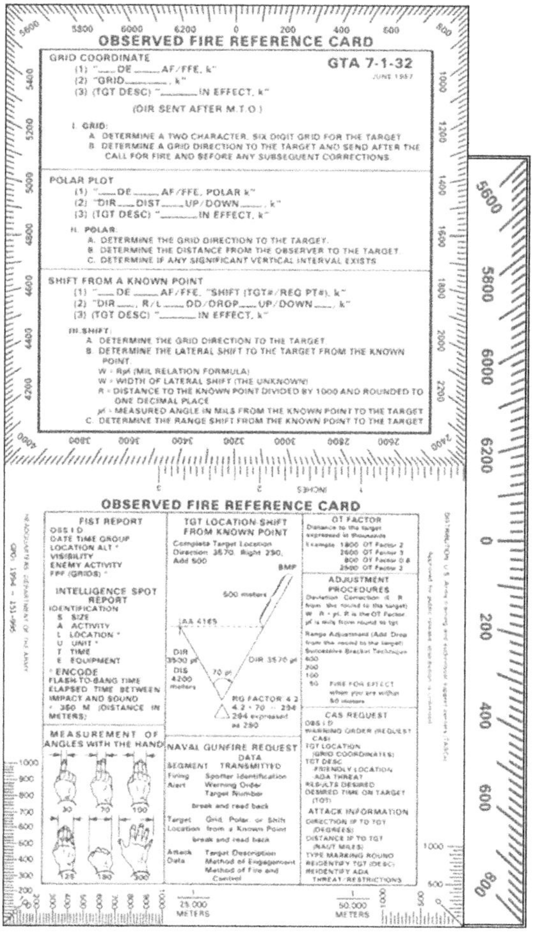

# OBSERVED FIRE REFERENCE CARD

**GTA 7-1-32**
JUNE 1987

GRID COORDINATE
(1) "___ DE ____ AF/FFE, k"
(2) "GRID_______, k"
(3) (TGT DESC) "______IN EFFECT, k"

(DIR SENT AFTER M.T.O.)

I. GRID:
A. DETERMINE A TWO CHARACTER, SIX DIGIT GRID FOR THE TARGET
B. DETERMINE A GRID DIRECTION TO THE TARGET AND SEND AFTER THE CALL FOR FIRE AND BEFORE ANY SUBSEQUENT CORRECTIONS.

POLAR PLOT
(1) "___ DE ____ AF/FFE, POLAR k"
(2) "DIR___DIST____UP/DOWN____, k"
(3) (TGT DESC) "______IN EFFECT, k"

II. POLAR:
A. DETERMINE THE GRID DIRECTION TO THE TARGET.
B. DETERMINE THE DISTANCE FROM THE OBSERVER TO THE TARGET.
C. DETERMINE IF ANY SIGNIFICANT VERTICAL INTERVAL EXISTS

SHIFT FROM A KNOWN POINT
(1) "___ DE ____ AF/FFE, "SHIFT (TGT#/REG PT#), k"
(2) "DIR____, R/L____ OD/DROP____UP/DOWN____, k"
(3) (TGT DESC) "______IN EFFECT, k"

III. SHIFT:
A. DETERMINE THE GRID DIRECTION TO THE TARGET.
B. DETERMINE THE LATERAL SHIFT TO THE TARGET FROM THE KNOWN POINT.
W = Rµ (MIL RELATION FORMULA)
W = WIDTH OF LATERAL SHIFT (THE UNKNOWN)
R = DISTANCE TO THE KNOWN POINT DIVIDED BY 1000 AND ROUNDED TO ONE DECIMAL PLACE
µ = MEASURED ANGLE IN MILS FROM THE KNOWN POINT TO THE TARGET
C. DETERMINE THE RANGE SHIFT FROM THE KNOWN POINT TO THE TARGET

INCHES

# OBSERVED FIRE REFERENCE CARD

**FIST REPORT**
OBS I D
DATE TIME GROUP
LOCATION ALT *
VISIBILITY
ENEMY ACTIVITY
FPF (GRIDS) *

**INTELLIGENCE SPOT REPORT**
IDENTIFICATION
S SIZE
A ACTIVITY
L LOCATION *
U UNIT *
T TIME
E EQUIPMENT
* ENCODE
FLASH-TO-BANG TIME
ELAPSED TIME BETWEEN IMPACT AND SOUND × 350 M (DISTANCE IN METERS)

**MEASUREMENT OF ANGLES WITH THE HAND**



**TGT LOCATION SHIFT FROM KNOWN POINT**
Complete Target Location
Direction 3570, Right 290, Add 500



**NAVAL GUNFIRE REQUEST**

| SEGMENT | DATA TRANSMITTED |
|---|---|
| Firing Alert | Spotter Identification<br>Warning Order<br>Target Number |
| | break and read back |
| Target Location | Grid, Polar, or Shift from a Known Point |
| | break and read back |
| Attack Data | Target Description<br>Method of Engagement<br>Method of Fire and Control |

**OT FACTOR**
Distance to the target expressed in thousands
Example: 1800 OT Factor 2
2600 OT Factor 3
800 OT Factor 0.8
2500 OT Factor 3

**ADJUSTMENT PROCEDURES**
Deviation Correction (L-R from the round to the target)
W = R × µ; R is the OT Factor; µ is mils from round to tgt
Range Adjustment (Add-Drop from the round to the target)
Successive Bracket Technique
400
200
100
50 FIRE FOR EFFECT when you are within 50 meters

**CAS REQUEST**
OBS I D
WARNING ORDER (REQUEST CAS)
TGT LOCATION (GRID COORDINATES)
TGT DESC
FRIENDLY LOCATION
ADA THREAT
RESULTS DESIRED
DESIRED TIME ON TARGET (TOT)

**ATTACK INFORMATION**
DIRECTION IP TO TGT (DEGREES)
DISTANCE IP TO TGT (NAUT MILES)
TYPE MARKING ROUND
REIDENTIFY TGT (DESC)
REIDENTIFY ADA THREAT/RESTRICTIONS

1/25,000 METERS
1/50,000 METERS

HEADQUARTERS, DEPARTMENT OF THE ARMY

# 9 Line MEDEVAC

| LINE ITEM | EXPLANATION |
|---|---|
| **1. Location of Pickup Site.** | Encrypt grid coordinates. When using *DRYAD Numeral Cipher, the same SET line* will be used to encrypt grid zone letters and coordinates. To preclude misunderstanding, a statement is made that grid zone letters are included in the message (unless unit SOP specifies its use at all times). |
| **2. Radio Frequency, Call Sign, Suffix.** | Encrypt the frequency of the radio at the pickup site, *not* a relay frequency. The call sign (and suffix if used) of person to be contacted at the pickup site may be transmitted in the clear. |
| **3. No. of Patients by Precedence.** | Report only applicable info & encrypt brevity codes. A = Urgent, B = Urgent-Surg, C = Priority, D = Routine, E = Convenience. (If 2 or more categories reported in same request, insert the word "*break*" between each category). |
| **4. Spec Equipment.** | Encrypt applicable brevity codes. A = None, B = Hoist, C = Extraction equipment, D = Ventilator. |
| **5. No. of Patients by Type.** | Report only applicable information and encrypt brevity code. If requesting MEDEVAC for both types, insert the word "*break*" between the litter entry and ambulatory entry: L + # of Pnt - Litter; A + # of Pnt - Ambul (sitting). |
| **6. Security Pickup Site (Wartime).** | N = No enemy troops in area, P = Possibly enemy troops in area (approach with caution), E = Enemy troops in area (approach with caution), X = Enemy troops in area (armed escort required). |
| **6. Number and type of Wound, Injury, Illness (Peacetime).** | Specific information regarding patient wounds by type (gunshot or shrapnel). Report serious bleeding, along with patient blood type, if known. |
| **7. Method of Marking Pickup Site.** | Encrypt the brevity codes. *A* = Panels, *B* = Pyrotechnic signal, *C* = Smoke Signal, *D* = None, *E* = Other. |
| **8. Patient Nationality and Status.** | Number of patients in each category need not be transmitted. Encrypt only applicable brevity codes. A = US military, B = US civilian, C = Non-US mil, D = Non-US civilian, E = EPW. |
| **9. NBC Contamination (Wartime).** | Include this line only when applicable. Encrypt the applicable brevity codes. N = nuclear, B = biological, C = chemical. |
| **9. Terrain Description (Peacetime).** | Include details of terrain features in and around proposed landing site. If possible, describe the relationship of site to a prominent terrain feature (lake, mountain, tower). |

Reference: FM 8-10-6, *Medical Evacuation in a Theater of Operations, pages 7-7 through 7-9.*

Made in the USA
Monee, IL
05 November 2026

67813166R00213